POLYMER SCIENCE AND TECHNOLOGY SERIES

COMPUTER SIMULATION OF POLYMERS

Editor: E. A. COLBOURN

Longman Scientific & Technical

POLYMER SCIENCE AND TECHNOLOGY SERIES
SERIES EDITORS: DR D M BREWIS AND PROFESSOR D BRIGGS

Published

D E PACKHAM (ed.), *Handbook of Adhesion*
I S MILES AND S ROSTAMI (eds), *Multicomponent Polymer Systems*
F R JONES (ed.), *Handbook of Polymer–fibre Composites*

Forthcoming

D M BREWIS AND B C COPE (eds.), *Handbook of Polymer Science*
H R BRODY (ed.), *Synthetic Fibres*
R N ROTHON (ed.), *Particulate Filled Polymer Composites*

Longman Scientific and Technical
Longman Group UK Limited
Longman House, Burnt Mill, Harlow,
Essex, CM20 2JE, England
and Associated Companies throughout the world.

Copublished in the United States with
John Wiley & Sons, Inc., 605 Third Avenue, New York, NY 10158

© Longman Group UK Limited 1994

All rights reserved; no part of this publication may be reproduced, stored in a retrieval system, or transmitted in any form or by any means, electronic, mechanical, photocopying, recording, or otherwise without either the prior written permission of the Publishers, or a licence permitting restricted copying in the United Kingdom issued by the Copyright Licensing Agency Ltd, 90 Tottenham Court Road, London, W1P 9HE

First edition 1994

ISBN 0 582 08374 5

British Library Cataloguing in Publication Data
A catalogue record for this book is available from the British Library

Library of Congress Cataloging-in-Publication Data
Computer simulation of polymers / editor, E. A. Colbourn.
    p.   cm.—(Polymer science and technology series)
Includes bibliographical references and index.
ISBN 0-470-23343-5
1. Polymers—Computer simulation.   I. Colbourn, E. A. (Elizabeth A.)
II. Series.
QD381.9.E4C68   1994
547.7'01'13—dc20   93-36416 CIP

Set by 16JJ in 10/12½pt Times

Printed and bound in Great Britain
by Bookcraft (Bath) Ltd

Contents

List of Contributors

Chapter 1 Molecular modelling of polymers 9. Description and application of torsion angle unit theory to predict polymer properties A J HOPFINGER AND M G KOEHLER 1

    1.1 Introduction 1
    1.2 Torsion angle unit (TAU) theory 5
    1.3 Estimation of intramolecular TAU physicochemical properties 9
    1.4 Estimation of intermolecular TAU physicochemical properties 17
    1.5 Results of applying TAU theory 23
    1.6 Discussion of TAU theory and results of QSPR studies 41
    References 44

Chapter 2 Molecular dynamics modelling of amorphous polymers J H R CLARKE AND D BROWN 46

    2.1 Introduction 46
    2.2 Models and methods 49
    2.3 Preparation of amorphous samples 57
    2.4 Mechanical 'experiments' on model polyethylene 63
    2.5 Conclusions 81
    Acknowledgements 82
    Appendix 1   The loose-coupling algorithm 82
    Appendix 2   Method 1 vs Method 2 86

	Appendix 3	Unbiased Monte Carlo sampling of chain configurations	88
	References		89

Chapter 3   Monte Carlo studies of collective phenomena in dense polymer systems   K BINDER   91

    3.1 Introduction: length and time scales   91
    3.2 Some details on 'coarse-graining' models for polymers   99
    3.3 Crossover in semi-dilute polymer solutions and the dynamics of polymer melts   100
    3.4 Unmixing in polymer blends   111
    3.5 Local ordering and chain stretching in block copolymer melts   121
    3.6 Final remarks   125
    References   127

Chapter 4   Molecular modelling of crystalline polymers
E A COLBOURN AND J KENDRICK   130

    4.1 Introduction   130
    4.2 Polymer morphology   131
    4.3 Molecular mechanics force fields   132
    4.4 Determination of force field parameters   136
    4.5 Simulations of crystal structure   145
    4.6 Computer simulation of poly(ethene) crystals   152
    4.7 Simulation of condis crystals   157
    4.8 Modelling studies of polyamide systems   157
    4.9 Conclusions   161
    References   162

Chapter 5   Computer simulation of polymer crystallization
G GOLDBECK-WOOD   165

    5.1 Introduction   165
    5.2 The microscopic level   166
    5.3 The macroscopic level   193
    References   197

Chapter 6	Molecular models for polymer deformation and failure Y TERMONIA	200
	6.1 Introduction	200
	6.2 Linear polymers	201
	6.3 Crosslinked polymers	211
	6.4 Conclusions	225
	References	226
Chapter 7	Monte Carlo simulations of the free energies and phase diagrams of macromolecular systems S KUMAR	228
	7.1 Introduction	228
	7.2 Chain chemical potentials from simulation	234
	7.3 Results and discussion	239
	7.4 Phase equilibrium behaviour	254
	7.5 Summary	258
	Acknowledgements	258
	References	259
Chapter 8	Computer simulation of polymer network formation B E EICHINGER AND O AKGIRAY	263
	8.1 Introduction	263
	8.2 Analytical theories of gelation	266
	8.3 Simulation techniques	282
	8.4 New results	294
	8.5 Conclusions	299
	References	300
Chapter 9	Computer simulations of biopolymers D J OSGUTHORPE AND P DAUBER-OSGUTHORPE	303
	9.1 Introduction	303
	9.2 Methods and techniques	307
	9.3 Examples of applications to peptides and proteins	317
	9.4 Conclusions and further directions	332
	References	333
Index		337

List of contributors

A J Hopfinger, Department of Chemistry and Department of Medicinal Chemistry and Pharmacognosy, M/C 781, University of Illinois at Chicago, Box 6998, Chicago, IL 60680, USA

M G Koehler, Department of Chemistry and Department of Medicinal Chemistry and Pharmacognosy, Box 6998, Chicago, IL 60680, USA (Permanent address: Allied-Signal Inc., Engineering Materials Research Center, 50 East Algonquin Road, Box 5016, Des Plaines, IL 60017-5016, USA)

J H R Clarke, Chemistry Department, UMIST, Manchester M60 1QD, UK

D Brown, Chemistry Department, UMIST, Manchester M60 1QD, UK

K Binder, Materialwissenschaftliches Forschungszentrum (MWFZ) und Institut für Physik, Johannes Gutenberg Universität Mainz, D-55099 Mainz, Staudinger Weg 7, Germany

E A Colbourn, Materials Division, Oxford Molecular, The Magdalen Centre, Oxford Science Park, Sandford-on-Thames, Oxfordshire OX4 4GA

J Kendrick, ICI Wilton Materials Research Centre, P.O. Box 90, Wilton, Middlesbrough, Cleveland TS6 8JE

G Goldbeck-Wood, H H Wills Physics Laboratory, University of Bristol, Tyndall Avenue, Bristol BS8 1TL, UK

Y Termonia, Central Research and Development, Experimental Station, E I du Pont de Nemours, Wilmington, DE 19880-0356, USA

S Kumar, Department of Materials Science and Engineering, Polymer Science Program, The Pennsylvania State University, University Park, PA, USA

B E Eichinger, BIOSYM Technologies, Inc., 9685 Scranton Road, San Diego, CA 92121-3752, USA

O Akgiray, BIOSYM Technologies, Inc., 9685 Scranton Road, San Diego, CA 92121-3752, USA

D J Osguthorpe, Molecular Graphics Unit and School of Chemistry, University of Bath, Claverton Down, Bath BA2 7AY, UK

P Dauber-Osguthorpe, Molecular Graphics Unit and School of Chemistry, University of Bath, Claverton Down, Bath BA2 7AY, UK

CHAPTER 1

Molecular modelling of polymers 9. Description and application of torsion angle unit theory to predict polymer properties

A J HOPFINGER AND M G KOEHLER

1.1 Introduction

1.1.1 History and background of computer-assisted molecular design

The application of computer-assisted molecular design (CAMD) approaches to the development of new polymeric materials is being actively explored.[1] The considerable interest in this 'hi-tech' approach to the development of polymer products arises jointly from the fundamental insight that can be realized, and the enormous gain in design efficiency that is possible from its use. The extension of CAMD approaches in materials science is a natural progression of the significant impact CAMD has had in the pharmaceutical industry. Virtually every major pharmaceutical company, worldwide, has a group of scientists who are doing computer-assisted drug design.[2-4] In terms of macromolecules, computer-assisted drug design has been directed to expand its focus and effort on modelling the interactions of ligands (potential drugs) with biological macromolecules–proteins and polynucleic acids. The conformations of the biological macromolecules have been determined mainly from X-ray crystallography, and, to a lesser extent, from NMR methods. X-ray crystallography and NMR spectroscopy will likely merge with current computer-assisted drug design methods over this decade to yield integrated experimental and computational approaches for the efficient and reliable design of new biopharmaceutical agents.

Unfortunately, the application of CAMD in polymer science is not as direct as it is in the pharmaceutical sciences. In the case of designing a drug, there is usually an isomorphic relationship between a change in the structure of a drug-candidate molecule and the corresponding biological response. Thus, one can deal with discrete molecular entities – the drug candidate, and even better, also its macromolecular receptor – in the molecular design process.

In the case of the design of polymeric materials the problem becomes more diffuse, less defined. What must be evaluated (modelled) is a volume element of the polymeric material that is of sufficient size so as to allow the estimation of key molecular properties that, at the least, correlate to, or are indicative of, the bulk properties of the material. Moreover, the polymer is not a distinct molecular entity unto itself. Issues of molecular weight, chemical defects and structural defects must be considered in even the most simple linear, synthetic polymers. In other words, there is not necessarily an isomorphic relationship between the physicochemical properties of a single polymer molecule and the bulk properties of a material derived from the polymer. Overall, the lack of an isomorphic molecule–property relationship makes polymer design more difficult than drug design.

1.1.2 Approaches to computer-assisted polymer design

The most comprehensive approach to the modelling of structural features of polymeric materials is computer simulation–molecular dynamics and/or Monte Carlo techniques.[5,6] Computer simulation of polymer structure permits estimation of average properties of polymer systems containing multiple polymer chains and other types of molecular entities such as solvent, plasticizer and/or crosslinking agents. At this time in the evolution of CAMD there are two significant drawbacks to the use of computer simulation of polymers in a practical design mode.

1. The time, effort and computer resources to do a polymer simulation are often so large as to negate the impact of the calculation on a design programme. It may simply be easier to make and test a targeted polymer system than to do the corresponding simulation.
2. It is often difficult to abstract the relevant information/molecular properties from a simulation that are indicative of the bulk property of interest. For example, what motions in molecular dynamics simulations correspond to the onset of the glass transition, T_g?

Molecular simulations on meaningful models of polymer materials have only been possible for a short time. Hence other approaches to predicting bulk properties, obviously of much less computational intensity, have been considered over the past years. The most straightforward of these approaches are models which are based upon the group additive property (GAP) concept.[7–9] The GAP concept assumes that some intrinsic contribution to any bulk property, $P_B(i)$, is associated with the ith structural group of the polymer. Most of the GAP models deal with homopolymers so that consideration of the structural groups composing the monomer unit is sufficient to define all contributions to the bulk property, P_B. P_B is simply taken to be the sum of the $P_B(i)$ composing the monomer,

$$P_B = \sum_{i=1}^{m} P_B(i) \qquad [1]$$

In Eqn [1] the index m corresponds to the number of structural groups in the monomer unit. Clearly, group additivity is often far from a reasonable assumption, and most GAP models include non-additive correction factors, $f_B(i)$, so that Eqn [1] becomes

$$P_B = \sum_{i=1}^{m} (P_B(i) + f_B(i)) \qquad [2]$$

The majority of polymer GAP models have been developed to predict one particular bulk property with a strong bias favouring the estimation of T_g.[7,10-13] However, VanKrevelen has undertaken the monumental task, with considerable success, of devising a family of homologous GAP models to estimate a wide variety of bulk properties of homopolymers for both the solution and solid-states (amorphous, semi-crystalline and crystalline).[7] While the mathematical formats of the VanKrevelen GAP models vary somewhat, they can be characterized by

$$P_B = \sum_{i=1}^{m} \frac{P_B(i) + f_B(i-1, i, i+1)}{M_W} \qquad [3]$$

where $P_B(i)$ is the weighing factor for the ith group for bulk property P_B, $f_B(i-1, i, i+1)$ is a non-additive correction factor for the ith group, and M_W is the monomer molecular weight. The introduction of M_W in the VanKrevelen GAP models means that the bulk polymer properties are estimated from a parameter set that scales to monomer molecular weight. The individual $P_B(i)$ and $f_B(i-1, i, i+1)$ are determined by fitting procedures using a training set of polymers for which the bulk property of interest has been measured.

The VanKrevelen GAP formalism, overall, has been quite successful. In particular, it has been reliable in predicting relative differences in bulk properties of polymers having structural homology. Moreover, GAP methods are quite easy to employ and interpret. The calculations are simple and fast, and the end-point bulk property well defined.

However, GAP methods have three serious limitations:

1. If the group parameters are not available for the polymer of interest, then a bulk property estimation is not possible. Moreover, there do not appear to be any overall guidelines to estimate group parameters for any of the more popular/general GAP formalisms.
2. The real world is not group-additive, and even the introduction of non-additive correction factors still leaves, ultimately, a chemical topological (graph) model, as opposed to a three-dimensional molecular

model. Consequently, those bulk properties that are highly dependent upon spatial behaviour on the molecular level may not be estimated accurately.
3. GAP models do not provide mechanistic information that can be used to provide an understanding of why a particular polymer exhibits a particular bulk property measure. This information can only come from higher level modelling.

There is a level of polymer modelling intermediate to GAP formalisms on the one hand, and molecular simulation on the other. This level (class) of molecular modelling can be described as three-dimensional static (3DS) modelling and encompasses conformational analyses of polymer chains,[14] polymer crystal packing calculations,[15] and statistical mechanics formalisms of which Flory's Rotational Isomeric State (RIS) theory[16] is most prominent. 3DS modelling studies actually represent the majority of polymer structure calculations over the past thirty years. However, 3DS studies have not often been used to predict bulk polymer properties with the exception of crystal structures. 3DS calculations have been used, in the main, to provide 'pictures' of polymer chain conformations under different conditions, usually ranging from the amorphous to crystalline solid-state, and/or dilute to dense solution states.

1.1.3 Overview of torsion angle unit theory

Given the current situation regarding approaches to the molecular modelling of polymers, with the goal of predicting properties, we asked ourselves if some hybrid technique could be devised which maximizes the advantages of each of the current techniques, but minimizes their respective drawbacks. One additional constraint was considered in formulating our answer – that computational power will increasingly grow so that the applications of large molecular simulations will become increasingly practical.

The remainder of this chapter describes our answer to the question posed above. We have developed a polymer modelling formalism we call torsion angle unit, TAU theory.[17,18] The essential features of TAU theory, which are expanded upon in the balance of this chapter, are:

1. The theory, in its current state of development, only treats homopolymers, and uses the structural repeat unit (SRU), which is the monomer in most cases, of a polymer, as the fundamental building block.
2. The SRU of the homopolymer is decomposed into its corresponding set of TAUs, and the molecular physicochemical properties of each TAU are taken from an existing table, or computed directly using

3DS modelling and/or molecular simulations. The net physicochemical property of the polymer is then taken as the sum of TAU contributions, that is, group additivity is assumed.
3. Step 2 is repeated for each member of a set of homopolymers for which the target bulk property (T_g, T_m, modulus, etc.) has been measured.
4. Step 3 leads to the generation of a structure–property relationship (SPR) table for the 'training set' of homopolymers. The data in the SPR table are used to formulate a quantitative structure–property relationship (QSPR). The measured bulk polymer properties are considered as dependent variables to the set of calculated physicochemical properties of the SPR table which, in turn, are derived from the TAU of the SRU.
5. The QSPR is generated by performing multi-dimensional linear regression analyses of the molecular physicochemical properties against the bulk polymer property measures. The preferred QSPR is that which maximizes the statistical significance of fit between a specific set of molecular physicochemical properties and the measured bulk polymer property.

1.2 Torsion angle unit (TAU) theory

1.2.1 The torsion angle unit

The key to being able to efficiently estimate physicochemical properties of polymers is the representation of the polymer in terms of torsion angle units, TAUs. A TAU is schematically defined as

$$G_i \xrightarrow{\theta_i} G_{i+1}$$

where G_i and G_{i+1} are structural groups connected by a bond about which the torsion angle θ_i occurs. The polymer is built up by connecting TAUs together such that the 'right' structural group of the ith torsion angle becomes the 'left' structural group of the $i + 1$ TAU. Figure 1.1 illustrates the structure of a linear polymer in terms of TAUs.

TAUs permit a global molecular property of a polymer system to be computed as a scalar sum of the individual molecular properties of the constituent TAUs. That is, we have formulated the estimation of global molecular properties in terms of a GAP model. The global molecular properties can be correlated to macroscopic properties of the system to hopefully yield a QSPR.

A major limitation of GAP models is that the requisite GAP parameters needed to make an estimation of a bulk property are not always available.

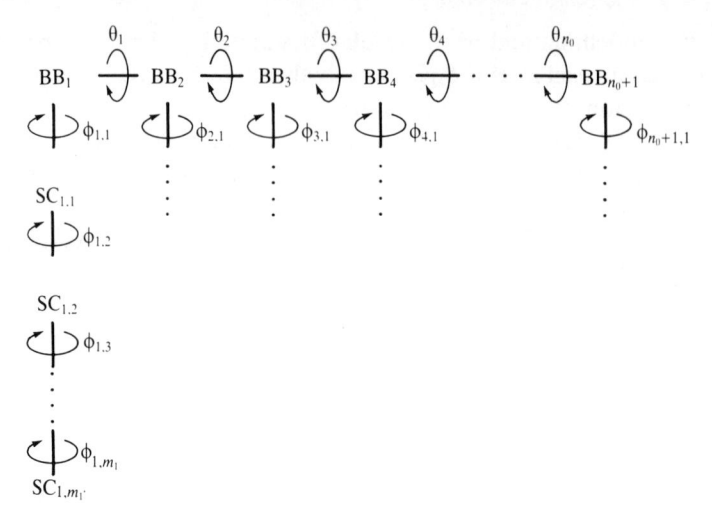

Figure 1.1 Structure of a linear polymer as formulated to define torsion angle units. BB refers to backbone TAUs and SC to sidechain TAUs.

Thus, while the calculation of the global molecular property is straightforward, it cannot be done because of the missing data. However, the formulation of a GAP model using TAUs permits the application of molecular modelling methods to estimate the GAP parameters associated with any TAU. Thus, one only needs to know, or develop, a scheme to compute the requisite group additive properties using molecular modelling to ensure being able to parameterize any TAU. Moreover, once the GAP properties are computed for a TAU they can be stored in a database and used whenever the TAU turns up in a study. Overall, the marriage of molecular modelling and GAP approaches, and their joint application to estimate TAU molecular properties, provides an open-ended, general approach to estimate bulk polymer properties.

It should be pointed out that the TAU described and used in this chapter is one-dimensional (one degree of conformational freedom) in torsion angle, θ_i, when bond lengths and angles are held constant. This representation of the TAU is not a restriction. Higher dimensional TAUs can be defined if needed. For example,

$$SG_{i-1} \xrightarrow{\theta_i} SG_i \xrightarrow{\theta_{i+1}} SG_{i+1}$$

defines a two-dimensional TAU. Obviously, the computational effort needed to compute the molecular properties increases with increasing torsion angle dimensionality. However, the coupling of structural units composing the polymer chain also increases which should enhance the quality of the estimation of molecular properties. The polymer chain can also be decomposed into arbitrary sequences of TAUs of different torsion angle dimensions.

In Figure 1.1 the G_is are abbreviated as BB_k or $SC_{k,j}$, which indicates whether the structural group is part of the backbone or a sidechain, respectively. That is, within the notation of Fig. 1.1

$$-BB_2 \overset{\theta_2}{\frown} BB_3-$$

is a representative backbone torsion angle unit, and

$$-SC_{1,1} \overset{\phi_{1,1}}{\frown} SC_{1,2}-$$

is a typical sidechain torsion angle unit. It is perhaps more informative to give an example of how the torsion angle units are derived from a monomer unit. Suppose

[monomer structure diagram]

is the monomer used in the example. Then the set of backbone torsion angle units is

[backbone torsion angle units diagram with $\theta_1, \theta_2, \theta_3, \theta_4, \theta_5, \theta_6$]

The only set of sidechain torsion angle units is

[sidechain torsion angle units diagram with $\phi_{3,0}, \phi_{3,1}, \phi_{3,2}$]

In the example above, a nine-dimensional problem, in terms of torsion angles, is reduced to nine one-dimensional problems by application of TAU theory. Of course this spatial decoupling process, which is what TAU theory achieves, is paid for at the expense of the accuracy and reliability of predicting the corresponding physicochemical properties of the polymer system. It is reiterated that loss in accuracy and reliability in estimating physicochemical properties can be 'controlled' by selection of TAUs of a higher torsion angle dimension than one.

1.2.2 Advantages of TAU theory in predictive polymer modelling

The one-dimensional TAU units, which will be the only TAUs discussed hereafter, represent relatively simple structures to use in both 3DS modelling and/or molecular simulation studies. In being able to make this statement TAU theory intrinsically overcomes two major drawbacks of GAP models, and one limitation of full-scale molecular simulation, to polymer property predictions. First, physicochemical molecular properties can be determined by prescribed molecular modelling schemes, described below, for any TAU. Thus, there is no parameterization limitation in TAU theory as is the case in GAP models. TAU theory provides an open-ended means to compute requisite physicochemical properties (the parameters).

Since TAU theory involves explicit three-dimensional geometries, that can be as complex as the user selects in terms of numbers of torsion angles, many of the non-additive properties inherent to polymer systems can be taken into account. Thus, TAU theory can overcome some of the shortcomings of GAP formalisms implicit to the additivity assumption.

The overall structural simplicity of TAUs makes the calculation of associated physicochemical properties quite rapid and straightforward. Thus, there are not the computational limitations inherent to molecular simulations of samples of polymer systems.

TAU theory does provide structure–property models that have more insight than GAP formalisms in terms of understanding mechanisms of bulk property behaviour. These TAU structure–property models are generally expressed as QSPRs. The QSPRs are derived from performing multi-dimensional regression analysis on a training set of polymers in which a bulk polymer property of interest is correlated against one, or more, TAU physicochemical properties. This is discussed in the next section. It is not clear at this time in the evaluation of polymer modelling to what extent the mechanistic information of a QSPR has as much insight or meaning compared to that possible from molecular simulations.

1.3 Estimation of intramolecular TAU physicochemical properties

1.3.1 Scaling intramolecular TAU calculations

All TAU calculations must be scaled with respect to the structure of the polymer. Only treatment of homopolymers has been developed to date. The SRU of the polymer, which is normally the monomer unit, serves to define the set of TAU units associated with a polymer chain. The TAUs are divided, as noted earlier, into backbone and sidechain TAUs. Backbone torsion angles of the ith BB–TAU are denoted by θ_i, while sidechain torsion angles of the SC–TAU are given as $\phi_{i,j}$, where j stands for the jth torsion angle of the sidechain attached to the ith backbone TAU. Using this notation the scaled backbone physicochemical TAU property for the entire SRU, F_{BB}, is given by

$$F_{BB} = \frac{\sum_{i=1}^{N_{BB}} f_{BB}(\theta_i)}{N_{BB}} \quad [4]$$

where $f_{BB}(\theta_i)$ is the physicochemical property for the ith backbone TAU and N_{BB} is the number of BB–TAU in the SRU.

In like fashion to Eqn [4] the physicochemical property of the ith sidechain for the monomer $F_{SC}(i)$ is given by

$$F_{SC}(i) = \frac{\sum_{i=1}^{N_{SC}(i)} f_{SC}(\phi_{i,j})}{N_{SC}(i)} \quad [5]$$

where $f_{SC}(\phi_{i,j})$ is analogous to $f_{BB}(\theta_i)$ and $N_{SC}(i)$ corresponds to N_{BB} in Eqn [4]. The total sidechain contribution for the monomer, F_{SC}, can be directly computed as

$$F_{SC} = \frac{\sum_{i=1}^{N_{BB}} F_{SC}(i)}{N_{BB}} \quad [6]$$

Three important components of TAU theory, expressed by Eqns [4]–[6], are (a) the set of TAU is defined by the SRU; (b) each physicochemical property is normalized/scaled against the number of backbone and/or sidechain torsion angles; and (c) physicochemical properties can be partitioned between backbone and sidechain structures as well as between individual sidechains of an SRU. TAU theory does not normalize/scale against SRU molecular weight, as does the VanKrevelen formalism.[7] Molecular weight within TAU theory, is, as discussed below, treated as a physicochemical property.

1.3.2 Conformational entropy, $s_{BB}(\theta_i)$ and $s_{SC}(\phi_{ij})$

The principal property used to quantitatively estimate intramolecular flexibility is the conformational entropy associated with TAUs. The estimation of $s_{BB}(\theta_i)$ is given by

$$Q \cong \sum_1 \exp(-E_1(\theta)/RT)\,\Delta\theta \qquad [7]$$

$$P_1 = \exp(-E_1(\theta)/RT)\,\Delta\theta/Q \qquad [8]$$

$$s_{BB}(\theta_i) = -R \sum_1 P_1 \ln P_1 \qquad [9]$$

where $E_1(\theta)$ is the conformational energy of the lth conformer state of the TAU generated by systematically changing θ in $\Delta\theta$ increments. That is, a one-dimensional conformational scan in θ is performed at $\Delta\theta$ resolution.[14] RT is the Boltzmann factor, with T normally set to 298 K.

Fixed valence geometry molecular mechanics (MM)[19] has been used to compute the conformational energies. In this approach $E_1(\theta)$ is given by

$$E_1(\theta) = \sum_u \sum_v \left[\frac{-A_{uv}}{r_{uv}^6(\theta)} + \frac{B_{uv}}{r_{uv}^{12}(\theta)} + \frac{\alpha Q_u Q_v}{\varepsilon r_{uv}(\theta)} \right] + T(\theta) + H_b(\theta) \qquad [10]$$

where u and v are members of $\geq 1,4$ atom-pair interactions, A_{uv} the atom-pair dispersion attraction coefficient, B_{uv} the atom-pair steric repulsion coefficient, Q_w the partial atomic charge on atom w, ε the molecular dielectric constant, set equal to 3.5 in this work,[14] and $r_{uv}(\theta)$ the distance between atom-pair u, v for θ. $T(\theta)$ is the intrinsic torsional barrier potential for the torsion unit bond, and $H_b(\theta)$ is a hydrogen-bonding potential, if appropriate.[14] The A_{uv}, B_{uv}, $T(\theta)$ $H_b(\theta)$ were taken from the sets proposed by Hopfinger,[14] and the Q_w were computed using CNDO/2[20] on a sterically allowable conformer state of the TAU. An analogous scheme is used to estimate $s_{SC}(\phi_{i,j})$.

The $E_1(\theta)$ and, therefore, $s_{BB}(\theta_i)$ are dependent upon the conformational potential energy function representation. This is also the case for $E_1(\phi_{ij})$ and $s_{SC}(\phi_{i,j})$. However, this dependence is relatively constant among reasonable sets of conformational potential energy functions. Moreover, scaling can be used to put the $s_{BB}(\theta_i)$ and $s_{SC}(\phi_{i,j})$ on the same energy base even if different potential functions are used. In the most direct scaling form, suppose $s_{BB}^A(\theta_i)$ is calculated using potential set A for torsion angle unit i, but we want $s_{BB}^B(\theta_i)$ using potential set B for i. Then,

$$s_{BB}^B(\theta_i) = \left[\frac{s_{BB}^B(\theta_r)}{s_{BB}^A(\theta_r)} \right] s_{BB}^A(\theta_i) \qquad [11]$$

where $s_{BB}^B(\theta_r)$ and $s_{BB}^A(\theta_r)$ are the entropies of some common reference TAU computed using the B and A potential sets, respectively. The ratio, $s_{BB}^B(\theta_r)/s_{BB}^A(\theta_r)$, can be used as a general conversion constant. Thus, the conformational entropy parameters become relatively independent of choice of potential function set. The principal reason that simple linear scaling works in this application is that although the absolute depth of conformational energy minima vary among potential function sets, the relative shapes of these energy wells are usually similar. The available set of TAU entropies are given as part of Table 1.1.

1.3.3 TAU masses

The mass of the TAU, and as a result the masses of the backbone and sidechain structures of the SRU, are used as explicit physicochemical properties in TAU theory. The TAU masses, $m_{BB}(\theta_i)$-backbone and $m_{SC}(\phi_{i,j})$-sidechain, are half the actual masses of the two backbone structural groups, one backbone and one sidechain structural group, or two sidechain structural groups composing the TAU. The word 'actual' is used because some of the TAUs can include an extra atom in one or both of the structural groups used to compute, for example, the entropy. The mass of the first backbone TAU in the SRU example given above,

$$O-\underset{H}{\overset{F}{C_1}} \overset{\theta_1}{-} \underset{H}{\overset{H}{C_2}}-C$$

should not include the left oxygen or the right carbon. That is, the mass of this TAU is

$$m_{BB}(\theta_1) = \tfrac{1}{2}[M(C_1) + M(F) + M(C_2) + 3M(H)] = 22$$

The oxygen and carbon are needed to adequately express the entropic behaviour of θ_1, but are not actual members of the TAU. On the other hand, for the third TAU of the SRU,

$$C-\underset{\underset{C}{|}}{\overset{H}{C_3}}\overset{\theta_3}{-}\bigcirc$$

the two carbons bonded to C_3 are deleted in the mass calculation, but the ring carbons and four of its protons are included in the estimation of the TAU mass. One of the carbons bonded to C_3 belongs to the second TAU and the other to the first sidechain TAU off C_3. A rule to use on deciding what terminal atoms to delete in computing a mass from a TAU that has been used to compute entropy is to see if the atom in question completes a single bond adjacent to the bond connecting the two structural groups forming the TAU. If this is the case, that atom should be deleted in estimating mass. The reason for using one-half the mass of each TAU is that each structural group of each TAU is used twice in defining the set of TAUs. The common structure group between adjacent TAUs establishes the group coupling inherent to this model. Actually, the results are not too sensitive to whether or not the 'actual' is half the mass

Table 1.1 TAUs and corresponding masses, m, entropies, s, and probe energies $\langle E_D \rangle - (CH_3)$, $\langle E_+ \rangle - (H^+)$ and $\langle E_- \rangle - (O^-)$.

TAU $-(X-\overset{\frown}{\underset{\smile}{}}-CH_2)-$ and $-[CH_2-\overset{\frown}{\underset{\smile}{}}-X]-$	m (a.m.u.)	s (cal K^{-1} mol^{-1})	$\langle E_D \rangle$ (kcal mol^{-1})	$\langle E_+ \rangle$ (kcal mol^{-1})	$\langle E_- \rangle$ (kcal mol^{-1})
$X = -CH_2-$	14.0	4.30	−1.39	0.45	−0.26
$-O-$	15.0	3.33	−0.86	−1.53	0.69
$-\overset{O}{\underset{\|}{C}}-$	21.0	0.78	−1.66	−2.19	−0.36
$-\overset{O}{\underset{\|}{C}}-O-$	29.0	2.17	−2.08	−2.69	−0.49
$-O-\overset{O}{\underset{\|}{C}}-$	29.0	2.17	−2.16	−2.55	0.30
$-S-$	23.0	2.28	−1.89	−0.89	−0.42
$-SO_2-$	39.0	0.68	−2.75	−3.25	−1.08
$-\overset{O}{\underset{\|}{C}}-NH-$	28.5	1.16	−2.16	−4.45	−4.16
$-NH-\overset{O}{\underset{\|}{C}}-$	28.5	1.30	−2.16	−4.22	−4.39
$-\langle O \rangle-$ (para)	45.0	0.67	−3.44	−0.96	−0.48
$-\langle O \rangle-$ (meta)	45.0	0.76	−3.18	−0.88	−0.51
$-CH(OH)-$	22.0	1.48	−1.83	−0.80	−1.88
$-CHF-$	24.0	1.69	−1.52	−0.81	−0.38
$-CHCl-$	31.0	0.80	−1.93	−0.90	−0.46
$-CF_2-$	32.5	1.48	−1.69	−1.27	−0.16
$-CCl_2-$	48.0	0.84	−3.00	−1.19	−0.20
$-CFCl-$	39.5	0.90	−2.20	−1.36	−0.38
$X = CH(R)$					
$R = -\overset{O}{\underset{\|}{C}}-O-$	23.6	2.38	−1.83	−2.93	0.66
$-CH_3$	14.0	3.06	−1.51	0.51	−0.37
$-CH_2-$	12.0	2.92	−1.44	0.46	−0.29
$-O-$	14.5	3.29	−0.89	−1.72	0.83
$-\overset{O}{\underset{\|}{C}}-$	18.3	1.17	−1.57	−2.33	−0.46
$-\langle O \rangle-$ (para)	34.3	1.00	−3.26	−1.09	−0.59
$-\langle O \rangle-$ (meta)	34.3	1.03	−3.31	−0.69	−0.45

Table 1.1 (continued)

TAU $-(X \frown CH_2)-$ and $-[CH_2 \frown X]-$	m (a.m.u.)	s (cal K^{-1} mol^{-1})	$\langle E_D \rangle$ (kcal mol^{-1})	$\langle E_+ \rangle$ (kcal mol^{-1})	$\langle E_- \rangle$ (kcal mol^{-1})
$R = -S-$	19.7	3.41	−1.75	−0.77	−0.49
$-SO_2-$	30.3	0.57	−2.58	−3.46	−1.37
$-\underset{\underset{O}{\parallel}}{C}-NH-$	23.3	1.46	−2.03	−2.67	−2.05
$-NH-\underset{\underset{O}{\parallel}}{C}-$	23.3	2.03	−1.95	−1.87	−2.43
$X = {>}C_bH \frown \underset{\underset{O}{\parallel}}{C}-$ (a)	20.5	0.68	−1.60	−1.39	−0.44
${>}C_bH \frown O-$	14.5	2.17	−0.92	−1.43	−0.67
${>}C_bH \frown \underset{\underset{O}{\parallel}}{C}-O-$	28.5	0.83	−1.83	−1.59	−0.27
${>}C_bH \frown O-\underset{\underset{O}{\parallel}}{C}-$	28.5	0.96	−1.70	−1.45	−0.56
${>}C_bH \frown CH_2-$	13.5	1.45	−1.26	+0.29	−0.40
${>}C_b(CH_3) \frown \underset{\underset{O}{\parallel}}{C}-O-$	27.5	0.40	−1.72	−2.89	−0.72
$-O \frown CH_3$ (b)	15.5	4.72	−0.95	−1.43	0.46
$CH_2 \frown CH_3$	14.5	4.33	−1.30	0.40	−0.31
$-\underset{\underset{O}{\parallel}}{C}-O \frown CH_3$	29.5	3.00	−2.13	−2.51	0.35
$-CH_2 \frown C(CH_3)X-$ and $-C(CH_3)X \frown CH_2-$					
$X = -CH_3$	28.0	1.22	−2.50	+0.77	−0.66
$-\underset{\underset{O}{\parallel}}{C}-O-$	28.3	0.76	−2.83	−1.89	−0.52
$-\underset{\underset{O}{\parallel}}{C}-$	23.3	1.17	−2.59	−1.58	−0.63

(continued)

Table 1.1 (continued)

TAU and [structures with X]	m (a.m.u.)	s (cal K⁻¹ mol⁻¹)	$\langle E_D \rangle$ (kcal mol⁻¹)	$\langle E_+ \rangle$ (kcal mol⁻¹)	$\langle E_- \rangle$ (kcal mol⁻¹)
X = —O—	46.0	0.65	−2.89	−1.09	0.17
—C(=O)—	52.0	0.39	−3.51	−2.69	−1.05
—O—C(=O)—O—	68.0	0.48	−4.09	−3.05	−0.23
—C(CH₃)₂—	59.0	0.57	−4.16	+0.53	−0.82
—C(=O)—C(=O)—	60.0	0.58	−3.78	−2.86	−0.65
—C(=O)—O—	60.0	0.58	−3.46	−2.98	−0.79
—CH(F)⌒CH(F)—	32.0	1.06	−1.36	−1.03	−0.52
—CF₂⌒CF(Cl)—	58.0	1.12	−2.18	−1.92	−1.08
⌒NH—C(=O)— (phenyl)	59.5	0.66	−4.08	−2.05	−2.54
C(=O)—NH— (phenyl)	59.5	0.60	−3.89	−2.60	−2.11
(phenyl)⌒NH—C(=O)—	59.5	0.58	−4.05	−2.07	−2.39
(phenyl)⌒C(=O)—NH—	59.5	0.53	−3.95	−2.62	−2.08
C(=O)—O⌒C(CH₃)₃	50.5	1.37	−3.52	−1.06	+0.28
CF₂⌒CF₂	31.0	0.91	−1.96	−2.13	−0.80

(a) These torsion angle units define backbone–sidechain branching units.
(b) These torsion angle units terminate sidechains.

of the TAU or if extra 'end' atoms are systematically attached to the structure groups composing TAUs used in the entropy calculations. We make the rules primarily for conceptual accuracy. A set of TAU masses is given in Table 1.1.

1.3.4 TAU principal moments of inertia

Mass-dependent physicochemical TAU properties complement the conformational entropies in characterizing bulk polymer transition behaviour. Entropy describes chain flexibility while mass-related properties take into account momentum features of chain-segments. Simple masses of the TAUs may not incorporate sufficient information regarding the behaviour of connected chain-segments acting through space. Thus, the principal moments of inertia, which include both geometric and mass components, of the TAU are put forward as an additional descriptor in developing TAU–QSPRs.

Principal moments of inertia, $(e^x_{BB}(\theta_i), e^y_{BB}(\theta_i), e^z_{BB}(\theta_i), e^t_{BB}(\theta_i))$-backbone and $(e^x_{SC}(\phi_{i,j}), e^y_{SC}(\phi_{i,j}), e^z_{SC}(\phi_{i,j}), e^t_{SC}(\phi_{i,j}))$-sidechain are dependent on TAU conformation. At this stage of development of TAU theory, the global minimum energy conformation of the TAU is used as the geometry for computing the principal moments of inertia. The global minimum energy conformation is determined using Eqn [10] and minimizing $E_1(\theta)$ for the lth TAU torsion angle. Table 1.2A contains the principal moments of inertia for a limited set of TAU.

Table 1.2A Principal moments of inertia of some TAU in their respective global minimum conformations. The X-direction corresponds to the direction of the G_i—G_{i+1} bond. Units are $(AMU - Å^2)$.

G_i—G_{i+1}	e^x	e^y	e^z
CH_2—CH_2	1.1	4.2	4.2
CH_2—O	0.7	3.4	3.4
CHCl—CH_2	2.7	15.5	17.1
CH_2—(C=O)—O	7.4	9.0	15.9
CH_2—(phenyl)	15.3	32.9	47.7
(phenyl)—(phenyl)	29.0	159.0	173.5
CH_2—(C=O)—(N—H)	7.4	9.1	15.9
CHF—CH_2	2.0	9.6	11.7
CH_2—CHMe	2.3	13.1	14.7
CH_2—(N—H)—(C=O)	6.7	10.9	17.3

Table 1.2B Dipole moment components and total dipole, in debyes, for some TAUs. The X-direction corresponds to the direction of the G_i—G_{i+1} bond.

G_i—G_{i+1}	ρ^x	ρ^y	ρ^z	ρ^t
CH_2—CH_2	0	0	0	0
CH_2—O	1.2	−1.2	−0.9	1.9
CHCl—CH_2	1.7	0	−0.7	1.8
CH_2—(C=O)—O	0.9	−0.9	−0.3	1.3
CH_2—⟨phenyl⟩	0.1	0.0	−0.2	0.2
⟨biphenyl⟩	0	0	0	0
CH_2—(C=O)·(N—H)	−3.7	0	−1.3	3.9

1.3.5 TAU dipole moments

The dipole moments, $(\rho^x_{BB}(\theta_i), \rho^y_{BB}(\theta_i), \rho^z_{BB}(\theta_i), \rho^t_{BB}(\theta_i))$-backbone and $(\rho^x_{SC}(\phi_{i,j}), \rho^y_{SC}(\phi_{i,j}), \rho^z_{SC}(\phi_{i,j}), \rho^t_{SC}(\phi_{i,j}))$-sidechain, are also computed using the global minimum energy conformation of the TAU. We have used the CNDO/2 method to compute the dipole moments which are given in Table 1.2B for a few TAUs.

By convention, TAU moment properties are defined in space such that the positive x-direction is along the torsion angle bond of the TAU and in the G_i to G_{i+1} direction,

$$G_i \overset{\theta_i}{\frown} G_{i+1} \rightarrow x(+)$$

Thus, $\rho^x_{BB}(\theta_i)$, for example, is the component of the dipole of the ith backbone TAU in the direction of the torsion angle bond. The superscript t in the dipole and principal moment of inertia descriptors refers to the total moment as given by way of example for the ith backbone dipole TAU,

$$\rho^t_{BB}(\theta_i) = \sqrt{[\rho^x_{BB}(\theta_i)]^2 + [\rho^y_{BB}(\theta_i)]^2 + [\rho^z_{BB}(\theta_i)]^2} \qquad [12]$$

Clearly, moment TAU properties are not additive since they are highly dependent upon molecular geometry.

1.3.6 Absolute global minimum conformational energy

Another physicochemical molecular property of a TAU, which can be determined during the estimation of the conformational entropy, is the global minimum conformational energy, $v^o_{BB}(\theta_i)$-backbone, and $v^o_{SC}(\phi_{i,j})$-sidechain of a TAU. One interpretation of the $v^o_{BB}(\theta_i)$ and $v^o_{SC}(\phi_{i,j})$ of a SRU is in terms of intramolecular cohesive energy density. That is, these physicochemical molecular properties reflect the intrinsic conformational stability of a polymer chain.

1.3.7 Second derivatives of global minimum energy conformations

The second derivatives about the global minimum energy conformation of a TAU, $v''_{BB}(\theta_i)$-backbone and $v''_{SC}(\phi_{i,j})$-sidechain, are measures of how easy/difficult it is to distort the TAU from its preferred conformation. Thus, these physicochemical molecular properties might be used to establish QSPRs related to bulk moduli properties. The $v''_{BB}(\theta_i)$ properties can be combined with the $v^o_{BB}(\theta_i)$ to yield TAU deformation energies

$$\Delta v^o_{BB}(\Delta \theta_i) = v^o_{BB}(\theta_i) + v''_{BB}(\theta_i) \Delta \theta_i^2 \qquad [13]$$

These deformation energies might also be used to construct QSPRs involving bulk moduli and/or deformation properties. Obviously, Eqn [13] is for backbone TAU and equivalent expressions can be written for sidechain TAUs.

1.4 Estimation of intermolecular TAU physicochemical properties

1.4.1 Scaling intermolecular TAU calculations

The intermolecular TAU physicochemical properties are also scaled against the number of backbone and/or sidechain TAUs in the SRU. In the case of explicit TAU–TAU pair interactions, the intramolecular scaling base is preserved by simply dividing the total TAU–TAU pair interaction property measure by two, and assigning each TAU half the measured value.

1.4.2 Intermolecular probe energies

The geometry for computing intermolecular probe energetics of a TAU is shown on Fig. 1.2.[18] The TAU is assumed to be in its global minimum energy conformation in order to facilitate the calculation. An intermolecular probe unit, p, is placed a perpendicular distance d^* from the geometric centre of the bond, C_i, between the two structure groups, G_i and G_{i+1}. The angular location of the probe relative to C_i is equivalent to θ_i and is initially taken to be arbitrary. However, d^* always corresponds to the minimum distance from C_i such that no steric violations occur between the probe and the atoms of the TAU. Once this distance, d^*, is identified by systematically increasing the radial distance, d, from C_i at

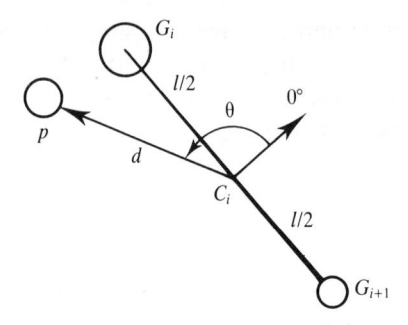

Figure 1.2 Geometry used to compute the intermolecular probe energies of TAUs. See text for definition of symbols.

Δd increments, the energy between the probe and the TAU is minimized with respect to d starting at d^*. The probe and the torsion angle unit energetics is taken to be the sum of the non-bonded steric and electrostatic components of a molecular mechanics force field as given by Eqn [14].

$$E_p(\theta_i, d) = \sum_{i=1}^{N} \left[\left(\frac{-A_{ip}}{r_{ip}^6} + \frac{B_{ip}}{r_{ip}^{12}} \right) + \left(\frac{Q_i Q_p}{\varepsilon r_{ip}} \right) \right] \quad [14]$$

In Eqn [14], θ_i is fixed, and $E(\theta_i, d)$ is minimized to $d_{\min}$ as a function of d. The first two terms in parentheses in Eqn [14] constitute the non-bonded steric potential function with the A_{ip} and B_{ip} parameters taken from the set proposed by Hopfinger. The last term in parentheses in Eqn [14] is the electrostatic potential with Q_x the partial charges on the atoms and probe. The Q_i were computed from CNDO/2[20] calculations. The molecular dielectric, ε, was set equal to 3.5.[14] Equation [14] is similar to Eqn [10] used to estimate the TAU conformational energy. However, Eqn [14] only involves interactions between the probe unit and TAU atoms, and hydrogen bonding and torsion potential terms are absent.

Equation [14] does not necessarily give proper accounting to intermolecular hydrogen bonding. However, the electrostatic term does very much dominate the favourable intermolecular energetics when atom i is a hydrogen bond donor/acceptor and the probe, p, is an acceptor/donor.

When $d_{\min}$, and correspondingly, $E_p(\theta_i, d_{\min})$, are determined for a given θ_i, θ_i is increased by $\Delta\theta_i$, and the energy minimization of $E_p(\theta_i, d_{\min})$ is repeated as a function of d. This cycle, in turn, is carried out over the $360°$ available to θ_i. The resulting set of $E_p(\theta_i, d_{\min})$ is then used to compute the average intermolecular energy, $\langle E_p \rangle_i$, for p which is taken to be a physicochemical property of the TAU.

$$\langle E_p \rangle_i = \frac{\sum_{j=1}^{N} E_p(\theta_i, d_{\min})_j \exp(-E_p(\Theta_i, d_{\min})_j / RT)}{\sum_{j=1}^{N} \exp(-E_p(\Theta_i, d_{\min})_j / RT)} \quad [15]$$

In the analyses reported here, $\Delta\theta_i = 30°$ so that $N = 12$ and T was set to room temperature, 298 K. In principle, the T in Eqn [15] should be treated as a variable in estimating bulk thermodynamic properties like T_g and T_m. However, conformational energy is independent of temperature for the fixed valence geometry approximation applied in this model. Moreover, since RT appears in the same functional form in the numerator and denominator of Eqn [15] it has a very modest effect upon $\langle E_p \rangle_i$. Lastly, the $\langle E_p \rangle_i$ are used as correlation, as opposed to thermodynamic, properties. The $\langle E_p \rangle_i$ are designed to reflect relative composite average values of intermolecular interactions. Thus, there is not a requirement for computational precision of these quantities, with respect to temperature, in order to yield significant QSPRs.

The major difficulty in using the formalism reported here for estimating intermolecular probe energies is the selection of a probe, p, which meaningfully reflects the dominant type of intermolecular interactions at play in the polymer system of interest. There does not seem to be any definite solution to this problem. Thus, it has been necessary to take advantage of the statistical nature of constructing a QSPR. This is the same approach that was successfully employed in developing quantitative structure–activity relationships, QSARs, using potential energy fields for a set of 2,4-diaminobenzylpyrimidine inhibitors of dihydrofolate reductase.[21] A probe to reflect non-bonded/dispersion interactions, a unified representation of a CH_3 group,[22] has been used in the intermolecular energy calculations. Probes to reflect electrostatic interactions with a positively charged group, a hydrogen atom with a unit positive charge, and an oxygen with a negative formal charge, have also been considered. The average interaction probe energies $\langle E_D \rangle$, $\langle E_+ \rangle$ and $\langle E_- \rangle$, respectively, of a TAU with these three probes have been determined and are reported in Table 1.1.

1.4.3 Explicit intermolecular TAU–TAU physicochemical properties

The geometry used to compute the TAU–TAU interactions is shown in Fig. 1.3. In Fig. 1.3 A_i is the G_i of one TAU and B_j the G_j of another TAU, and l_A, l_B and c_A, c_B are the same l and c, respectively, in Fig. 1.2. θ_i, chosen arbitrarily over θ_j, is identical to the TAU torsion angle, but refers to rotation of TAU–B about TAU–A. The inter TAU–TAU space, in terms of d and θ, is explored in the identical manner as done in the probe scheme described above (1.4.2). However, there are two additional geometric constraints followed by one degree of geometric freedom and relaxation. TAU–A and TAU–B are each frozen into their respective global intramolecular minimum energy conformations. In addition, the

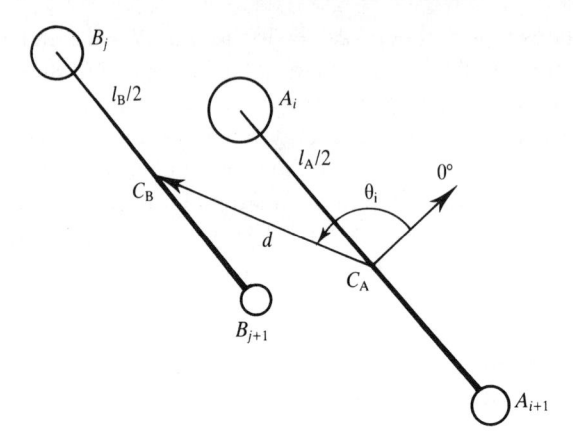

Figure 1.3 The geometry of the inter TAU–A–TAU–B energy calculations.

TAU bond vectors, l_A and l_B, are held parallel to one another over the scan of (θ, d)-space. However, after the set of apparent inter TAU–TAU energy minima have been identified, each apparent minimum, within 5 kcal/mole of the apparent global minimum, is then rigorously minimized with respect to the six degrees of intermolecular freedom between the TAUs. The intermolecular potential energy between TAU–A and TAU–B is computed using Eqn [10], less $T(\theta_i)$ and $H(\theta_i)$, and having u and v span the atoms of TAU–A and TAU–B. From this analysis of the TAU–A and TAU–B interaction the following molecular physicochemical properties can be determined:

$v^\circ(AB)$ the potential energy of inter TAU–A–TAU–B global minimum.

$d^\circ(AB)$ the inter TAU–A–TAU–B distance for the global energy minimum.

$\dfrac{\partial^2 v^\circ(AB)}{\partial d^\circ(AB)^2}$ the 'force constant' for movement from the global inter TAU minimum in the d direction.

l_{AB} the length of the larger of TAU–A or TAU–B in the direction of the TAU torsion angle bond.

$\dfrac{\partial^2 v^\circ(AB)}{\partial l_{AB}^2}$ the force constant for movement of TAU–B, relative to TAU–A, in the direction of l_{AB}.

These molecular physicochemical properties for some TAU–TAU pairs are given in Table 1.2C.

Table 1.2C Inter TAU-A–TAU-B molecular physicochemical properties. See text for definition of properties.

$(A_1 - A_2)$	$(B_1 - B_2)$	$v^o(\mathbf{AB})$ (kcal/mole)	$d^o(\mathbf{AB})$ (Å)	$\dfrac{\partial^2 v^o(\mathbf{AB})}{\partial d^2(\mathbf{AB})}$ kcal/mole Å²	l_{AB} Å	$\dfrac{\partial^2 v^o(\mathbf{AB})}{\partial l^2_{AB}}$ kcal/mole Å²
CH_2-CH_2	CH_2-CH_2	−1.67	4.6	0.25	1.3	0.30
CH_2-O	CH_2-O	−1.93	4.4	0.50	1.1	0.90
CH_2-CHF	CH_2-CHF	−1.89	4.7	0.45	1.3	0.55
CH_2-CHCl	CH_2-CHCl	−2.19	5.0	0.65	1.3	1.00
CH_2-CONH	CH_2-CONH	−5.62	4.2	3.15	2.3	1.45
⌬—CH_2	⌬—CH_2	−3.05	4.5	0.60	4.2	0.60
⌬—CONH	⌬—CONH	−5.80	4.3	3.20	5.1	1.30
⌬—NH—CO	⌬—NH—CO	−5.89	4.3	3.00	4.9	1.35
CH_2—NH—CO	CH_2—NH—CO	−5.50	4.2	3.25	2.3	1.45
CH_2—CHMe	CH_2—CHMe	−1.90	4.9	0.60	1.3	0.85
CH_2—CH⌬	CH_2—CH⌬	−3.27	5.2	0.75	1.3	1.05

(continued)

Table 1.2C (continued)

$(A_1 - A_2)$	$(B_1 - B_2)$	$v^o(AB)$ (kcal/mole)	$d^o(AB)$ (Å)	$\dfrac{\partial^2 v^o(AB)}{\partial d^2(AB)}$ kcal/mole Å²	l_{AB} Å	$\dfrac{\partial^2 v^o(AB)}{\partial l_{AB}^2}$ kcal/mole Å²
CH_2-CMe_2	CH_2-CMe_2	−2.39	5.1	0.65	1.3	1.10
CH_2-CH_2	CH_2-O	−1.85	4.5	0.30	1.3	0.55
CH_2-CH_2	NHCO—C$_6$H$_5$	−2.48	4.9	0.40	4.9	0.40
CH_2-CH_2	CH_2-CHMe	−1.85	4.7	0.50	1.3	0.45
NHCO—C$_6$H$_5$	$CONH-CH_2$	−5.70	4.3	3.10	5.1	1.05
NHCO—C$_6$H$_5$	CH_2-CMe_2	−1.98	4.8	0.45	5.0	0.35
CH_2-CHMe	CH_2-CMe_2	−2.10	5.2	0.40	1.3	0.35
$CONH-CH_2$	CH_2-CMe_2	−2.25	4.8	0.45	2.3	0.40
$CONH-CH_2$	CH_2-CHMe	−1.97	4.7	0.45	2.3	0.40
CH_2-CH_2	CH_2-NHCO	−2.15	4.2	0.40	2.3	0.35
CH_2-CH_2	$NHCO-CH_2$	−2.22	4.2	0.40	2.3	0.35
CH_2-CH_2	CH_2-CONH	−2.10	4.2	0.40	2.3	0.35
CH_2-CH_2	$CONH-CH_2$	−2.12	4.2	0.40	2.3	0.35
CH_2-CONH	$CONH-CH_2$	−4.35	4.5	1.76	2.3	0.80

1.5 Results of applying TAU theory

1.5.1 Estimation of the glass transition temperature, T_g

Table 1.3 contains one of the sets of polymers investigated[7] along with the observed T_gs, conformational entropies, masses and the three average intermolecular probe energies. This set of polymers was taken from the work of VanKrevelen[7] in order to evaluate how well TAU theory could be applied to a set of polymers of diverse structure. In addition, it was of interest to see if T_g estimation by TAU theory was as accurate as the VanKrevelen GAP model.

Multidimensional linear least-square regression analysis has been used to construct the correlation equation representing each QSPR.[17,18,23,24] A variety of functional representations of the selected independent variables have been considered for the QSPR. The optimum functional representation, based upon the statistical significance of fit for the regression analyses explored, has been found to contain only linear terms in the independent variables

$$T_x = \alpha S_{BB} + \beta M_{BB} + \sum [\sigma_i S_{SC}(i) + r_i M_{SC}(i)] + \varepsilon \bar{E}_D + \rho \bar{E}_+ + \Delta \bar{E} + \omega \quad [16]$$

In Eqn [16] the subscript x of T_x refers to the application of this equation to construct T_g ($x = g$) and T_m ($x = m$) [the crystal to melt transition temperature] QSPRs. The independent variables S_{BB}, M_{BB}, $S_{SC}(i)$, $\bar{E}_D$, $\bar{E}_+$ and $\bar{E}_-$ are the backbone entropy and mass, the sidechain entropy and the probe energies, respectively, as defined in Eqns [4] and [5], e.g. $S_{BB} = F_{BB}$ in Eqn [4].

In many cases, two observed T_gs are reported for a particular polymer. The T_g value which best fits the QSPR equation was used in each QSPR analysis. In the initial investigation to predict T_g, in which intermolecular energetics were neglected, the following QSPR was established for 30 of the polymers in Table 1.3:

$$\left. \begin{array}{l} T_g = -35.18 S_{BB} + 1.55 M_{BB} - 18.26 S_{SC} + 1.35 M_{SC} + 327.30 \\ N = 30 \quad R = 0.929 \quad SD = 17.9 \quad F = 39.5 \end{array} \right\} \quad [17]$$

In Eqn [17] N is the number of compounds, R the correlation coefficient, SD the standard deviation of fit and F the statistical significance of fit. The entropy terms in Eqn [17] account for more than 70% of the variance in the T_g values. Thus, an initial thought was to replace the non-entropy terms in Eqn [17] with the intermolecular energy probe descriptors. This resulted in the following QSPR equation:

$$\left. \begin{array}{l} T_g = -21.46 S_{BB} - 9.83 S_{SC} - 17.24 \bar{E}_D - 29.97 \bar{E}_+ - 18.12 \bar{E}_- + 263.84 \\ N = 35 \quad R = 0.912 \quad SD = 20.2 \quad F = 28.8 \end{array} \right\} \quad [18]$$

Table 1.3 Polymers, corresponding entropy, mass moment and intermolecular probe energy descriptors, and the observed and calculated (using Eqn [20]) T_g values as well as the differences, ΔT_g, in observed and calculated values.

No.	Polymer	S_{BB} (cal mol^{-1} K^{-1})	M_{BB} (a.m.u.)	S_{SC} (cal mol^{-1} K^{-1})	M_{SC} (a.m.u.)	$\bar{E}_D$ (kcal mol^{-1})	$\bar{E}_+$ (kcal mol^{-1})	$\bar{E}_-$ (kcal mol^{-1})	T_g(obs.) (K)	T_g(calc.) (K)	ΔT_g (K)
1	—CH$_2$—O—	3.33	15.0	—	—	−0.86	−1.53	0.69	188/243	220.9	22.1
2	—(—CH$_2$)$_2$—O—	3.65	14.7	—	—	−1.04	−0.87	0.37	206/246	221.5	−15.5
3	—(—CH$_2$)$_3$—O—	3.82	14.5	—	—	−1.13	−0.54	0.22	195/228	210.6	17.4
4	—(—CH$_2$)$_4$—O—	3.91	14.4	—	—	−1.18	−0.34	0.12	185/194	204.7	−10.7
5	—(—CH$_2$—)—	4.30	14.0	—	—	−1.39	0.45	−0.26	143/250	176.9	−33.9
6	—CH$_2$—CH(CH$_3$)—	1.93	14.0	—	—	−1.51	0.51	−0.37	238/299	249.0	−11.0
7	—CH$_2$—CH(ϕ)—	1.00	34.3	—	—	−3.26	−1.09	−0.59	353/380	334.3	18.7
8	—CH$_2$—CH(F)—	1.70	24.0	—	—	−1.52	−0.81	−0.38	253/314	307.8	6.2
9	—CH$_2$—CF$_2$—	1.48	32.5	—	—	−1.69	−1.27	−0.16	238/286	325.6	−39.6
10	—CH(F)—CH(F)—	1.06	32.0	—	—	−1.36	−1.03	−0.5	323/371	329.4	−6.4
11	—CH$_2$—CH(Cl)—	0.80	31.0	—	—	−1.93	−0.90	−0.46	247/354	341.1	12.9
12	—CF$_2$—CF(Cl)—	1.12	58.0	—	—	−2.18	−1.92	−1.08	318/373	377.0	−4.0
13	—O—C(=O)—⌬—C(=O)—O—(CH$_2$)$_2$—	1.96	38.4	—	—	−2.59	−2.10	−0.22	346	351.2	−5.2
14	—⌬—CH$_2$—⌬—O—C(=O)—⌬—C(=O)—O—	0.58	56.5	—	—	−3.77	−2.00	−0.36	393/420	397.2	−4.2
15	—⌬—C(CH$_3$)$_2$—⌬—O—C(=O)—⌬—C(=O)—O—	0.53	63.5	—	—	−4.13	−1.26	−0.53	414/423	381.1	32.9

#	Structure										
16	—CH₂—C(CH₃)— COOCH₃	0.76	28.3	1.70	28.5	−2.38	−2.29	−0.41	378	353.8	24.2
17	—CH₂—C(CH₃)— COOCH₂CH₃	0.76	28.3	2.29	23.7	−2.15	−1.79	−0.56	338	319.3	18.7
18	—CH₂—C(CH₃)— COO(CH₂)₁₅CH₃	0.76	28.3	3.95	15.7	−1.59	−0.14	−0.34	288	258.6	29.4
19	—CH₂—C(CH₃)— COOC(CH₃)₃	0.76	28.3	0.89	39.0	−2.72	−1.93	−0.42	380	373.3	6.7
20	—CH₂—C(CH₃)— COO—C₆H₅	0.76	28.3	0.49	43.7	−2.71	−2.41	−0.69	385	393.9	−8.9
21	—CH₂—C(CH₃)₂—	1.22	28.0	—	—	−2.50	0.77	−0.66	198/243	272.6	−29.6
22	—CH₂—CH(OH)—	1.48	22.0	—	—	−1.83	−0.80	−1.88	343/372	332.1	10.9
23	—CH₂—CH(C₂H₅)—	2.92	12.0	2.89	14.0	−1.36	0.40	−0.32	228/249	218.2	9.8
24	—CH₂—CH(C₃H₇)—	2.92	12.0	3.36	14.0	−1.37	0.41	−0.31	221/287	213.0	8.0
25	—CH₂—CH(OCH₃)—	3.29	14.5	3.45	15.0	−0.91	−1.58	0.36	242/260	242.2	−0.2
26	—CH₂—CH(OCH₂CH₃)—	3.29	14.5	3.28	14.7	−0.97	−1.20	0.27	231/254	234.2	−3.2
27	—CH₂—CH— COOCH₃	2.38	23.6	1.92	29.0	−1.91	−2.49	0.35	279/282	318.1	−36.1
28	—CH₂—CH— COOCH₂CH₃	2.38	23.6	2.44	24.0	−1.77	−1.95	0.05	251	283.3	−32.3

(*continued*)

Table 1.3 (continued)

No.	Polymer	S_{BB} (cal mol^{-1} K^{-1})	M_{BB} (a.m.u.)	S_{SC} (cal mol^{-1} K^{-1})	M_{SC} (a.m.u.)	$\bar{E}_D$ (kcal mol^{-1})	$\bar{E}_+$ (kcal mol^{-1})	$\bar{E}_-$ (kcal mol^{-1})	T_g(obs.) (K)	T_g(calc.) (K)	ΔT_g (K)
29	—CH$_2$—CH— | COOC(CH$_3$)$_3$	2.38	23.6	1.10	39.5	−2.25	−2.13	0.33	314	305.3	8.7
30	—CH$_2$—CH— | COO(CH$_2$)$_3$CH$_3$	2.38	23.6	3.19	20.0	−1.66	−1.26	−0.04	219	252.1	−33.1
31	—(—CH$_2$)$_{10}$— ...	3.40	23.4	—	—	−1.84	−0.54	−0.31	268/298	240.5	27.5
32	—NH— ...	0.59	59.5	—	—	−3.97	−2.34	−2.25	411/428	438.4	−27.4
33	—NH—(CH$_2$)$_6$—NHC—(CH$_2$)$_4$—	3.27	18.8	—	—	−1.65	−1.16	−1.58	318/330	277.2	40.8
34	—CH$_2$—CH— | C$_6$H$_{15}$	2.92	12.0	3.83	14.0	−1.38	0.43	−0.29	208/228	207.8	0.2
35	—CH$_2$—CH— | O—(CH$_2$)$_5$CH$_3$	2.92	12.0	3.86	14.3	−1.28	0.02	−0.21	196/223	216.9	6.1

The relative magnitudes of the regression coefficients of Eqn [17] suggest that none of the physicochemical molecular properties (descriptors) can be deleted without compromising the integrity of the QSPR. Moreover, a comparison of Eqns [17] and [18] suggests that the intermolecular probe energy terms do not fully compensate for the deletion of the mass terms.

The decision to delete mass descriptors was based upon the marginal statistical significance of these terms in Eqn [17], and not upon a conceptual understanding of the glass transition process. Conceptually, conformational entropy is associated with the thermodynamic component of the glass transition process, while the mass terms represent an attempt to account for the kinetic nature of the glass transition. Equation [18] only contains thermodynamic descriptors, and its limited significance, relative to Eqn [17], indicates that mass terms are needed in the QSPR. Thus, a multidimensional linear regression analysis was carried out using all seven descriptors (S_{BB}, M_{BB}, S_{SC}, M_{SC}, $\bar{E}_D$, $\bar{E}_+$ and $\bar{E}_-$) for the 35 compounds whose T_g values are reported in Table 1.3.

$$T_g = -25.3 S_{BB} - 7.66 S_{SC} + 1.40 M_{BB} - 0.14 M_{SC} - 0.77 \bar{E}_D - 31.6 \bar{E}_+ \\ - 24.4 \bar{E}_- + 275.17 \\ N = 35 \quad R = 0.959 \quad SD = 14.9 \quad F = 44.0$$ [19]

An analysis of the coefficients of the independent variables (physicochemical molecular properties) in Eqn [19] indicates that the normalized mass of the sidechain TAUs, and the intermolecular dispersion energy, make minimal contributions to the variance of T_g over the set of polymers given in Table 1.3. Thus, a multidimensional linear regression analysis was carried out without these two terms. This led to the most significant QSPR for T_g,

$$T_g = -27.3 S_{BB} - 10.1 S_{SC} + 1.07 M_{BB} - 29.3 \bar{E}_+ - 15.1 \bar{E}_- + 288.83 \\ N = 35 \quad R = 0.954 \quad SD = 15.6 \quad F = 58.9$$ [20]

Figure 1.4 contains a plot of observed versus predicted T_gs based upon Eqn [20].

Two families of homologous polymers have also been analysed to predict their respective sets of T_gs within the framework of Eqn [16]. These are the acrylates (I) and methacrylates (II)

$$\begin{array}{cc} -CH_2-\underset{|}{CH}- & -CH_2-\underset{|}{C(CH_3)}- \\ O=C-O-X & O=C-O-X \\ I & II \end{array}$$

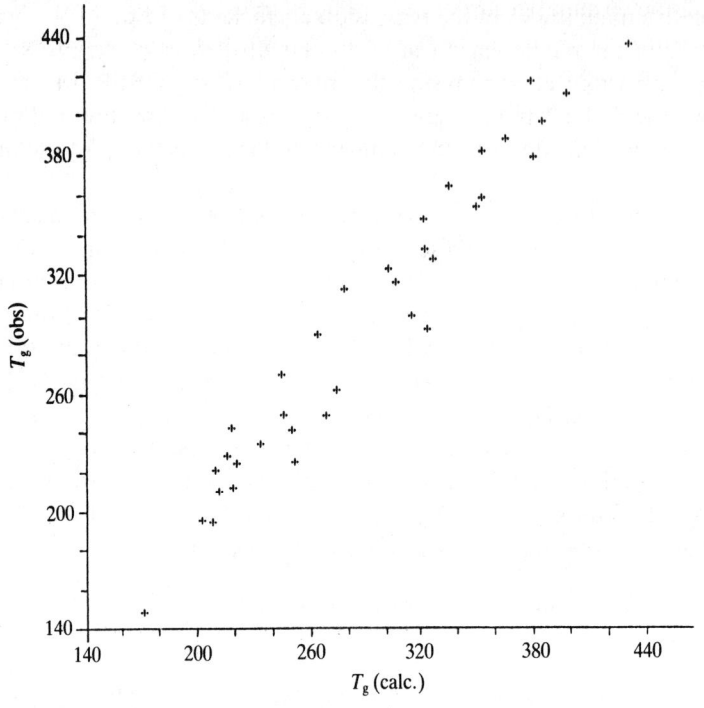

Figure 1.4 Plot of observed versus calculated Eqn [20] T_g values for the polymers in Table 1.3.

Table 1.4. The database of polyacrylate T_gs used to generate Eqn [21], including predicted T_gs and entropy and mass TAU measures.

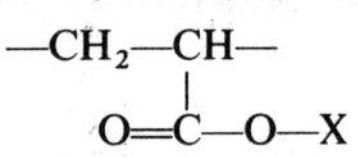

Poly-X-acrylates

—CH$_2$—CH—
         |
     O=C—O—X

X	T_g(obs.)[a] (K)	T_g(pred.) (K)	ΔT_g (K)	S_{SC}	M_{SC}
Methyl	281	272.9	8.1	1.88	29.5
Ethyl	251	225.6	25.4	3.30	21.8
Isopropyl	268	261.2	6.8	2.21	43.5
n-Butyl	219	208.9	10.0	3.80	17.9
iso-Butyl	230	258.6	−28.6	2.31	26.1
s-Butyl	253	245.5	7.5	2.70	26.2
t-Butyl	314	299.5	14.5	1.06	50.5
2-Ethylhexyl	188	221.9	−33.9	3.42	14.3
3,3,5-Trimethylcyclohexyl	288	300.1	−12.1	1.00	84.0
2-Methoxyethyl	224	221.4	2.6	3.43	18.4
2-Ethoxyethyl	223	215.3	7.7	3.61	17.5
2-Butoxyethyl	201	209.0	−8.0	3.80	16.5

[a] Taken from Ref. 25.

in which the terminal group of the sidechain unit has been varied. Unlike the measures of T_g in Table 1.3, the T_gs for both the acrylates and methacrylates have all been measured under identical conditions.[25] Thus, the T_gs in Tables 1.4 and 1.5 should, on average, be more accurate than those in Table 1.3.

A very important simplification is possible for both these classes of polymers within the TAU formalism. Since the backbone structure is constant over each homologous series, and because each TAU is decoupled from all other TAUs in the generation of entropy and mass descriptors, contributions from the backbone in Eqn [16] can be neglected since they are constant over the set of homologues. Moreover, the contributions from the backbone to sidechain TAUs are also constant within each class of polymer. Hence, we need only consider $O{=}\overset{|}{C}{-}O{-}X$ contributions to Eqn [16] for estimating T_g. Obviously, this simplification

Table 1.5. The database of polymethacrylate T_gs used to generate Eqn [22], including predicted T_gs and entropy and mass TAU measures.

	Poly-X-methacrylates $-CH_2-C(CH_3)-$ $O{=}\overset{|}{C}{-}O{-}X$				
X	T_g(obs.)[a] (K)	T_g(pred.) (K)	ΔT_g (K)	S_{SC}	M_{SC}
Methyl	378[b]	359.2	18.8	1.88	29.5
Ethyl	338	287.5	50.5	3.30	21.8
Isopropyl	354	343.8	10.2	2.21	43.5
n-Butyl	293	262.2	30.8	3.80	17.9
iso-Butyl	321	337.4	−16.4	2.31	26.1
s-Butyl	333	317.9	15.1	2.70	26.2
t-Butyl	380	401.9	−21.9	1.06	50.5
n-Hexyl	268	253.6	14.4	3.97	16.6
n-Octyl	253	249.5	3.5	4.05	15.9
Isodecyl	232	256.0	−24.0	3.92	15.7
Lauryl	208	246.0	−38.0	4.12	15.4
Tridecyl	227	244.5	−17.5	4.15	15.2
Stearyl	235	242.9	−7.9	4.18	14.9
Phenyl	385	382.8	2.2	1.42	37.0
Cyclohexyl	377	404.0	−27.0	1.04	63.5
3,3,5-Trimethylcyclohexyl	398	407.7	−9.7	1.00	84.0
Isobornyl	443	417.8	25.2	0.80	85.5
2-Hydroxyethyl	328	323.3	4.7	2.59	25.7
Glycidyl	319	329.6	−10.6	2.46	22.2
Dimethylaminoethyl	291	293.7	−2.7	3.18	24.2

[a] Taken from Ref. 25.
[b] The value used in Table 1.3 to construct Eqn [20] is 399 K from the two values 266/399 reported in Ref. 7. This was done for self-consistency in the data set of observed T_gs given in Table 1.3.

is not possible if the goal is to predict T_gs for two or more homologous families of polymers simultaneously.

Table 1.4 lists the set of poly-X-acrylates and their respective measured T_gs used to carry out the QSPR analysis. This table also contains the estimated T_g values, the differences, ΔT_g, in observed and estimated values and the values for S_{SC} and M_{SC}, which have been used as independent QSPR variables, and derived from the entries in Table 1.1. The QSPR is

$$T_g = -33.59 S_{SC} - 0.045 M_{SC} + 337.40$$
$$N = 12 \quad R = 0.885 \quad SD = 12.8 \quad F = 16.3$$
[21]

Table 1.5 reports the same data as Table 1.4 but for poly-X-methacrylates. The QSPR for the methacrylate polymers is

$$T_g = 50.01 S_{SC} + 0.082 M_{SC} + 450.76$$
$$N = 20 \quad R = 0.943 \quad SD = 16.6 \quad F = 67.9$$
[22]

Equation [22] is, interestingly, considerably more significant than Eqn [21], and moderately more significant than Eqn [20]. One would have expected both Eqns [21] and [22] to be more significant than Eqn [20] because Eqns [21] and [22] are derived from sets of homologous polymers, as opposed to structurally diverse polymers. Also, the measured T_gs of the acrylates and methacrylates should be more precise, on average, than the observed T_gs of Table 1.3. The lower significance of Eqn [21] may be a statistical artifact in that the range of T_g values is less than the databases used to construct Eqns [20] and [22]. This leads to a more demanding fit to the observed T_g values than in Eqns [20] and [22] to yield equivalent statistical significance.

The two major outliers for the acrylates are X = *iso*-butyl and X = 2-ethylhexyl. In both cases the T_g are overestimated so that molecular flexibility is underestimated. It is not possible to assign this lack of flexibility to entropy or mass TAU descriptor representation. The ethyl derivative is the major outlier for the methacrylates with its T_g underestimated.

1.5.2 Estimation of the crystal to melt transition temperature, T_m

It is generally held that the same interactions at play in governing the T_g process also control other bulk polymer transition processes, including the crystal to melt transition temperature, T_m. The *absolute* magnitudes of the molecular interactions control the temperature regime of a particular process, while the *relative* strengths of the molecular interactions define the type of transition process than can occur. Hence, there is no reason to believe that an equation like Eqn [16] cannot be used to construct a QSPR for T_m.

VanKrevelen reports a set of homopolymers for which T_ms have been measured.[7] Many of these polymers have also been included in the data set used to construct Eqn [20], the best TAU-theory T_g-QSPR. These polymers have been used to construct a T_m-QSPR. The optimum QSPR equation to estimate T_m contains the same set of molecular properties as Eqn [16],

$$T_m = -32.6 S_{BB} - 22.1 S_{SC} - 2.51 M_{BB} - 50.5 \bar{E}_+ - 109.8 \bar{E}_- + 493.7$$
$$N = 30 \quad R = 0.907 \quad SD = 31.6 \quad F = 22.4 \quad [23]$$

However, it is clear from the R, SD and F values that Eqn [23] is inferior for predicting T_m as compared with Eqn [20] for predicting T_g. The set of polymers, corresponding correlation descriptors and the observed, predicted and differences in T_m values are reported in Table 1.6. An inspection of the observed, predicted and difference T_m values suggests that there is no pattern to the errors in estimating T_m. Indeed, only 11 predicted T_m values are within ± 20 K of the observed T_m measurements. Eight predicted T_m values are more than ± 50 K from the measured values. Figure 1.5 is a plot of observed versus predicted T_m values, and reflects the relatively high random scatter inherent to the predicted and observed T_m values.

1.5.3 Comparison of T_m and T_g molecular mechanisms

There has been a long-standing interest in polymer science as to the extent of intra- versus intermolecular interactions in governing T_g and T_m. Intuition has favoured a larger intermolecular contribution to T_m than to T_g. Considerable interest, and debate, has also centred around what specific types of intra- and/or intermolecular interactions, e.g. dispersion, electrostatic, hydrogen-bonding, etc. are most critical to T_g and T_m processes.

Equations [20] and [23] permit mechanistic inspections of the T_g and T_m processes. Moreover, since Eqns [20] and [23] were both developed for a largely common set of homopolymers, these equations can be compared to one another in terms of transition mechanisms. Of course, such interpretations are predicated upon (a) the validity of TAU-theory, (b) the significance/correct selection of QSPR physicochemical molecular properties and (c) the extent to which the polymer database reflects general features inherent to T_g and T_m processes.

Table 1.7 contains a summary of the various contributions of each of the physicochemical molecular properties in Eqns [20] and [23] to T_g and T_m. The total intramolecular ($S_{BB} + S_{SC} + M_{BB}$) and total intermolecular ($\bar{E}_+ + \bar{E}_-$) contribution to T_g and T_m are also reported. The

Table 1.6 Polymers used to construct the QSPR given by Eqn [23]. The descriptor terms have the same meaning as in Table 1.3. T_m is the crystal-melt transition temperature.

No.	Polymer	S_{BB} (cal mol^{-1} K^{-1})	M_{BB} (a.m.u.)	S_{SC} (cal mol^{-1} K^{-1})	M_{SC} (a.m.u.)	$\bar{E}_D$ (kcal mol^{-1})	$\bar{E}_+$ (kcal mol^{-1})	$\bar{E}_-$ (kcal mol^{-1})	T_m(obs.) (K)	T_m(calc.) (K)	ΔT_m (K)
1	—CH$_2$—O—	3.33	15.0	—	—	−0.86	−1.5	0.69	333/473	349.3	−16.3
2	—(CH$_2$)$_2$—O—	3.65	14.7	—	—	−1.04	−0.87	0.37	335/349	341.4	−6.4
3	—(CH$_2$)$_3$—O—	3.82	14.5	—	—	−1.13	−0.54	0.22	308	336.2	−28.2
4	—(CH$_2$)$_4$—O—	3.91	14.4	—	—	−1.18	−0.34	0.12	308/333	334.3	−26.4
5	—(CH$_2$)—	4.30	14.0	—	—	−1.39	0.45	−0.26	410	324.5	85.5
6	—CH$_2$—CH(CH$_3$)—	1.93	14.0	—	—	−1.51	0.51	−0.37	385/481	410.7	70.3
7	—CH$_2$—CH(ϕ)—	1.00	34.3	—	—	−3.26	−1.09	−0.59	498/523	495.0	3.0
8	—CH$_2$—CH(F)—	1.70	24.0	—	—	−1.52	−0.81	−0.38	473	460.9	12.1
9	—CH$_2$—CF$_2$—	1.48	32.5	—	—	−1.69	−1.27	−0.16	410/511	445.8	−35.8
10	—CH$_2$—C(CH$_3$)$_2$—	1.22	28.0	—	—	−2.50	0.77	−0.66	275/317	417.4	−100.4
11	—CH$_2$—CH(Cl)—	0.80	31.0	—	—	−1.93	−0.90	−0.46	485/583	485.9	−0.9
12	—CF$_2$—CF(Cl)—	1.12	58.0	—	—	−2.18	−1.92	−1.08	483/533	527.4	5.6
13	—CH$_2$—C(CH$_3$)— COOCH$_3$	0.76	28.3	1.70	28.5	−2.38	−2.29	−0.41	433/473	521.2	−48.3
14	—CH$_2$—CH(OCH$_3$)—	3.29	14.5	3.45	15.0	−0.91	−1.58	−0.36	417/423	393.5	23.5
15	—CH$_2$—CH(OCH$_2$CH$_3$)—	2.92	12.0	2.89	14.0	−1.36	0.40	−0.32	359	319.8	39.2
16	—CH$_2$—CH— COO(CH$_2$)$_3$CH$_3$	2.38	23.6	3.19	20.0	−1.66	−1.26	−0.04	275/317	354.8	−37.8
17	—(CH$_2$)$_{10}$—O—C(=O)—⌬—C(=O)—O—	3.40	23.4	—	—	−1.84	−0.54	−0.31	396/411	385.7	10.3

#	Structure									
18	—CH$_2$—CCl$_2$—	0.84	48.0	—	—	−3.00	−1.36	−0.38	463/483	456.5	6.5
19	—CF$_2$—CF$_2$—	0.91	31.0	—	—	−1.96	−2.13	−0.80	292/672	581.8	90.2

20	—CH$_2$—CH—
	    |
	   CH$_2$—CH$_3$
		2.92	12.0	2.89	14.0	−1.36	0.40	−0.32	379/415	319.8	59.2

21	—CH$_2$—CH—
	    |
	  (CH$_2$)$_5$—CH$_3$
		2.92	12.0	3.83	14.0	−1.38	0.43	−0.29	235	294.2	−59.2

22	CH$_2$—CH—
	   |
	   O—C(=O)—O—(CH$_2$)$_2$—CH$_3$
		2.38	23.6	2.91	21.5	−1.71	−1.55	0	388/435	371.2	16.8

23	—(CH$_2$)$_2$—NHC(=O)—C$_6$H$_4$—C(=O)NH—	1.56	38.0	—	—	−2.72	−2.47	−2.59	728	756.8	−28.8
24	—(CH$_2$)$_2$—C(=O)—O—C$_6$H$_4$—O—(CH$_2$)$_{10}$—	3.43	21.3	—	—	−1.72	−0.48	−0.10	338	363.9	−25.9

25	—CH$_2$—CH—O—
	    |
	  (CH$_2$)$_9$—CH$_3$
		3.29	14.5	4.02	14.2	−1.23	−0.18	−0.05	280	276.2	3.8

26	—(CH$_2$)$_2$—O—C(=O)—C$_6$H$_4$—C(=O)—O—	2.35	34.3	—	—	−2.39	−1.67	−0.23	533/537	440.9	92.1

(continued)

Table 1.6 (continued)

No.	Polymer	S_{BB} (cal mol^{-1} K^{-1})	M_{BB} (a.m.u.)	S_{SC} (cal mol^{-1} K^{-1})	M_{SC} (a.m.u.)	$\bar{E}_D$ (kcal mol^{-1})	$\bar{E}_+$ (kcal mol^{-1})	$\bar{E}_-$ (kcal mol^{-1})	T_m(obs.) (K)	T_m(calc.) (K)	ΔT_m (K)
27	—(CH$_2$)$_4$—O—C(=O)—O—	3.24	21.5	—	—	−1.75	−1.10	−0.24	332	416.3	−84.3
28	—NH—(CH$_2$)$_6$—NHC(=O)—(CH$_2$)$_4$—C(=O)—	3.27	18.8	—	—	−1.65	−1.16	−1.58	523/545	572.2	−49.2
29	—(CH$_2$)$_2$—NHC(=O)—C$_6$H$_4$—C(=O)—NH—	2.39	29.2	—	—	−2.30	−1.20	−1.40	606/613	557.0	49.0
30	—(CH$_2$)$_{10}$—O—C(=O)—(CH$_2$)$_8$—C(=O)—O—	3.87	17.0	—	—	−1.53	−0.17	−0.27	344/358	363.4	−19.4

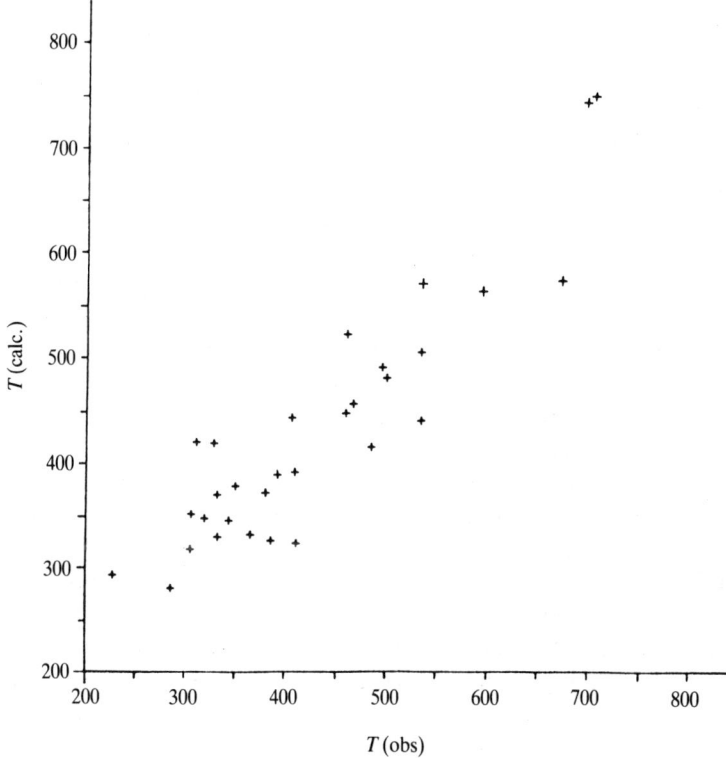

Figure 1.5 Plot of observed versus calculated Eqn [23] T_m values for the polymers in Table 1.6.

Table 1.7. The relative contributions of S_{BB}, S_{SC}, M_{BB}, the total intramolecular contribution, $\bar{E}_+$, $\bar{E}_-$, and the total intermolecular contribution to the range in T_g (Eqn [20]) and T_m (Eqn [23]). R_T is the range in temperature, F_I the intra- or intermolecular fraction of the range covered by the variable, and F_T the total fraction of the range covered by the variable.

Molecular property	T_g			T_m		
	R_T	F_I	F_T	R_T	F_I	F_T
S_{BB}	102.9	0.536	0.310	115.4	0.409	0.413
S_{SC}	34.0	0.177	0.102	151.3	0.182	0.064
M_{BB}	55.1	0.287	0.166	115.5	0.409	0.143
Total intramolecular	192.0	1.000	0.578	282.2	1.000	0.350
$\bar{E}_+$	95.5	0.683	0.288	163.6	0.312	0.203
$\bar{E}_-$	44.4	0.317	0.134	360.1	0.688	0.447
Total intermolecular	139.9	1.000	0.422	523.7	1.000	0.650

column headed by R_T refers to the contribution of the property in terms of the total range in T_g and T_m that is spanned by the computed range in the property. The column of F_i refers to the relative fractional weight of a property in explaining the range in T_g and T_m with respect to the total intramolecular or total intermolecular contribution. The F_T column contains the fractional weight of each property, and the respective intra- and intermolecular total contributions to T_g and T_m.

Several conclusions/inferences can be drawn from Table 1.7. First, from the F_T total contribution entries, T_g is governed, on an approximate 60/40 ratio, by intramolecular interactions. Conversely, T_m is controlled on 65/35 ratio by intermolecular contributions. Second, backbone entropy and mass terms account for over 80% of the intramolecular contributions to T_g. Sidechain entropy accounts for less than 20% of the intramolecular behaviour on T_g. The positive intermolecular probe energy contribution plays a 70/30 dominant role over negative intermolecular probe energy in the intermolecular contributions to T_g. This may be the result of a biased data set of polymer structures. Third, backbone entropy and mass dominate in an 80/20 ratio over sidechain entropy in contributing to T_m. This is the same behaviour as seen for T_g, but backbone mass is more significant in T_m than in T_g. Lastly, the negative probe energy term dominates the positive probe energy term in a 70/30 ratio for intermolecular contributions to T_m. This finding may further indicate a biased data set of polymers, but does bring into question the possible significance of these unexpected intermolecular electrostatic energy findings. The analyses of more, and/or different, polymer systems will have to be done before these findings are validated.

1.5.4 Estimation of polymer–polymer phase behaviour

It has long been of interest to be able to predict the phase behaviour of a mixture of two or more polymers. Of particular concern has been the identification of the conditions necessary for blending of a mixture into a single phase.

In this section we discuss some initial studies of the phase behaviour of a UV binary polymer mixture where U is nylon 3Me6T,

$$\left[-NHCO-\bigcirc-CONH-CH_2-CMe_2-CH_2-CHMe-CH_2-CH_2-\right]$$

and V is nylon 66,

$$\left[-CH_2-CH_2-CH_2-CH_2-CONH-CH_2-CH_2-CH_2-CH_2-CH_2-CH_2-NHCO-\right]$$

using TAU theory. The basic mixing equation governing UV binary mixing is the famous Flory–Huggins expression[26,27]

$$\frac{\Delta G_m}{k_B T} = \frac{\phi_U}{N_U} \ln \phi_U + \frac{(1-\phi_U)}{N_V} \ln(1-\phi_U) + \phi_U(1-\phi_U)\chi \quad [24]$$

where ΔG_m is the free energy change per 'segment' associated with random mixing of the polymer chains, ϕ_U is the volume fraction of polymer chain for U and V, respectively. Finally, χ is the Flory–Huggins segment–segment interaction parameter[28,29] defined as

$$\chi = \frac{1}{k_B T}[E(UV) - \tfrac{1}{2}(E(UU) + E(VV))] \quad [25]$$

The key to estimating polymer–polymer phase behaviour in a binary UV-system is being able to estimate $E(UV)$, $E(UU)$ and $E(VV)$. These three quantities are the UV, UU and VV segment–segment interaction energies. The meaningful determination of these interchain energies for amorphous binary mixtures has been elusive, and may be best attacked by a QSPR formalism. TAU theory offers a general approach to computing relative, and unscaled, χ values provided a pseudo-topological/geometrical scheme to minimize and fully search the segment–segment interactions can be devised. The inter TAU–TAU physicochemical properties, and their scheme of determination, is the foundation for computing χ. The relative χ value for a given UV system is arrived at by individually minimizing $E(UV)$, $E(UU)$ and $E(VV)$ as a function of linear alignment of the SRUs with respect to equal numbers of torsion angles in each of the two interacting SRUs. This is straightforward for UU and VV interactions, but normally requires the 'smaller' SRU to have added TAUs to match the larger SRU in UV interactions. The total interaction energy is minimized with respect to a translation of one (the larger) SRU past the other SRU such that TAU groups are deleted from one side of the moving SRU and added to the other end so that a one-to-one TAU–TAU interchain relationship is always maintained. The minimum energy alignment configuration, and corresponding set of inter TAU–TAU physicochemical properties, is given in Table 1.8 for the nylon 3Me6T(U)-nylon 66(V) binary polymer mixture system. Both first and second nearest-neighbour TAU–TAU interactions are considered in estimating $E(UV)$, $E(UU)$ and $E(VV)$. Equation [13] is the form of the theoretical expression for estimating the inter TAU–TAU energies. However, detailed geometric modelling of interTAU–TAU energies suggests that the form of Eqn [13] be modified to

$$E(\alpha\beta) = \sum_\alpha \sum_\beta \left[v^o(\alpha\beta) - \frac{1}{2}\frac{\partial^2 v^o(\alpha\beta)}{\partial l_{\alpha\beta}^2} l_{\alpha\beta} \right] M_j^{-1} \quad [26]$$

Table 1.8. Minimum energy alignment and corresponding inter TAU–TAU physicochemical properties from Table 1.2C, for the nylon 3Me6T(U)-nylon 66(V) system.

Alignment

U $\quad$ $\{-CH_2-CHMe-CH_2-CH_2-NHCO-\langle\bigcirc\rangle-CONH-CH_2-CMe_2-\}$

V $\{-CH_2-CH_2-CH_2-CH_2-CONH-CH_2-CH_2-CH_2-CH_2-CH_2-$
$\qquad\qquad -CH_2-CHMe-CH_2-$
$\qquad\qquad\qquad\qquad -CH_2NHCO-\}$

Part A First nearest-neighbour (NN_1) inter TAU–TAU interaction

TAU-A	TAU-B	$v^o(AB)$ (kcal/mole)	$\dfrac{\partial^2 v^o(AB)}{\partial l_{AB}^2}$	l_{AB} (Å)
CH_2-CHMe	$\leftrightarrow CH_2-CH_2$	−1.85	0.45	1.3
$CHMe-CH_2$	$\leftrightarrow CH_2-CH_2$	−1.85	0.45	1.3
CH_2-CH_2	$\leftrightarrow CH_2-CH_2$	−1.67	0.30	1.3
CH_2-NHCO	$\leftrightarrow CH_2-CONH$	−6.91	1.50	2.3
$NHCO-\langle\bigcirc\rangle$	$\leftrightarrow CONH-CH_2$	−5.85	1.25	4.9
$\langle\bigcirc\rangle-CONH$	$\leftrightarrow CH_2-CH_2$	−2.48	0.40	4.9
$CONH-CH_2$	$\leftrightarrow CH_2-CH_2$	−2.10	0.35	2.3
CH_2-CMe_2	$\leftrightarrow CH_2-CH_2$	−1.85	0.85	1.3
CMe_2-CH_2	$\leftrightarrow CH_2-CH_2$	−1.85	0.85	1.3
CH_2-CHMe	$\leftrightarrow CH_2-CH_2$	−1.85	0.45	1.3
$CHMe-CH_2$	$\leftrightarrow CH_2-NHCO$	−1.90	0.35	2.3

Part B Second nearest-neighbour (NN_2) inter TAU–TAU interactions

TAU-A	TAU-B	$v^o(AB)$ (kcal/mole)	$\dfrac{\partial^2 v^o(AB)}{\partial l_{AB}^2}$	l_{AB} (Å)
CH_2-CHMe	$\leftrightarrow CH_2-CH_2$	−1.85	0.45	1.3
	CH_2-NHCO	−1.90	0.35	2.3
$CHMe-CH_2$	$\leftrightarrow CH_2-CONH$	−1.93	0.35	2.3
	$NHCO-CH_2$	−1.96	0.35	2.3
CH_2-CH_2	$\leftrightarrow CH_2-CH_2$	−1.67	0.25	1.3
	$CONH-CH_2$	−2.12	0.35	2.3
CH_2-NHCO	$\leftrightarrow CH_2-CH_2$	−2.15	0.35	2.3
	CH_2-CH_2	−2.15	0.35	2.3
$NHCO-\langle\bigcirc\rangle$	$\leftrightarrow CH_2-CH_2$	−2.48	0.40	4.9
	CH_2-CH_2	−2.48	0.40	4.9
$\langle\bigcirc\rangle-CONH$	$\leftrightarrow CH_2-CH_2$	−2.73	0.40	4.9
	CH_2-CONH	−5.90	1.05	4.9

Table 1.8 (continued).

TAU-A	TAU-B	$v^o(AB)$ (kcal/mole)	$\dfrac{\partial^2 v^o(AB)}{\partial l_{AB}^2}$	l_{AB} (Å)
CONH—CH$_2$ ↔	CH$_2$—CH$_2$	−2.12	0.35	2.3
	CONH—CH$_2$	−5.50	1.45	2.3
CH$_2$—CMe$_2$ ↔	CH$_2$—CH$_2$	−1.85	0.85	1.3
	CH$_2$—CH$_2$	−1.85	0.85	1.3
CMe$_2$—CH$_2$ ↔	CH$_2$—NHCO	−2.45	0.35	2.3
	CH$_2$—CH$_2$	−1.85	0.85	1.3
CH$_2$—CHMe ↔	NHCO—CH$_2$	−2.09	0.35	2.3
	CH$_2$—CH$_2$	−1.85	0.45	1.3
CHMe—CH$_2$ ↔	CH$_2$—CH$_2$	−1.85	0.45	1.3
	CH$_2$—CH$_2$	−1.85	0.45	1.3

where α and β can be U or V, v^o is the inter TAU–TAU interaction energy, M_j is the larger number of TAUs in the SRU of α or β, and $l_{\alpha\beta}$ is the longer of the two SRUs. The sums are over the inter TAU–TAU interactions for the alignment. $E(\alpha\beta)$ in Eqn [26] has been normalized to a per TAU per chain scale so that $E(\alpha\beta)$ can be used with all other TAU physicochemical properties in constructing QSPRs. The values of $E(UV)$, $E(UU)$ and $E(VV)$ are given in Table 1.9A for the nylon 3Me6T(U)-nylon 66(V) system. The corresponding unscaled value of χ, as a function of T, is included as part of Table 1.9B. It is important to note that χ is negative, suggesting that single phase blending is predicted which is experimentally observed.[30]

The value of χ cannot be scaled without using experimental data from at least one binary system of a homologue family with respect to U and/or V. Scaling requires a temperature and composition which is at the interface of the one and two phase regimes for a binary system which blends, $\Delta G_m = 0$. The onset of a single glass transition temperature, T_g, in a binary polymer mixture of a particular composition, ϕ_U^*, is often used to define this interface point. If C_1 is the needed scaling constant, then Eqn [25] is modified to

$$\chi = \frac{C_1}{k_B T}[E(UV) - \tfrac{1}{2}(E(UU) + E(VV))] \qquad [27]$$

Substitution of Eqn [27] into Eqn [24] and solving for C_1 at T_g and ϕ_U^* yields

$$C_1 = \left[\frac{-k_B T_g}{E(UV) - \tfrac{1}{2}(E(UU) + E(VV))}\right]\left[\frac{\ln \phi_U^*}{N_U(1 - \phi_U^*)} + \frac{\ln(1 - \phi_U^*)}{N_V \phi_U^*}\right] \qquad [28]$$

For the nylon 3Me6T(U)-nylon 66(V) system $T_g = 347$ K for a 50:50

Table 1.9 Summary of polymer–polymer blend profile of nylon 3Me6T(U)-nylon 66(V) system using TAU theory.

Part A The segment–segment interaction energies

Interaction	NN$_1$ kcal/mole SRU	NN$_1$ kcal/mole SRU/TAU	NN$_2$ kcal/mole SRU	NN$_2$ kcal/mole SRU/TAU	Total kcal/mole SRU	Total kcal/mole SRU/TAU
E(UU)	−16.43	−2.054	−7.15	−0.894	−23.58	−2.948
E(VV)	−15.30	−1.391	−6.26	−0.569	−21.56	−1.961
E(UV)	−15.08	−1.371	−12.60	−1.145	−27.68	−2.516

Part B χ value and C_1 constants

Variable	per/SRU	per(SRU/TAU)
χ	$-5.11/RT$	$-0.0615/RT$
C_1	-4.5×10^{-3}	-1.12

Part C ϕ_U versus T(blend) for $N_U = 67$ and $N_V = 105$. T(blend) corresponds to $\Delta G_m = 0$

ϕ_U	0	0.2	0.4	0.5	0.6	0.8	1.0
T(blend) (K)	$-\infty$	290.5	338.9	351.2	362.9	331.2	$-\infty$

mixture ($\phi_U^* = 0.5$) and $N_U = 67$ and $N_V = 105$.[30] Thus, for these values and the computed E(UV), E(UU) and E(VV) given in Table 1.9B $C_1 = 1.12$. In principle, C_1 should be a constant within a homologue family of binary polymer mixtures. That is, small structural changes in U and/or V should not alter C_1. Hence, χ, as given by Eqn [27], can be estimated directly from E(UV), E(UU) and E(VV), and *predictions* can be made on polymer–polymer phase behaviour for the homologue family of polymers to U and V. We are currently investigating this approach to studying the phase behaviour of nylon binary mixtures.

Once C_1 is available, it is also possible to determine the (T, ϕ_U) interface space for which $\Delta G_m = 0$. This leads to an equation for the blending temperature, T(blend) given by

$$T(\text{blend}) = \frac{C_1 \phi_U (1 - \phi_U)[\tfrac{1}{2}(E(UU) + E(VV)) - E(UV)]}{k_B \left[\dfrac{\phi_U}{N_U} \ln \phi_U + \dfrac{(1 - \phi_U)}{N_V} \ln(1 - \phi_U) \right]} \quad [29]$$

The T(blend) versus ϕ_U are reported in Table 1.9C and are plotted in Fig. 1.6. It is clear from Fig. 1.6 that TAU theory provides a phase diagram very much the same as seen experimentally in binary polymer mixtures. However, it remains to be seen how far TAU theory can go to predicting

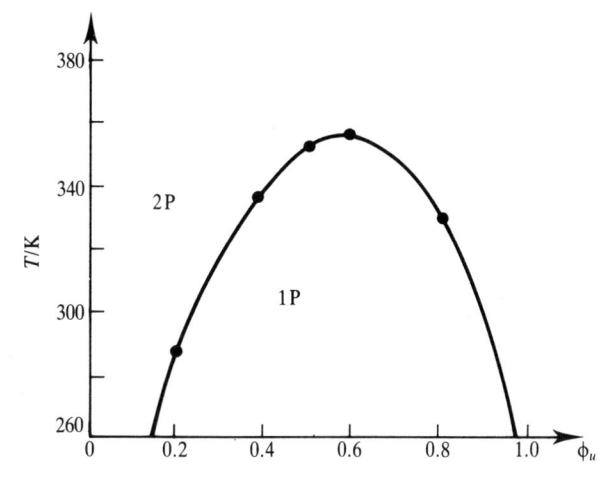

Figure 1.6 A plot of ϕ_U versus T for the nylon 3Me6T(U)-nylon 66(V) system with $N_U = 67$ and $N_V = 105$. The curve corresponds to the $(\phi_U, T(\text{blend}))$-space, $\Delta G_m = 0$. One phase (1P) is under the curve while two separate phases are elsewhere (2P).

polymer–polymer phase behaviour over a structural range of homologous binary systems.

1.6 Discussion of TAU theory and results of QSPR studies

1.6.1 T_g and T_m QSPR

Equation [20] suggests that it is possible to meaningfully estimate T_gs in terms of conformational entropy, mass distribution in an SRU and electrostatic intermolecular interactions. However, these molecular properties do not contribute uniformly to the specification of T_g. Thus, multidimensional linear regression T_g equations were generated for all possible combinations of molecular properties in Table 1.2. The minimum number of molecular properties was sought which yielded $R > 0.92$, $SD < 20$ and $F > 50.0$. It was found that an equation involving only S_{BB}, E_+ and E_- could satisfy these conditions.

$$\left. \begin{array}{l} T_g = -35.3 S_{BB} - 29.9 \bar{E}_+ - 29.3 \bar{E}_- + 321.7 \\ N = 35 \quad R = 0.923 \quad SD = 18.2 \quad F = 59.5 \end{array} \right\} \quad [30]$$

One might dismiss Eqn [20] in favour of Eqn [30] for estimating T_gs as well as inferring mechanisms of molecular action in the T_g process. That is, one might not consider mass or sidechain conformational entropy as important. This, however, is dangerous to do in that the size of our T_g database is small – 35 compounds. In addition, the upper T_g values in the training set are low relative to the T_gs of many of the high temperature engineering plastics which have been synthesized.

It is also important to remember that the QSPRs for the analog series of polyacrylates and the polymethacrylates (see Eqns [21] and [22]) account for the variations in T_g solely in terms of sidechain entropy and mass. Many more polymers need to be considered in generating a reliable universal T_g QSPR. However, it is probably fair to say that S_{BB}, $\bar{E}_+$ and $\bar{E}_-$ are the dominant molecular properties correlating with T_g.

There is no obvious explanation why T_m cannot be significantly correlated against some combination S_{BB}, M_{BB}, S_{SC}, M_{SC}, $\bar{E}_D$, $\bar{E}_+$ and $\bar{E}$. If polymers in Table 1.3 having an absolute difference in predicted and observed T_ms greater than 85 K are deleted in constructing a QSPR, the following correlation is achieved:

$$\left.\begin{array}{l} T_m = -28.9 S_{BB} - 17.5 S_{SC} - 1.54 M_{SC} - 37.4 \bar{E}_+ + 459.4 \\ N = 26 \quad R = 0.959 \quad SD = 24.7 \quad F = 45.9 \end{array}\right\} \quad [31]$$

Equation [31] is quite an improvement over Eqn [23]. Unfortunately, the four polymers deleted in construction of Eqn [23] do not share any common features which might explain why it is more difficult to accurately predict T_m as compared to T_g. The four deleted compounds are polyethylene, poly(isobutylene), polytetrafluoroethylene, and poly(ethylene terephthalate). Thus, it is not possible to propose what additional molecular features need to be considered in order to successfully predict T_ms.

One surprising finding is that T_m is predicted to decrease as the normalized mass of the backbone SRU increases in both Eqns [23] and [31]. This is the case since the regression coefficients are negative. The opposite is true for T_g. Intuition suggests that both T_g and T_m should increase with normalized SRU mass. Since the regression coefficients for M_{BB} have a small absolute value, the mass terms make relatively small contributions to specifying T_g and T_m. Hence, this discrepancy may not be meaningful and an artefact of the statistical fit.

A comparison of the regression coefficients of Eqns [20] and [23] suggests that sidechain entropy and the electrostatic intermolecular energies are more important (the absolute values of the coefficients are larger) for specifying T_m than T_g. The reference point is the backbone entropy whose regression coefficient does not change very much between Eqns [20] and [23]. In particular, the coefficient of E_+ changes by more than a factor of seven. It is not obvious why these specific changes are seen, but intuition would support intermolecular interactions being more important in maintaining a crystal than a glass. Moreover, sidechain entropy in a crystal might reflect the onset of disruption of both intrachain and interchain order since sidechains are in more intimate contact in a crystal than in a glass. This could explain the increased significance of S_{sc} in Eqn [23] as compared with Eqn [20].

A probable reason for not being able to predict T_ms as well as T_gs may be due to the anisotropic geometric environment inherent to a polymer crystal, in contrast to the orthotropic medium of a polymer glass. The molecular properties we have computed using one-dimensional TAUs model the orthotropic environment of a glass better than the anisotropic geometry surrounding a TAU in a polymer crystal. There is no allowance in the current model to consider the effects of direction-dependent intermolecular interactions. Specific, periodic lattice interactions which stabilize the crystal are not accounted for in the current model. We have seen in previous polymer crystal structure calculations[15] that the electrostatic interactions play a crucial role in specifying lattice geometry, especially the setting angles of the chains. This finding from lattice packing calculations can be taken as evidence that larger TAUs, which build anisotropic molecular interactions into the model, are probably required in order to generate significant T_m QSPRs.

In so far as the data sets are large enough to impart self-consistency in the regression equations, the QSPRs reported here might be used to resolve which of two, or more, reported transition temperatures for a polymer is likely to be correct. That has, in fact, been done in the derivation of QSPR equations. When two observed T_g, or T_m, values are given, the observed value which maximizes the regression fit is used. Thus, TAU theory may be of use in resolving conflicting experimental measurements.

1.6.2 TAU theory and polymer-polymer phase behaviour

The application of the inter TAU-TAU physicochemical properties (see Table 1.2C) to estimate segment-segment interaction energies, provides reasonable results for the nylon 3Me6T-nylon 66 binary mixture. Minimization of the interchain SRU-SRU energy by optimum linear alignment of the SRUs, in terms of first and second nearest-neighbour inter TAU-TAU energies, appears to be an efficient, yet realistic, means of estimating chain-chain interactions in amorphous polymer mixtures. It is interesting to note in the nylon 3Me6T-nylon 66 system that it is the second nearest-neighbour energy term that makes $E(UV)$ more stable than $[1/2E(UU) + E(VV)]$.

The number of inter TAU-TAU pairs needed in interchain applications, in the most general case, is N^2 as compared to N TAUs for intrachain QSPR development. This increase in computational requirements needs to be kept in mind when undertaking a project.

1.6.3 Summary statement

Overall, it would seem at this time in the evolution of TAU theory that it provides an open-ended means to generate a large number of physicochemical molecular properties to use in devising a QSPR. The investigator, however, must be clever in selecting the appropriate set of molecular properties, and in interpreting the 'meaning' of the resultant QSPR.

References

1. Roe R J (ed.) *Computer Simulations of Polymers*, Prentice-Hall, Englewood Cliffs, NJ, 1990.
2. Hopfinger A J *J. Med. Chem.* **28**: 1133 (1985).
3. Perun T J and Propst C L (eds) *Computer-Aided Drug Design*, Marcel Dekker, New York, 1989.
4. Bristol J A (ed) *Annual Reports in Medicinal Chemistry*, Vol. 26, Academic Press, New York, Section VI, p. 259, 1991.
5. McCammon J A, Harvey S C *Dynamics of Proteins and Nucleic Acids*, Cambridge University Press, New York, 1987.
6. Mezei M, Mehrotra P K, Beveridge D L, *J. Am. Chem. Soc.* **107**: 2339 (1985).
7. VanKrevelen D W *Properties of Polymers*, Elsevier, New York, 1976.
8. Hansch C, Fujita T *J. Am. Chem. Soc.* **86**: 1616 (1964).
9. Bondi A *Physical Properties of Molecular Crystals, Liquids, and Gases*, 2nd edn, McGraw-Hill, New York, 1966.
10. Wiff D R, Altieri M S, Goldfarb I J *J. Polym. Sci. Polym. Phys. Ed.* **23**: 1165 (1985).
11. Askadskii A A, Slonimskii G A *Polym. Sci. USSR* **13**: 2158 (1971).
12. Barton J M *J. Polym. Sci., Part C* **30**: 573 (1970).
13. Lee W A and Knight G J *Br. Polym. J.* **2**: 73 (1970).
14. Hopfinger A J *Conformational Properties of Macromolecules*, Academic Press, New York, 1973.
15. Tripathy S K, Hopfinger A J, Taylor P L *J. Phys. Chem.* **85**: 1371 (1981).
16. Flory P J *Statistical Mechanics of Chain-Molecules*, Wiley, New York, 1969.
17. Hopfinger A J, Koehler M G, Pearlstein, R A, Tripathy S K *J. Polym. Sci.: Part B: Polym. Phys.* **26**: 2007 (1988).
18. Koehler M G, Hopfinger A J *Polymer* **30**: 116 (1989).
19. Allinger N L *J. Am. Chem. Soc.* **99**: 8127 (1977).
20. Pople J A, Beveridge D C *Approximate Molecular Orbital Theory*, McGraw-Hill, New York, 1970.
21. Hopfinger A J *J. Med. Chem.* **26**: 990 (1983).
22. Brant D A, Flory P J *J. Am. Chem. Soc.* **87**: 663, 2791 (1965).
23. Wold S, Hellber S, Dunn W J III, *Acta Pharmacol. Toxicol.* **52**: 158 (1983).
24. Dunn W J III, Wold S *Bioorg. Chem.* **9**: 505 (1980).
25. Rohm and Hass Company, Technical Reference Report CM-20, Philadelphia, PA (1976).
26. Flory P J *J. Chem. Phys.* **10**: 51 (1942).

27. Huggins M *J. Phys. Chem.* **46**: 151 (1942); *J. Am. Chem. Soc.* **64**: 1039 (1942).
28. deGennes P-G *Scaling Concepts in Polymer Physics*, Cornell University Press, Ithaca, NY, 1979.
29. Flory P J *Principles of Polymer Chemistry*, Cornell University Press, Ithaca, NY, 1953.
30. Ellis T S *Macromolecules* **22**: 742 (1989).

CHAPTER 2

Molecular dynamics modelling of amorphous polymers

J H R CLARKE AND D BROWN

2.1 Introduction

Polymers present a particularly demanding challenge to computer simulation. Unlike molecular liquids, for which many properties are determined predominantly by short range interactions and the important fluctuations occur over short times, a full description of amorphous polymers requires consideration of length scales ranging from the size of monomer units to the end-to-end length of chains. In addition the thermally driven exploration of the random coil structures can span many decades of time. Local conformational fluctuations, for instance, may occur in the nanosecond regime for fairly flexible chains, whereas fluctuations of the end-to-end distance in a polymer melt can be of the order of seconds.[1,2]

One way to diminish the problem is to simplify the model so that it is computationally more tractable. For instance, one may sacrifice all the chemical detail in the chain and use a freely-jointed 'bead' model for the chain interactions; here the bead represents a statistical unit of the chain. One may further simplify the situation by restricting the chain to a lattice, which again increases the computational efficiency. Elsewhere in this book there are discussions of some of the elegant results that have been obtained using these approaches. The problem with these simplifications is that there exists no satisfactory *ab initio* method of identifying the 'bead' with any structural unit characteristic of a real chain.

In the past five or ten years there has been a growing interest in the more detailed atomistic dynamical modelling of polymers in continuous space using energy minimization[3] and molecular dynamics simulation,[4,5] particularly for dense amorphous polymers where there is strong competition between entropic effects and intermolecular interactions. Atomistic molecular dynamics offers the opportunity to explore time dependent properties, but these simulations are limited by current computing resources to times of order nanoseconds. Full 'brute force' equilibrium molecular dynamics simulations of relatively high molecular weight polymer melts, where the chains explore a significant amount of the available configurational space, are just not feasible.

So what contributions can be made to understanding polymers using atomistic modelling? One area where the technique promises to be very useful is in the study of polymer glasses in order to improve our understanding of, for example, their mechanical and barrier properties. In polymer glasses the diffusional and large scale configurational fluctuations are completely frozen out and the properties are determined by the short range van der Waals interactions moderated by the connectivity and flexibility of the chains.

The glass transition arises from the failure of a system to achieve full thermodynamic equilibrium during continuous cooling. It is a kinetic phenomenon determined by the temperature dependence of characteristic structural relaxation processes in the material. What one then observes, for instance, as a polymer melt is cooled depends on the time scale of the experiment. If it is performed sufficiently slowly then changes in properties will be observed, first as diffusional motions and then as conformational motions are frozen out. For some materials these effects are sufficiently distinguished that they give rise to separate liquid–rubber and rubber–glass transitions. If the cooling is performed rapidly, as in a molecular dynamics experiment, then the material is already configurationally arrested on the time scale of the experiment and only the conformational motions will be frozen out. This is illustrated schematically in Fig. 2.1, where we indicate the temperature dependences of the relaxation times for interchange of *gauche* and *trans* conformational states in a simple hydrocarbon chain as compared to that of the terminal relaxation time for a typical dense polymer.[2] *Trans-gauche* interconversion times can be obtained from simulations using a method previously described;[6] the form of the curve is a fit to high temperature data obtained for model polyethylene.

Even for real polymers the glass transition is usually associated primarily with the freezing out of conformational fluctuations. Although simulation cooling rates are much higher (of order 10^{12} K s^{-1}) than could be achieved in even the fastest laboratory splat-quenching experiments and hence molecular dynamics simulations will not access the same spectrum of conformational fluctuations, nevertheless the process is well defined. There is much to do in the area of understanding scaling laws that will enable such short time data to be mapped onto the much longer time scales familiar in the laboratory.

There have been several approaches to modelling polymers using molecular dynamics.[7] One of the first applications was to study the mechanical properties of ordered and amorphous polymer chains suspended

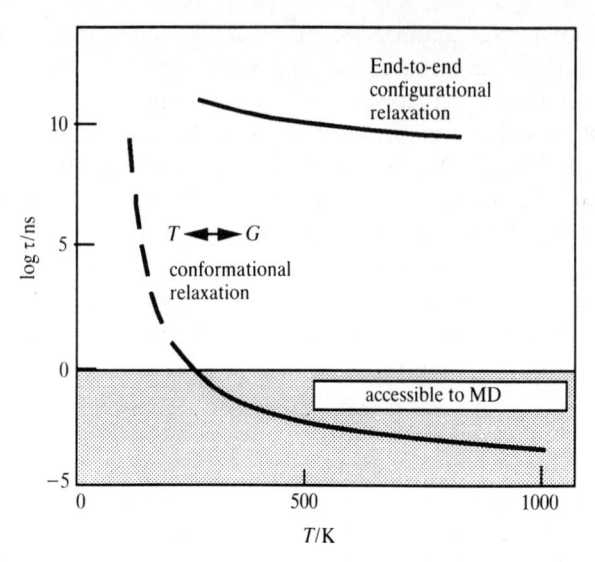

Figure 2.1 Illustrating the temperature dependences of characteristic relaxation times for a polyethylene chain with $\sim 10^3$ monomers. See text for details.

between two parallel surfaces producing a lamina-type model.[8] More recently a method has been proposed for modelling high molecular weight polymers in which the length of a single continuous chain is many times that of the unit cell.[9] Other simulations have been restricted to short chains,[10] much less than the entanglement length (by 'entanglement' here we do not imply a geometric definition but merely that beyond this length that there are topological constraints on the polymer which are characterized by very long relaxation times).

We shall here concentrate attention on a continuous space model of a linear polymer in which the van der Waals interactions are simplified by ignoring some of the chemical detail but retaining the essential features of connectivity and restricted flexibility in a fairly realistic way. The model bears a resemblance to polyethylene but the main objective is not to reproduce the properties of any particular material. The degree of polymerization is 1000, which is several times the expected entanglement length.

We shall show that the general form of the results bears a striking resemblance to those of laboratory studies on time scales many orders of magnitude slower. We shall show how the glass transformation can be characterized and will describe one recent application showing how the stress–strain properties of the polymer glass can be related to its configurational properties.

Although the interest here is primarily with molecular dynamics simulations mention should be made of the recent emergence of promising new Monte Carlo techniques for modelling chain molecules at high densities.[11] Although currently limited to rather short chains, these methods might be adapted for the study of the equilibrium properties of melts and solutions. It is not clear, however, how they can be used for systematic studies of non-equilibrium states such as glasses where time is an essential variable.

2.2 Models and methods

2.2.1 A model for polymer interactions

Having identified a polymer which we wish to study the choice of 'atomistic' model for the simulations should be approached with a degree of caution. The level of sophistication required in the model depends very much on the questions being asked; for instance, if we wish to replicate as accurately as possible the properties of a particular polymer then we could bring together as much information as possible from experiment and quantum mechanical calculations to develop a 'good' force field. It must be remembered, however, that we have only a limited knowledge of inter- and intramolecular forces in complex molecules and no model force field will correctly reproduce all properties.

Here we shall be concerned with an alternative approach where the emphasis is on understanding phenomena rather than reproducing experimental data. In this case the aim is to use the simplest monomer model which incorporates all the essential features of inter- and intramolecular interactions. An example of this kind of approach is the identification of functional groups which are expected to maintain a fairly rigid structure (e.g. CH_3 and CH_2 groups) in a polymer and the representation of them as single interaction sites. This seems to be reasonable for amorphous polymers[4,10] although there is reason to suspect, however, that these simplified potentials may not correctly reproduce phenomena in crystalline polymers where the delicate balance of close packed interactions requires a more detailed description of the monomer.[12]

The essential properties that any polymer model should exhibit are connectivity, chain flexibility and, in the parlance of polymers, long range van der Waals interactions. The connectivity may be linear (as for the models discussed here) or may involve branches and/or crosslinks. The flexibility of chains will be limited by both 'chemical' forces along the chain and local van der Waals interactions between chains which restrict torsional and bond angle motions.

In the polyethylene-like polymer discussed here each of the N monomer units is treated as a single site and is given a mass corresponding to that of a CH_2 group. Neighbouring sites on the chain are connected together by rigid bonds of length $b_0 = 0.153$ nm. Flexibility of the chains is limited by incorporating a harmonic valence angle potential, $\Phi(\theta)$, and a torsional potential, $\Phi(\alpha)$, into the model. $\Phi(\theta)$ is of the form

$$\Phi(\theta) = \tfrac{1}{2}k_\theta(\cos\theta - \cos\theta_0)^2 \qquad [1]$$

where $k_\theta = 520$ kJ mol^{-1} and $\theta_0 = 112.813°$. We use the cosine of the angle in the harmonic potential for computational convenience but it is possible to express $\Phi(\theta)$ directly in terms of angle displacements[13] using the same value of k_θ. For small displacements there is very little difference in the two potentials. In the absence of precise experimental data the choice of k_θ is in any case somewhat arbitrary.

The torsional potential restricting internal rotations about a bond in the chain can be parametrized in terms of the dihedral angle, α, formed by this bond and the two adjacent bonds. A widely used parametrization is that due to Ryckaert and Bellemans[14] but the form used here is that due to Steele[15] and is given below:

$$\Phi(\alpha)/\text{J mol}^{-1} = C_0 + C_1 \cos\alpha + C_2 \cos^2\alpha + C_3 \cos^3\alpha \qquad [2]$$

where $C_0 = 8832$, $C_1 = 18\,087$, $C_2 = 4880$ and $C_3 = -31\,800$. This parametrization is based on a wide range of *ab initio* data for *n*-butane. Finally, non-bonded monomer interactions, i.e. those between sites separated by at least three others, are represented by a Lennard–Jones (LJ) 12-6 potential:

$$\Phi(|\mathbf{r}_{ij}|) = 4\varepsilon\left\{\left(\frac{\sigma}{|\mathbf{r}_{ij}|}\right)^{12} - \left(\frac{\sigma}{|\mathbf{r}_{ij}|}\right)^{6}\right\} \qquad [3]$$

with $\varepsilon/k_B = 57$ K and $\sigma = 0.428$ nm. The value of ε has been used previously by Maréchal[16] and Jorgensen *et al.*[17] and along with the value given for σ we find these parameters give a reasonable fit to the density of real polyethylene, as extrapolated from the data of Richardson *et al.*[18] in the temperature range from 500 K to 600 K. The LJ 12-6 potential was truncated at $r_c = 2.5\sigma$ and the appropriate long range corrections were made to the potential energy and the virial at each step according to the density and assuming $g(r) = 1$ for $r > r_c$.

2.2.2 Molecular dynamics for polymers

A characteristic of small samples of dense polymers under constant volume conditions is that the average pressure tensor is often anisotropic with substantial differences between the on-diagonal components and, in addition, non-zero off-diagonal terms. It is very unlikely that a polymer system will be able to adjust its configuration in the time available, so that the properties obtained cannot be expected to bear much resemblance to those that might be obtained in a laboratory experiment carried out at the usual conditions of constant hydrostatic pressure. It is desirable therefore to use a technique which allows pressure anisotropies to be relaxed out. Once implemented such a method has the added advantage of being easily adapted to measure the response of such systems to externally applied pressure fields as we shall show later.

For the majority of atomic and 'small' molecule systems at equilibrium in the $(N, \mathbf{P}, T)$ ensemble ($\mathbf{P}$ is the pressure tensor) it is widely accepted that the most rigorous approach is to use the controlled pressure technique proposed by Rahman and Parrinello (RP)[19] in conjunction with the Nosé–Hoover thermostat.[20,21] However, the choice of method must take careful account of the material we wish to study, how it is modelled and any external perturbations which we wish to apply. For polymers the Berendsen loose-coupling controlled pressure MD technique[22] is a good compromise. Although the theoretical basis of this method has been criticized[21,23] in practice it has been found[23] that to within statistical uncertainties first order properties are the same as those obtained by more rigorous approaches.

2.2.2.1 Loose-coupling controlled pressure dynamics for polymers

We outline here a method[24] which utilizes weak coupling of an external tensorial pressure field, $\mathbf{P}_0$, to the system through a simple feedback loop. It is assumed that provided the coupling is loose enough it will have an insignificant effect on the first order properties of the system. The implementation of this 'loose-coupling' method is discussed in Appendices 1 and 2.

The coupling is implemented by allowing the matrix $\mathbf{h}$, made up from the basis vectors, a, b and c which determine the shape of the primary dynamics cell, to respond to imbalances between the internally measured pressure tensor and an externally applied pressure tensor. We have modified the original method to ensure that $\mathbf{h}$ remains symmetric throughout by defining the equation for the rate of change of the $\mathbf{h}$ matrix with time to be

$$\dot{\mathbf{h}} = \frac{\mathbf{P} - \mathbf{P}_0}{M} \quad [4]$$

where M is a constant and $\mathbf{P}$ is the internally measured pressure tensor, which in this case is defined in an 'atomic' frame of reference, i.e. the momentum is localized at the positions of the CH_2 sites, and hence is symmetric:

$$\mathbf{P} = \frac{1}{V} \sum_{i=1}^{N} \left[\frac{1}{m_i} \mathbf{p}_i \mathbf{p}_i + \mathbf{r}_i \mathbf{f}_i \right] \qquad [5]$$

Although Eqn [5] is formally correct great care has to be taken in a simulation using periodic boundaries when calculating **P**. Exact details are given in Appendix 1. The volume, V, is calculated from the determinant of the matrix **h**,

$$V = 8 \times \det \mathbf{h} = 8 \times \mathbf{a} \cdot (\mathbf{b} \times \mathbf{c}) \qquad [6]$$

The factor of 8 appears in our calculations because the origin for the basis vectors is the centre of the primary cell.

In keeping with the loose-coupling method[22] and other schemes[19,25] a simple proportional scaling of coordinates is used to minimize local disturbances. So defining a set of scaled coordinates, **s**, by

$$\mathbf{s} = \mathbf{h}^{-1}\mathbf{r} \qquad [7]$$

differentiation gives the following equation of motion for the sites

$$\dot{\mathbf{r}}_i = \mathbf{h}\dot{\mathbf{s}}_i + \dot{\mathbf{h}}\mathbf{s}_i = \frac{\mathbf{p}_i}{m_i} + \dot{\mathbf{h}}\mathbf{h}^{-1}\mathbf{r}_i \qquad [8]$$

The motion is thus seen to be split into two contributions which are integrated separately, that due to the momenta and that resulting from the change in shape and size of the cell. The 'fast' motions due to the momenta are dealt with in the usual way using a 'leapfrog' algorithm incorporating an iterative scheme to maintain the constraints whereas a simple first order Taylor expansion is considered sufficient to integrate the equation for the relatively 'slow' motion of the box

$$\mathbf{h}(t + \Delta t) = \mathbf{h}(t) + \left[\frac{\mathbf{P} - \mathbf{P}_0}{M} \right] \Delta t \qquad [9]$$

It can then simply be shown that to first order the motion of the box results in a scaling of the position of a site

$$r_i(t + \Delta t) = \mathbf{h}(t + \Delta t)\mathbf{h}^{-1}(t)\mathbf{r}_i(t) = \mathbf{H}\mathbf{r}_i(t) \qquad [10]$$

The use of the term constant pressure to describe these simulations is somewhat of a misnomer. In common with most of the other schemes the pressure is not a constant of the motion and indeed the pressure fluctuations can be larger for a second order method than at constant volume. The simulations merely maintain the mean value of the pressure close to some predetermined value. For this reason we prefer to use the term 'controlled pressure molecular dynamics'.

Although the method is less rigorous than the alternative Rahman–Parrinello (RP) technique[19] it does have at least one important practical advantage. This is that the pressure imbalance is coupled to the first derivative of the basis vectors rather than to the second derivative; this means that motions of the box are overdamped and so there is little tendency for an unphysical oscillatory response to changes in the applied pressure. For this reason this method comes into its own when the non-equilibrium properties of dense highly viscoelastic systems are of interest. Typically the response to uniaxial tension or compression provides useful information concerning the material properties of the system, e.g. Young's moduli, Poisson's ratio and yield stresses. With care this method allows such experiments to be performed in a well-controlled manner. The ease of implementation of the loose coupling method is also an advantage as it is compatible with the simple second order integration algorithms much in use in MD.

2.2.2.2 Choice of the coupling constants M and τ_T

In this section we consider criteria that can be used in choosing values for these two coupling constants. There are two competing factors which have to be taken into account. Firstly, to minimize the disruption to the system it is desirable to reduce the coupling of the system with the external pressure and temperature 'baths' as much as possible, and this can be achieved by using larger values of M and τ_T. Since an important application of the method is, however, to measure the time dependent response to external pressure and/or temperature perturbations M and τ_T must not be so large as to couple or interfere with the physical phenomena being investigated on the time scale of interest. No unique universal values of these parameters can be given as they will depend on many factors, e.g. the system being studied, the applied conditions, the rate of change of those conditions, etc. For this reason we have devised some procedures that can be used as a guide in the choice of M and τ_T.

The first procedure is designed to explore the limits of the 'stability envelope' of the controlled pressure algorithm. To do this a change, $\Delta \mathbf{P}_0$, is made within one time step to the external pressure of a chosen start configuration. Starting from this same configuration several short simulations are then performed with different values of M in an attempt to find the minimum value for which the algorithm is stable. Generally we have found that 100 steps is enough to establish whether the algorithm is going to break down (indicated in our case by a failure of the constraints routine to find a solution within a hundred iterations) or produce wildly oscillating pressures (and hence densities). Once a minimum has been established a different value for $\Delta \mathbf{P}_0$ is chosen and the procedure repeated.

In this particular case we have chosen to change the pressure isotropically, i.e. $\Delta \mathbf{P}_0 = \Delta P_0 \mathbf{1}$ and in the range from 10^3 to 10^6 bar, though there are many other possible alternatives. Potentially the stability can be affected by the value of τ_T used; however, in practice we have found values as widely spaced as $\tau_T = \Delta t = 5$ fs and $\tau_T = \infty$ have no affect on the stability envelope within the precision with which we have obtained the minimum values of M.

In Fig. 2.2 the applied pressure change has been plotted against the minimum values of M that produced stable results. For convenience the values of M have been normalized by an arbitrarily chosen value $M_0 = 5.25 \times 10^6$ Pa s m^{-1}. Note also that both axes are on a logarithmic scale. The two sets of data shown in Fig. 2.2 were obtained using the two methods of satisfying the bond constraints that are outlined in Appendix 1. It can be seen that the application of a second constraint loop to correct the bond lengths following the change in cell shape and size has the effect of increasing the stability range of the algorithm at the lower values of $\Delta \mathbf{P}_0$. At the higher values, though, there appears to be little or no advantage in using the extra constraint loop. It should be pointed out that instantaneous pressure changes of order 10^6 bar are extreme by most standards and are considered here not for any physical reason but just as a severe test of the robustness of the algorithm.

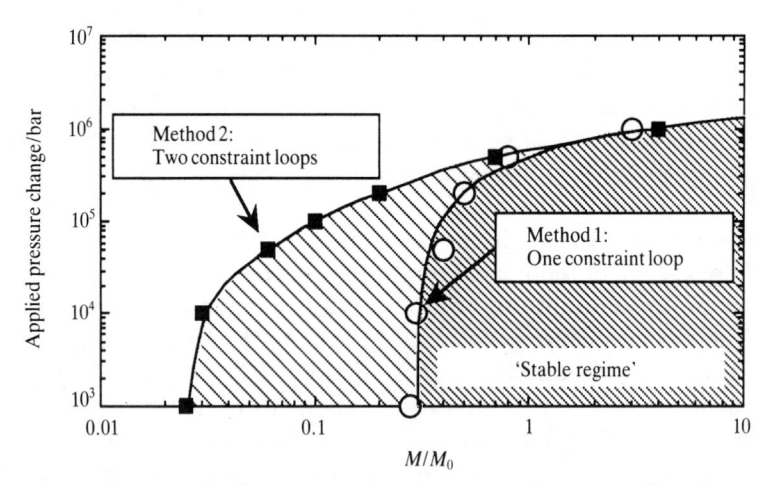

Figure 2.2 The 'stability envelope' for the pressure loose-coupling algorithm. The points represent the minimum values of M/M_0 ($M_0 = 5.25 \times 10^6$ Pa s m^{-1}) for which the algorithm produced stable results for 100 time steps following an instantaneously applied isotropic pressure change. The two sets of data arise from the two different methods of implementing the algorithm discussed in Appendices 1 and 2. From Brown and Clarke.[24]

Having obtained the range of possible values of M that can be used it remains to test the degree to which they cause coupling between the internal pressure and that of the external 'bath'. To do this we have calculated the root mean square deviations in the pressure from the mean, $\langle \Delta P^2 \rangle^{1/2}$, over a certain time interval for a series of equilibrium simulations starting from an identical configuration but using different values of M. We then define the 'coupling index' for the pressure, C_P^1, as

$$C_P^1(M) = 1 - \frac{\langle \Delta P^2 \rangle_M^{1/2}}{\langle \Delta P^2 \rangle_{M=\infty}^{1/2}} \qquad [11]$$

where the $M = \infty$ result is just that obtained at constant volume. Note that $C_P^1 = 1$ for complete coupling and 0 for no coupling. The results of 1 ps simulations are given in Fig. 2.3. They were obtained using the same initial starting conditions and the fact that correlation times for the pressure are quite short in this system, ~ 0.2 ps, should ensure a reasonably good comparison. Once again calculations have been made for both variants of the algorithm. The results in Fig. 2.3 were obtained with $\tau_T = \infty$. Although other values of τ_T were used the differences were not

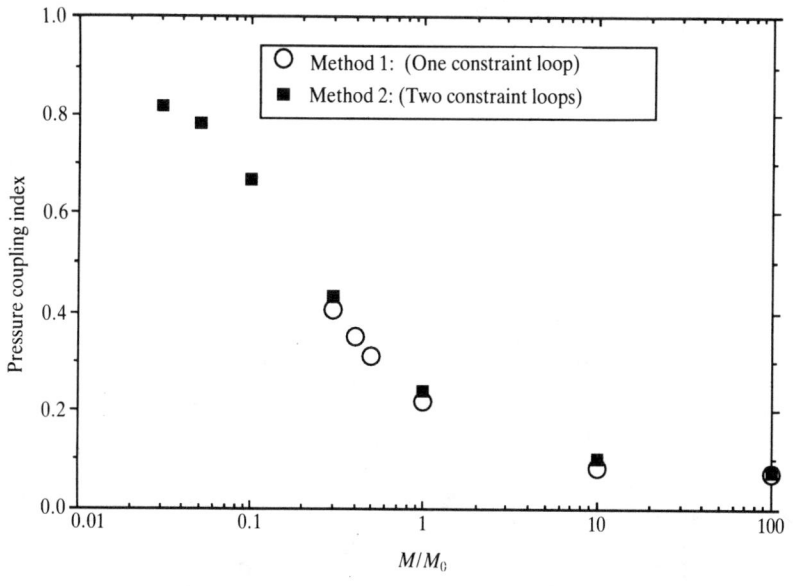

Figure 2.3 The pressure coupling index parameter (see Eqn [11]) plotted as a function of M. Results were obtained from a series of short (1 ps) simulations starting from the same initial configuration of an $N = 1000$ site linear chain model dense amorphous polymeric system at $T = 300$ K and an applied isotropic pressure of 1 bar. From Brown and Clarke.[24]

perceptible on the scale of Fig. 2.3. This is simply a reflection of the fact that for this particular system at the conditions used most of the fluctuations in the pressure come from the force terms in the pressure tensor rather than from the kinetic ones.

To assess the effect of the magnitude of τ_T on the coupling between the internal temperature and the 'external' temperature a coupling index, analogous to that for the pressure, has been defined for the temperature as

$$C_T^I(\tau_T) = 1 - \frac{\langle \Delta T^2 \rangle_{\tau_T}^{1/2}}{\langle \Delta T^2 \rangle_{\tau_T = \infty}^{1/2}} \quad [12]$$

The results for C_T^I from another set of simulations of duration 1 ps, this time varying τ_T and keeping M fixed at ∞, are given in Fig. 2.4. There are no problems with algorithmic stability in this case as in the limit $\tau_T = \Delta t$ the loose coupling method simply returns a form of constant temperature algorithm often termed *ad hoc* rescaling of momenta which has been discussed extensively in the literature.[26] The fact that C_T^I is only of order 0.5 for $\tau_T = \Delta t$ again emphasizes the dominant influence of the forces for this particular system. Even on a logarithmic scale the temperature coupling index soon decays to very small values such that for $\tau_T = 100\Delta t$ there is barely any significant difference from that of the

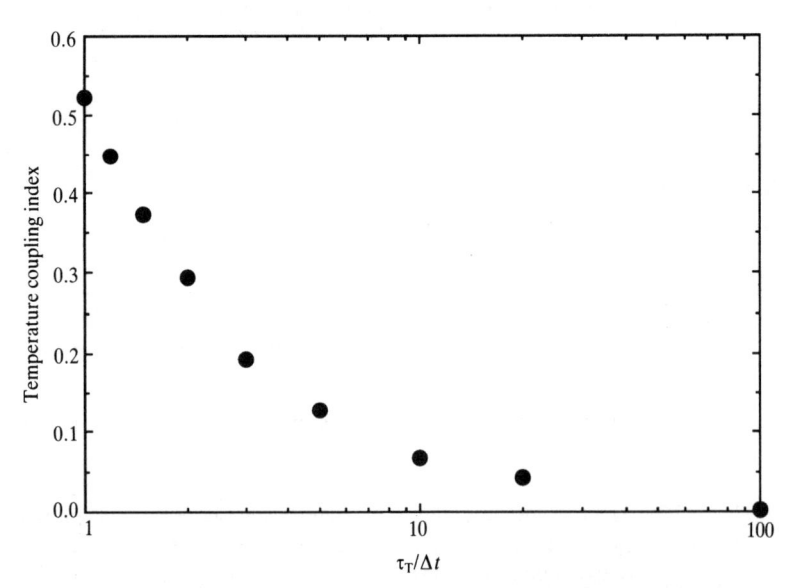

Figure 2.4 The temperature coupling index parameter (see Eqn [12]) plotted as a function of τ_T. Results were obtained in an analogous way as for Fig. 2.3. From Brown and Clarke.[24]

infinitely large result. It was also verified that C_T^l was rather insensitive to the choice of M.

2.3 Preparation of amorphous samples

The importance of this part of the polymer simulation cannot be over-emphasized and it is worth devoting some space to issues that it raises. One of the basic problems is, as stated before, that relaxation times in dense polymers are so long that it is not really feasible fully to sample the equilibrium state of, for instance, an entangled polymer melt using molecular dynamics. Instead we have to devise methods of preparing individual samples of the amorphous polymer from the correct distribution and then average the measured properties over as many independent samples as possible. Most of the methods so far used involve growing the polymer chain using some kind of random walk recipe and then locally relaxing the structures using molecular mechanics (at 0 K) or molecular dynamics (at finite temperature).

The use of periodic boundary conditions also presents some problems. The traditional requirement is that the size of the primary simulation cell is large enough to prevent molecules interacting with images of themselves through the periodic boundaries. For an amorphous polymeric system this criterion could be relaxed somewhat so that the cell size is greater than the expected end-to-end distance of a polymer molecule. Nevertheless for a degree of polymerization of 1000 we would have to include over 30 chains and a total of 30 000 monomers in the simulation which, although feasible, would be a very expensive calculation. A more radical, although controversial, approach is to use a cell which is only larger than the correlation lengths important to the phenomenon being studied. Here we are mainly interested in the mechanical properties of amorphous polymers at low strains ($< 100\%$). We do not anticipate any large N-dependence of these properties so we use just one chain of 1000 monomers to form a dense amorphous polymeric system through the replicative properties of periodic boundaries. The primary chain spans many neighbouring cells. The model is therefore one of a monodisperse polymer entangled with replicas of itself. A two-dimensional schematic diagram of this model is shown in Fig. 2.5. The effects of boundary conditions in this kind of model have yet to be fully evaluated but we can expect that one important condition might be the size of the unit cell in relation to the correlation length of the chain. For small N there is no doubt that the model gives a poor representation of bulk behaviour, particularly for less flexible polymer chains, but as N becomes larger we expect it to be an increasingly better approximation to a dense amorphous system.

Figure 2.5 A two-dimensional schematic diagram of the polymer model in which a single chain is replicated by the periodic boundaries.

2.3.1 Chain growth including excluded volume

This method has been used to generate linear chains sequentially site-by-site incorporating realistic excluded volume and intramolecular potentials. The method could be adapted in a straightforward way for branched chains and is reasonably efficient at moderate densities when a simple Monte-Carlo acceptance criterion is applied successively at the addition of each monomer in the chain although this does produce a biased distribution of chain configurations.

For the linear polyethylene chain a new proposed site is generated by choosing an intended dihedral angle (α) using a random number generator. The coordinates of the next site, r_{i+1}, are calculated using the equilibrium bond length b_0, the equilibrium valence angle θ_0 and the coordinates of the three previous sites. The total energy change, $\Delta\Phi$, resulting from the introduction of this site is then determined by summing all new interactions (using the minimum image convention) and adding to this the new contribution to the dihedral angle energy $\Phi(\alpha)$. The probability of this energy change is calculated from the Boltzmann factor, $\exp(-\Delta\Phi/k_B T)$, and a comparison between this value and a random number between 0 and 1 is then made. The new site is accepted only if the random number is less than the Boltzmann factor, otherwise a new proposed site is generated. Should repeated trials[50] fail to find an acceptable position for the new site the chain is shortened by one site and the process starts afresh by attempting to generate a new r_i and if successful a new r_{i+1}. If again a suitable r_{i+1} cannot be found then the chain is shortened this time by two sites and so on until a route circumventing the obstruction is found. The process stops when all N sites have been placed.

With this method it is not really practical to generate configurations at high densities. For the polyethylene model used here a reasonable choice for the initial density was found to be $0.5\,\mathrm{g\,cm^{-3}}$. A more fundamental problem is that even at this low density there is an increasing discrimination against *trans* states as the growth proceeds. As sites are added the effective volume available to subsequent ones diminishes. Towards the end the growing chain 'sees' a very different environment than at the start. Examination of the distribution of conformers along the polyethylene chain shows that there is a monotonic decrease in the %*trans* from 68% at the start of the growth to 52% at the end.

It was found that these initial biases disappear once the chain is dynamically relaxed[9] so that the method appears to be considered justified at least for this linear chain model. The effect may be a more serious problem for chains where monomer motions are subject to a higher degree of steric hindrance.

2.3.2 Growth followed by the introduction of excluded volume

This method produces a uniform although still biased distribution of conformational states.[9] It utilizes the hypothesis introduced by Flory[27] that the configurational properties of chains in a pure melt should be the same as those in a theta solvent due to screening of all long range van der Waals interactions. This is the basis of Flory's analysis of chain structure and is a fundamental premise underpinning the rotational isomeric state (RIS) theory approximation.[28] There is strong evidence in support of Flory's hypothesis[29] from experimental measurements.

Growth proceeds at finite temperature in a very similar way to the self-avoiding random walks except that successive torsion angles in a chain occur with a probability that is related to immediately adjacent torsion angles only (the so-called 'pentane effect' in the case of linear polyethylene). All longer range excluded volume effects are ignored.

Having grown the chain the complete set of van der Waals interactions must now be added. This is done within the confines of a periodic system and the chosen size of the unit cell defines the density. The choice of the initial density is not so critical. It can be the expected final relaxed density or even an experimental value. The choice can, however, affect the rate of dynamical relaxation.

An unavoidable side-effect of phantom chain growth (PCG) at melt densities is that there will be a large number of overlaps between sites. In principle, energy minimization could be used to remove the high energy contacts but since this effectively means quenching the system to zero Kelvin it could lead to a shift in the carefully prepared distribution of conformers. It is preferable to maintain the system at the desired temperature using dynamical relaxation.

During the initial stages of this relaxation it is necessary to use an altered LJ potential which moderates the forces resulting from strong overlaps. The most satisfactory results have been obtained using a 'truncated force' potential in which the short range force is constrained to be constant below a critical separation r_{tr}, i.e.

$$\frac{-d\Phi_{mLJ}(r)}{dr} = F_{tr}\left(=\frac{-d\Phi_{LJ}(r_{tr})}{dr}\right) \quad \text{for } r \leqslant r_{tr} \quad [13]$$

The full definition of the resulting modified potential is then

$$\Phi_{mLJ}(r) = \Phi_{LJ}(r) \quad \text{for } r > r_{tr} \quad [14]$$

$$\Phi_{mLJ}(r) = \Phi_{LJ}(r_{tr}) + (r_{tr} - r)F_{tr} \quad \text{for } r \leqslant r_{tr} \quad [15]$$

r_{tr} must be sufficiently small so that only a few pairs will be within this distance in the equilibrium distribution (which at this stage is unknown), but not so small that the large magnitude of F_{tr} causes breakdown of the algorithm. In practice a useful procedure is to reduce the value systematically from $\sim 0.90\sigma$ to $\sim 0.70\sigma$ during the course of the initial relaxation (~ 3 ps). The transition from the modified potential to the full LJ potential can then be made and so far this approach has not failed at least for the polyethylene model. The sequence of pair distribution functions shown in Fig. 2.6 clearly illustrates the effectiveness of this method for introducing excluded volume.

2.3.3 Dynamic relaxation of samples

This is carried out most conveniently using controlled pressure molecular dynamics. The characteristics of this relaxation as a function of the growth conditions and temperature have been fully discussed elsewhere.[9] Properties such as the density and the fractions of different conformers are useful indicators of the extent of relaxation. Studies of the preparational history dependence[9] at melt temperatures show that these properties decay in a few hundred picoseconds for the polyethylene model samples obtained by PCG. Much longer times are required for full relaxation of samples grown by the self-avoiding random walk as is shown in Figs. 2.7 and 2.8 which show the time variation of the densities and the fraction of *trans* conformers (since we use a continuous torsional potential

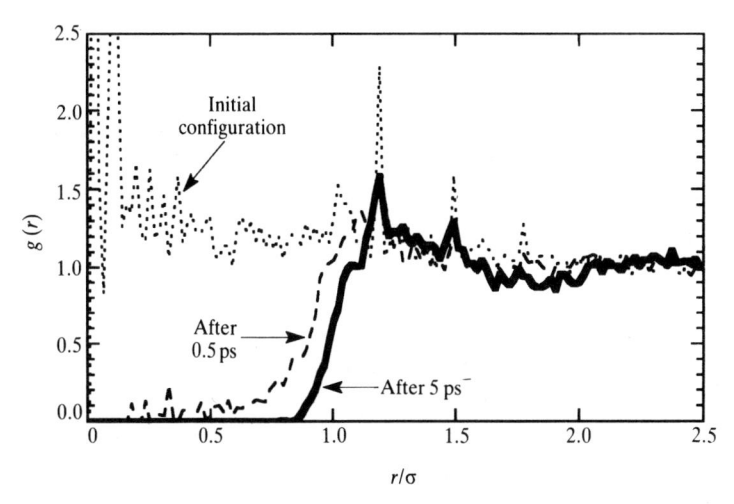

Figure 2.6 Typical radial distribution functions, $g(r)$, obtained at various times following the introduction of excluded volume using the 'truncated force' potential on chains grown using PCG (see text for details). From McKechnie et al.[9]

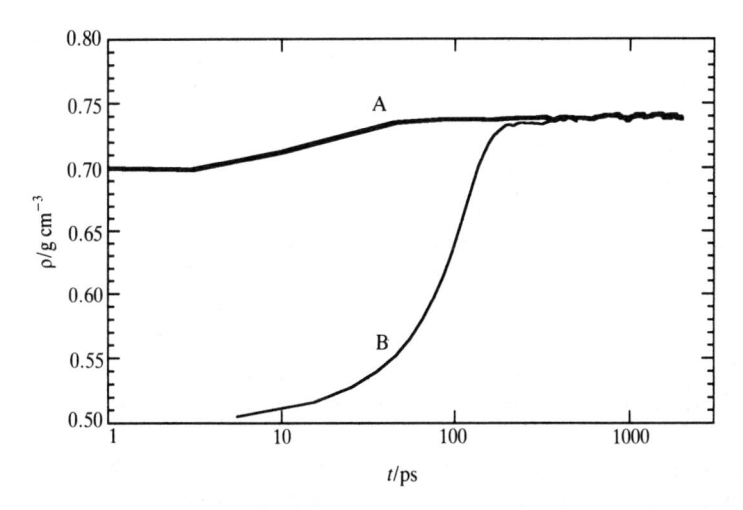

Figure 2.7 The average variation of the density during controlled pressure relaxation at 1 bar for configurations initially obtained by (A) phantom chain growth and (B) self-avoiding random walks at 500 K. From McKechnie et al.[9]

a dihedral angle is deemed to be in the *trans* state if $-60° < \alpha < +60°$) for the two methods above.

One point worth emphasizing is that once the excluded volume effects are included it is not possible using molecular dynamics significantly to

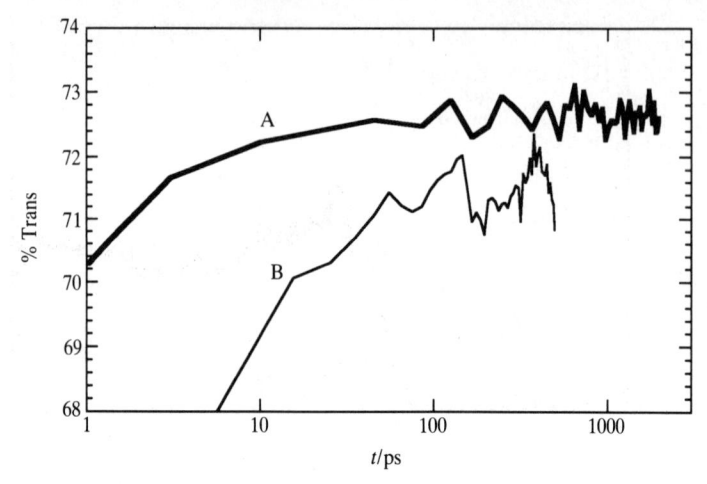

Figure 2.8 The average variation of the percentage of dihedral angles in the *trans* state during controlled pressure relaxation at 1 bar for configurations initially obtained by (A) phantom chain growth and (B) self-avoiding random walks at 500 K. From McKechnie et al.[9]

alter the overall configuration of a chain which is longer than its entanglement length due to the very long associated relaxation times. This characteristic can actually be turned to advantage in model studies of polymer glasses (see later).

Although the distribution of conformational states is guaranteed to be uniform along the chain in PCG, it is seen from the results shown in Fig. 2.8 that the overall fraction of *trans* states in the initial configurations is in fact below that in the relaxed melt configurations. The reason for this is that the growth procedure for both PCG and the self-avoiding random walk, as mentioned above, sample a biased distribution of chains as shown by Batoulis and Kremer.[30] This turns out to be not a problem at 500 K since the discrepancy is relaxed out quickly during the initial stages of the molecular dynamics (see Fig. 2.8).

On the basis of Flory's hypothesis we would expect that unbiased phantom chain sampling (or RIS calculations using the correct statistical weights) should return the same conformational statistics as observed in the melt where there are interactions between all non-bonded monomer pairs. Details of a suitable Monte Carlo scheme are given in Appendix 3 and we have used this method to verify the hypothesis for our model by comparing the conformational distributions in melts composed of short linear chains using molecular dynamics with those obtained using unbiased phantom chain growth; the fractions of *trans* conformers agree well in the two cases. The data also extrapolate smoothly to the results shown in Fig. 2.8 for the $N = 1000$ chains.[31]

2.4 Mechanical 'experiments' on model polyethylene

One of the advantages of controlled pressure MD is that control of the applied pressure tensor P_0 can be used to impart strain to a sample as a function of time in much the same way as in laboratory experiments. As illustrated in Fig. 2.9, control of appropriate components of the pressure tensor can be used to produce uniaxial tension, compression or shear, to take just three examples.

Of course very high rates (equivalent to frequencies in the GHz region) must be used in order to observe significant changes in the short simulation times that are currently available to us. Nevertheless, as we shall show below, the results obtained show a striking similarity to those of laboratory experiments performed on time scales many orders of magnitude slower.

For the experiments described below five independent samples were generated and allowed to relax for 500 ps at 500 K at an applied isotropic pressure of 1 bar. Samples at different temperatures were then obtained

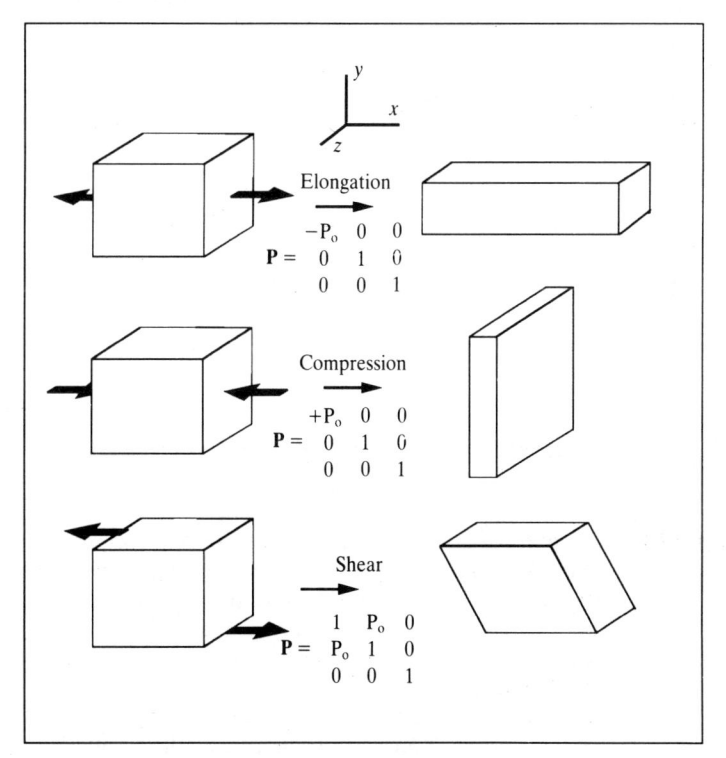

Elongation
$$P = \begin{pmatrix} -P_0 & 0 & 0 \\ 0 & 1 & 0 \\ 0 & 0 & 1 \end{pmatrix}$$

Compression
$$P = \begin{pmatrix} +P_0 & 0 & 0 \\ 0 & 1 & 0 \\ 0 & 0 & 1 \end{pmatrix}$$

Shear
$$P = \begin{pmatrix} 1 & P_0 & 0 \\ P_0 & 1 & 0 \\ 0 & 0 & 1 \end{pmatrix}$$

Figure 2.9 Examples of mechanical deformations that can be induced by an applied control pressure P_0 using the loose-coupling algorithm.

by cooling or heating at a rate of 1 K ps^{-1} to the desired temperature under isotropic controlled pressure conditions (1 bar) followed by subsequent periods of relaxation of order 1 ns. Using this procedure additional samples were generated at 600, 400, 300, 200, 100 and 10 K.

The mean densities obtained at the various temperatures are shown in Fig. 2.10. As the temperature is lowered from 500 K the density is seen to increase with a gradual decrease in the thermal expansivity of the polymer (obtained from the slope of the plot) towards values typical of amorphous solids. Although it is not easy to make useful comparisons with data on real polyethylene since the laboratory material is very difficult to prepare in a completely amorphous state, our data for 500 K and 600 K suggest that the model polymer has a lower expansivity than the laboratory material. This may be a deficiency of the model since the chain interaction parameters were fitted only to the density at 500 K.

The decrease in expansivity as the sample is cooled is exactly as expected upon glass formation; the relaxation in density cannot keep up with the

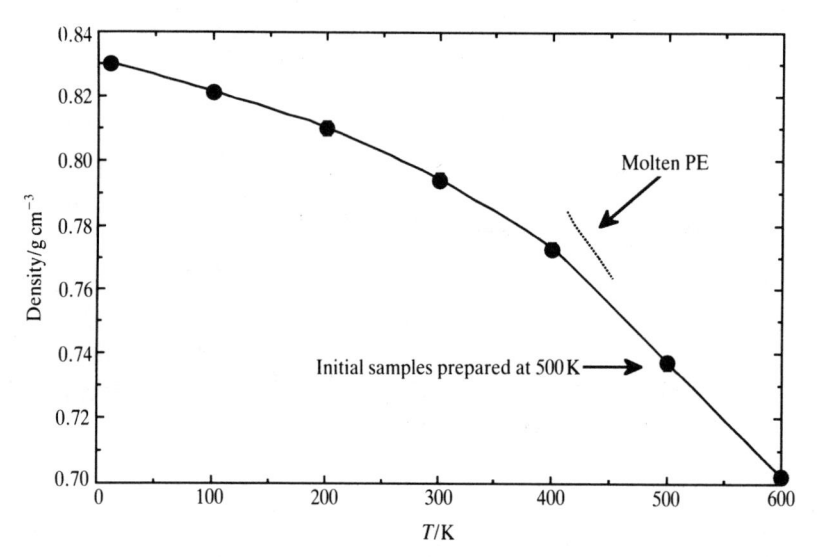

Figure 2.10 The average density of the polymer model at a pressure of 1 bar as a function of temperature. The initial samples were grown and relaxed at 500 K using the method described in the text. Samples at other temperatures were prepared by heating or cooling at a rate of 1 K ps^{-1} (see text for details). The Lennard–Jones potential parameters were adjusted to give a density at 500 K which corresponds approximately to the extrapolation of experimental data for molten polyethylene (dotted line).[18] The spread in the results from the five independent samples is less than the size of the symbols. From Brown and Clarke.[4]

rate of removal of kinetic energy as the sample is cooled and it falls progressively further out of equilibrium. As we shall discuss later, the apparent broad nature of the transformation range can be explained in terms of the very rapid effective cooling rate – amounting to about 10^{11} K s^{-1} – in the simulation experiment.

2.4.1 Uniaxial tension experiments

2.4.1.1 Non-linear behaviour

We describe here the results of simulations in which the above prepared samples were subjected to a gradually increasing uniaxial tension by changing the y component of the applied pressure tensor, $\mathbf{P}_{yy}^0$, at a constant rate

$$\frac{d\mathbf{P}_{yy}^0}{dt} = -\dot{\tau} \qquad [16]$$

where the tension application rate, $\dot{\tau}$, used was either 5 bar ps^{-1} or 1 bar ps^{-1} and the minus sign accounts for the fact that tension is a negative pressure. Employing two different values for $\dot{\tau}$ is useful in gauging the extent to which the measured properties are rate dependent.

We shall be interested primarily in the strain induced in the sample so the applied tension is best considered as a control variable which produces a change in the strain. The response is given by the measured tension, i.e. $-\mathbf{P}_{yy}$, within the sample. In these experiments then both the strain and the measured tension are dependent variables. The method is preferable to direct control of the strain since there is no way *a priori* of predicting how the shape or density of the sample will respond to a change in the external conditions. Studies have recently been reported of polymer deformation using energy minimization at constant volume,[5] a restriction which imposes a value of 0.5 for Poisson's ratio. This is equivalent to applying a complex pressure field which makes interpretation and comparison with real systems more difficult.

With small samples of about 1000 monomers experiments were continued until the sample had extended by about 50–100% of its original length. Extensions beyond about 100% were not possible without violating the truncation radius criterion for the site–site potential due to the contraction in the transverse direction.

The primary information which results from these tension experiments is the response of the **h** matrix, defining the size and shape of the primary cell, and that of the measured pressure tensor, **P**. These together allow us to elucidate the stress vs. strain behaviour. In general the small system sizes used means that there is a relatively complex and fluctuating response of the shape and dimensions of the system to the applied tension.

From the **h** matrix we can calculate the length of the primary cell, L, parallel to the direction of the applied tension and the cross-sectional area, A, perpendicular to it. From A an effective width, W, can also be defined:

$$W = A^{1/2} \qquad [17]$$

Using L and W, the effect of the applied tension can be resolved into the nominal strains parallel and perpendicular to the direction of application as in the case of a laboratory experiment:

$$\gamma_L = \frac{L - L_0}{L_0} \qquad [18]$$

$$\gamma_W = \frac{W - W_0}{W_0} \qquad [19]$$

Here the zero suffix denotes the equilibrium value and is defined in this case as the average over a time interval of about 100 ps just prior to the application of tension.

Figure 2.11 shows the average response of five independent samples each relaxed at six temperatures from 10 K to 500 K. The results for the measured tension, i.e. $-\mathbf{P}_{yy}$, are plotted against the percentage extension ($100 \times \gamma_L$) for experiments carried out with the two tension application rates of 1 and 5 bar ps^{-1}. There is clearly a wide range of behaviour observable in the model system; at low temperatures the material can support the tension up to strains of $\sim 20\%$ before undergoing yield and at progressively higher temperatures there is a gradual change in behaviour until at 500 K it is unclear from Fig. 2.11 whether there is any elastic response at all.

Plotting the extension against time demonstrates the difference between elastic and viscous behaviour as shown in Fig. 2.12. On a log–log scale an elastic response should have an asymptotic slope of 1 at low strains for a system with a well-defined Young's modulus, E,

$$-\mathbf{P}_{yy} = E\gamma_L \qquad [20]$$

Alternatively if the response to the applied tension is viscous, i.e.

$$-\mathbf{P}_{yy} = \eta_e \dot{\gamma}_L \qquad [21]$$

where η_e is an elongational viscosity coefficient, then it is easy to show that for the experiment we perform here that the strain would increase quadratically in time and hence give a slope of 2 on Fig. 2.12.

Both types of behaviour are evident, confirming the trend from elasto-plastic to viscous response as the temperature is increased. The estimated extensional viscosity at 500 K from the simulations is of the order of 0.01 Pa s. Although this is much lower than the equilibrium extensional

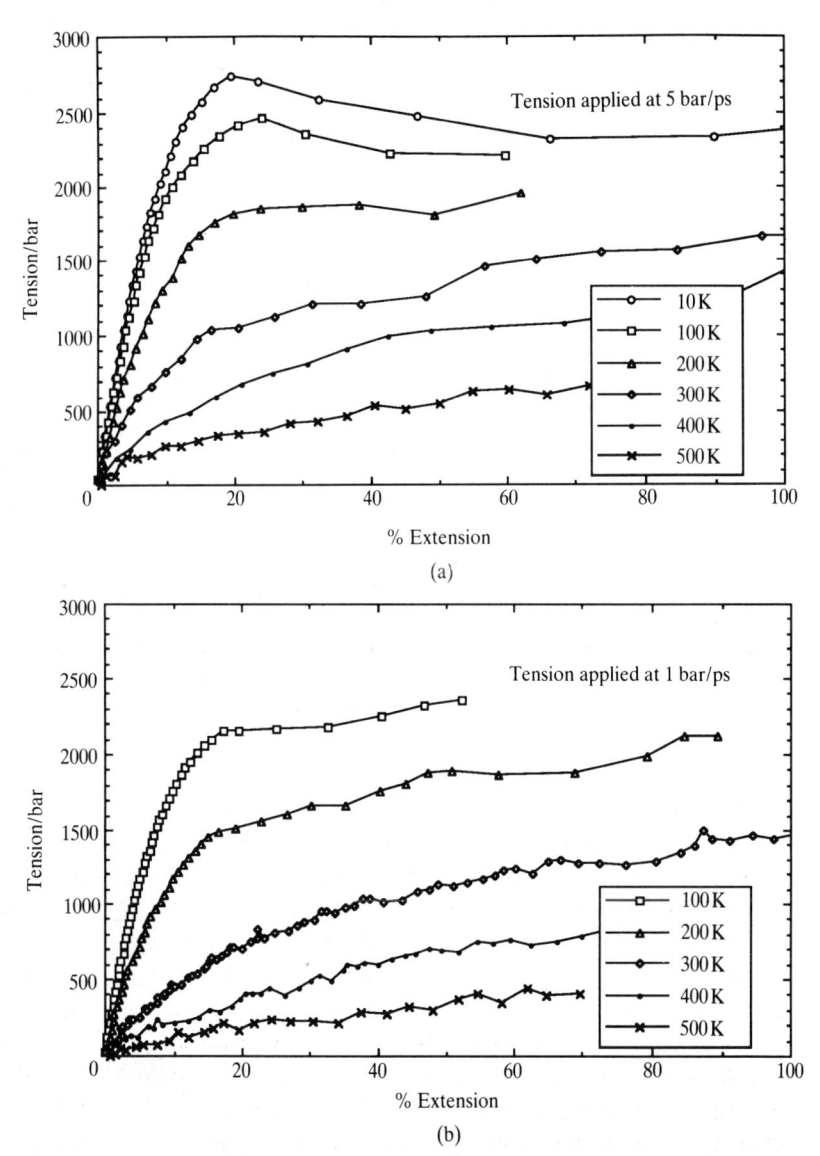

Figure 2.11 The measured tension ($-P_{yy}$) as a function of percentage extension ($\gamma_L \times 100$) for tension applied at (a) 5 bar ps^{-1} and (b) 1 bar ps^{-1}. The data at each temperature represent the average behaviour over five independent samples. From Brown and Clarke.[4]

viscosity of polyethylene it is known that the viscosity does decrease significantly with increasing strain rate.[32] At the extension rates used in our simulations, $10^7 \rightarrow 10^9$ s^{-1}, we expect the behaviour to be similarly non-Newtonian.

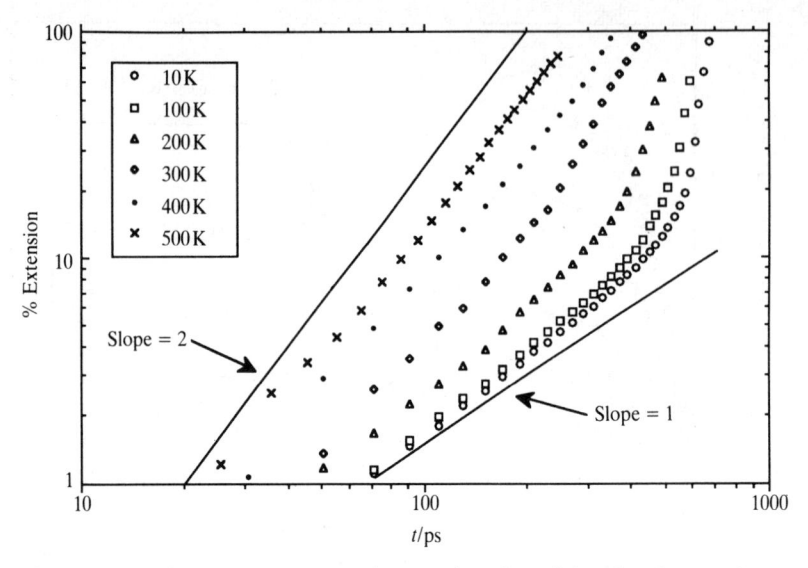

Figure 2.12 The percentage extension as a function of time for the samples subjected to a tension application rate of 5 bar ps^{-1}. On the log–log plot a slope of 1 indicates an elastic response to the applied tension, whereas a slope of 2 is that expected of a viscous material. From Brown and Clarke.[4]

In the laboratory it is not easy to measure the true tension in a macroscopic sample due to the inhomogeneity in the distribution of stress and strain[33] which accompanies necking and cold-drawing. The more usual plot in this case is of the load versus extension in which yield is often accompanied by a noticeable load drop. If we take the product of the measured tension, P_{yy}, and the effective cross-sectional area, A, then a pseudo-load can be calculated. In Fig. 2.13 this load has been plotted for all the data obtained at the two different tension application rates. For our systems the dimensions of the cell are only of order 30 Å so the loads obtained appear small, $\sim 10^{-9}$ N, in comparison to laboratory values. The load drop is quite evident in the plot for the tension applied at 5 bar ps^{-1} at a temperature of 200 K and below. Brown and Ward[34] suggest that there is a drop in true tension at the yield point; this is supported by our data at 5 bar ps^{-1} but not by that for the lower tension application rate.

The observed values of the yield stress and strain are much larger than those typically observed in the laboratory. In qualitative terms we can understand this as the result of the slow response of the internal structure as compared to the time scale of the perturbation. In comparing Figs. 2.11(a) with 2.11(b) and Figs. 2.13(a) with 2.13(b) it is evident that there

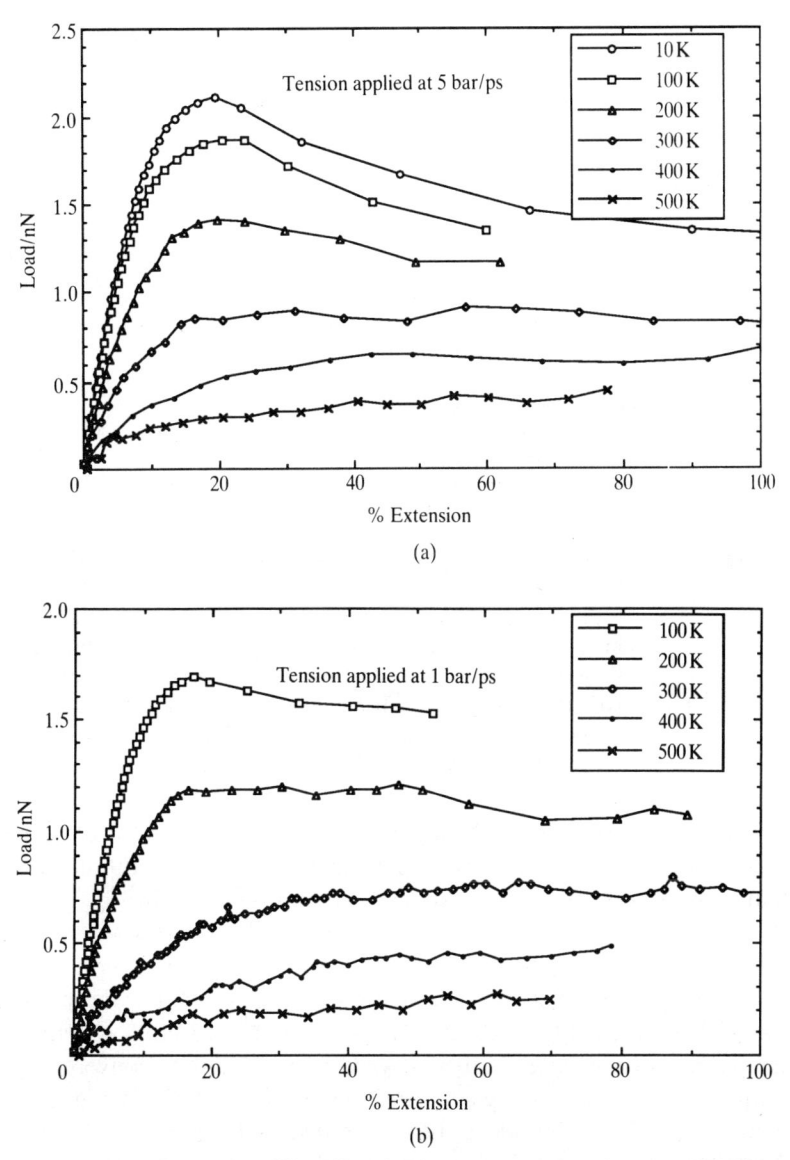

Figure 2.13 The load as a function of percentage extension ($\gamma_L \times 100$) for tension applied at (a) 5 bar ps^{-1} and (b) 1 bar ps^{-1}. From Brown and Clarke.[4]

is a decrease in the values of both the yield stress and the yield strain as the tension application rate (or strain rate) is decreased, behaviour which is analogous to that of real polymeric materials.[35,36]

One working definition of the yield stress in the laboratory is the true stress at the maximum observed load.[37] For convenience we have chosen

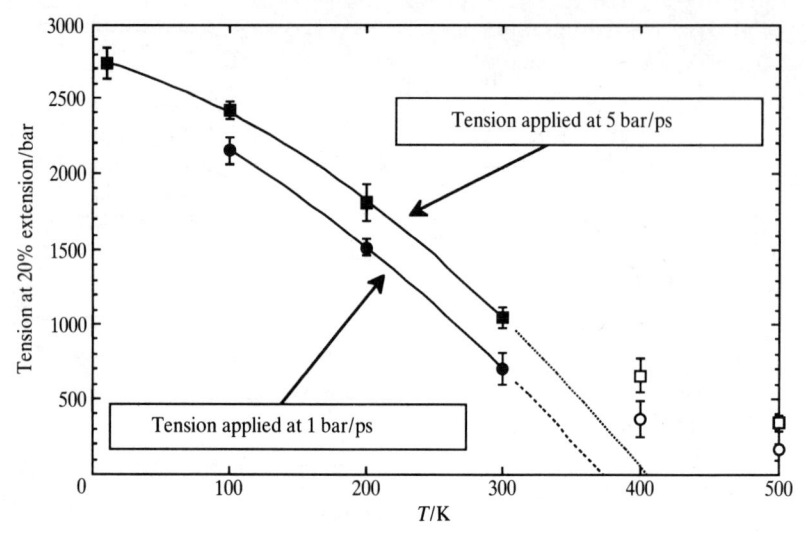

Figure 2.14 Tension at 20% extension ('yield stress') as a function of temperature. Squares and circles refer to tension application rates of 5 bar ps^{-1} and 1 bar ps^{-1} respectively. Open symbols indicate data for which no discernible yield was observed; these points are excluded from the curve fits and extrapolations to zero tension. The error bars shown are the standard deviations in the results for the five independent samples. From Brown and Clarke.[4]

to define the yield stress as the measured tension at a strain of 20% which corresponds closely to observed maxima in the load for those samples that show a maximum. The resultant values are plotted in Fig. 2.14 and the behaviour is very similar to that found in real systems where the yield stress decreases approximately linearly with increasing temperature.[38]

Our data cover a very wide temperature range which may account for the slight non-linearity. It has also been shown in laboratory experiments that extrapolating the data to zero yield stress results in convergence to a temperature close to the glass transition temperature.[38] If we ignore the points above 300 K for which there is no discernible yield point our data extrapolate to zero yield stress at around the same temperature where there is a change in expansivity (see Fig. 2.10). As for laboratory measurements there is a dependence on the rate of application of the tension with the lower rate leading to consistently lower values of the yield stress and hence a lower extrapolated temperature of zero yield stress.

We can gain some insight into the nature of the yield process by examining the response of the system densities to the applied tension at the different temperatures. At high temperatures (400–500 K) where the flow process is predominantly viscous there is a hardly perceptible change in density during the extension of the samples (see Fig. 2.15).

In contrast at low temperatures there is a noticeable dilation effect as the tension is applied and the density decrease continues until just beyond the yield point. Once the material yields the density remains relatively constant as plastic flow takes place. What is interesting is that for a given tension application rate this apparent critical density for yield and plastic flow is independent of temperature over the range 10–300 K. We should not conclude from this that yield is purely associated with a reduction in density; it is worth recalling the well known experimental result that yield can also occur in amorphous polymers under compression.

The decrease in density under extension at the lower temperatures is entirely consistent with the typical values of Poisson's ratio (μ) for amorphous polymeric solids which are generally in the range 0.3 to 0.4.[39–42] Indeed our estimates of Poisson's ratio from the extensional and contractile strains

$$\mu = \lim_{\gamma_L \to 0} \frac{-\gamma_C}{\gamma_L} \qquad [22]$$

give values of about 0.41 at the lowest temperatures.

2.4.1.2 Extensional modulus

The extensional and shear moduli are important properties which quantify the mechanical response of a polymer in the linear (small strain) regime. In the laboratory generally a dynamic technique is used in which an oscillating small amplitude strain is applied to a sample at a certain frequency and the components of the (complex) moduli are then determined from the amplitude and phase lag of the resultant stress. As a function of decreasing temperature a significant change in the storage (real) modulus usually accompanies the transition to glassy type behaviour. Although our experiments are carried out in a different manner it is still possible to obtain an estimate of the moduli from the initial slopes of the tension vs. extension plots.

Young's modulus is defined as the ratio of stress to strain in the limit of zero strain. It is, however, difficult to obtain precise values of this quantity from the experiments in our case due to the increasingly poor statistics of the points at low extensions. As a compromise we have determined values of the modulus from a least squares fit of the tension versus extension data for those points having an extension less than 5%. This procedure is somewhat arbitrary, but only at the very lowest temperatures can an obvious non-linear trend be seen in these plots (not shown).

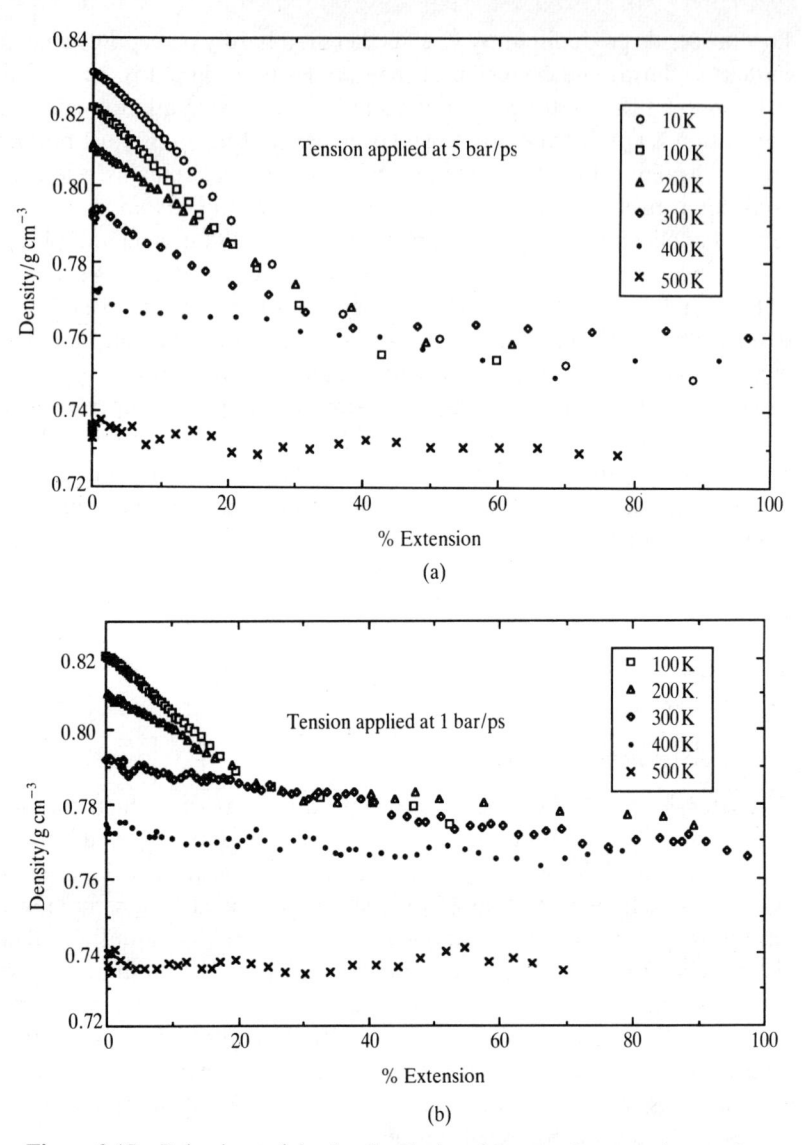

Figure 2.15 Behaviour of the density during the extension experiments for tension application rates of (a) 5 bar ps^{-1} and (b) 1 bar ps^{-1}. Note the distinct density decrease as the extension approaches the yield point at the lower temperatures. This contrasts with post yield plastic flow, and the viscous flow observed at high temperatures. From Brown and Clarke.[4]

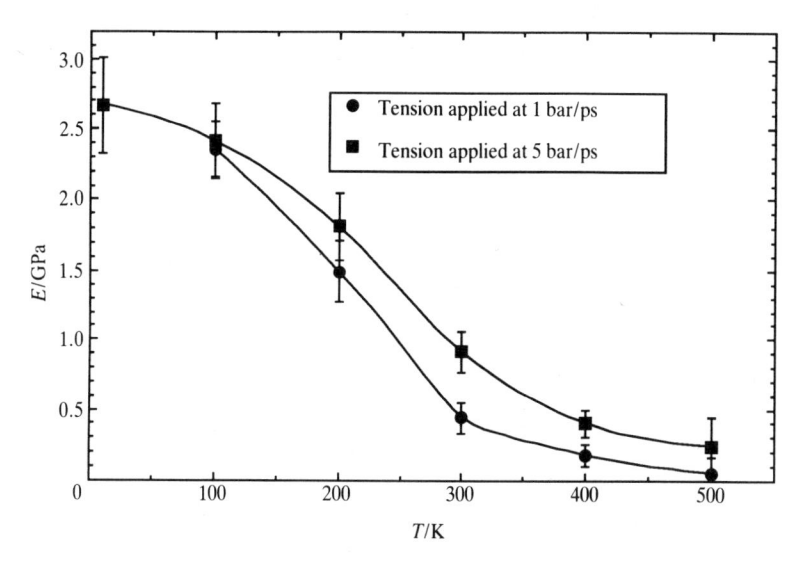

Figure 2.16 Extensional (Young's) modulus obtained from small (<5%) strain behaviour (see text) as a function of temperature. The squares and circles refer to the data obtained at tension application rates of 5 bar ps^{-1} and 1 bar ps^{-1} respectively. From Brown and Clarke.[4]

In Fig. 2.16 the average moduli obtained at the different temperatures from the experiments at both applied tension rates are shown. To give an indication of the errors involved the standard deviations evaluated from the five samples are also shown. The convergence of the moduli at the lowest temperatures is expected as the material becomes more solid like. At higher temperatures lower values of the modulus occur for the lower tension application rate, emphasizing the viscoelastic nature of the response.

The results of our very rapid experiments should best be compared with high frequency experimental data, but at low temperatures the moduli are relatively insensitive to frequency. The actual values of the modulus at low temperatures are of the same order as those measured in the laboratory ($\sim 10^9$ Pa[43]). At higher temperatures (400 K and 500 K) the moduli fall to relatively small values although the decrease is much smaller than the three or four orders of magnitude decrease that is seen in the laboratory. It is, however, unlikely that we would be able to measure such a large decrease due to statistical problems.

2.4.2 The glass transformation

There is clear evidence of a glass transformation in the region of 350–400 K both from the mechanical experiments and from the behaviour

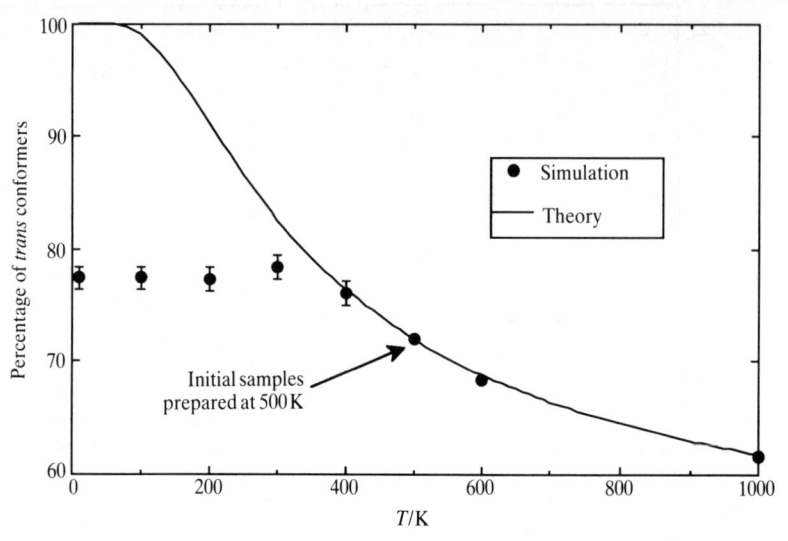

Figure 2.17 The percentage of *trans* conformers as a function of temperature for the unperturbed polymer samples prepared either by cooling or heating those generated at 500 K. The theoretical curve is an Arrhenius fit of the data for the equilibrium constant at 500 K and 1000 K. The error bars shown are the standard deviations in the results for the five independent samples. From Brown and Clarke.[4]

of the density. Direct observation of the freezing out of conformational transitions in this temperature range provides strong supporting evidence. This is shown in Fig. 2.17 which gives the temperature dependence of the fraction of *trans* conformers averaged over the five samples of unperturbed chains at the various temperatures. We define a dihedral angle to be *trans* if it lies between $\pm 60°$, otherwise the angle is in one of the two *gauche* states. The solid line in the plot is an extrapolation to lower temperatures made on the basis of fitting the data at 500 K and 1000 K to the form

$$\frac{\langle X_T \rangle}{\langle X_G \rangle} = A \exp\left(\frac{\Delta \Phi}{RT}\right) \quad [23]$$

where X denotes the fraction of *trans* (T) or *gauche* (G) states and A and $\Delta \Phi$ are the adjustable parameters. The actual values used for the curve shown were $A = 1.01$ and $\Delta \Phi / R = 463$ K. The nominal difference in energies between the *gauche* and *trans* wells for the dihedral angle potential is equivalent to 530 K but this takes no account of the excluded volume effects which, for example, largely prevent sequences of the type G^+G^- from occurring. The low value of 463 K obtained simply implies an overall enhancement of the *trans* fraction.

Figure 2.17 confirms that on the time scale of about a nanosecond the torsional degrees of freedom rapidly fall out of equilibrium when the temperature is reduced below ~ 400 K. Indirectly the same phenomenon has been previously inferred[4] from the behaviour of the end-to-end distance of short $N = 10$ and $N = 20$ alkane-like chains. The temperature at which this occurs correlates reasonably well with the temperature of the glass transformation indicated in our case by the behaviour of the modulus and disappearance of the yield stress. This is in accord with the torsional degrees of freedom being the dominant modes of relaxation in these systems.

At first sight it would appear that the characteristics of the glass transformation observed in the simulated polymer are quantitatively quite different to what is observed on macroscopic time scales in the laboratory. The transition is taking place at a much higher temperature than one would expect, and from the behaviour of the modulus the transition region seems to extend over a very wide range of temperature (~ 200 K). We can only estimate T_g for real polyethylene as a result of the experimental difficulties in realizing the pure amorphous polymer in the laboratory. However, values found in the literature are in the range ~ 150–300 K.[44,45]

Both of the above characteristics can be explained in terms of the very rapid cooling rates employed in simulations and they have nothing to do with the details of the model. Firstly we should note that similar behaviour has been observed in simulations of glass formation for the Lennard–Jones fluid using cooling rates of the order 10^{12} s^{-1}.[46] Properties such as the expansivity showed a gradual transformation from liquid to solid-like values over a temperature range of about half the temperature of the mid-point of the relaxation T_r. Also the value of T_r is high compared to that expected from a comparison with simple fluids on laboratory time scales.[47]

We can turn to laboratory data on glass-forming liquids for more compelling evidence for the above conclusions. For example, Brillouin scattering measurements have been used to probe the sound velocity at frequencies of about 10 GHz in cooling experiments on the glass-forming liquid $2Ca(NO_3)_2 \cdot 3KNO_3$.[48] The high frequency longitudinal compliance and the adiabatic compressibility derived from the measurements both show a smearing out of transition to high temperatures whilst the more familiar static measurements show a sharp transition at the low end of the transformation range.

In both the fast quenching and the high frequency probe experiments the explanations for these phenomena are the same. Firstly, as a sample is cooled, structural relaxation rates begin to lag behind the rate of perturbation of the system at higher temperatures so the transformation starts well above the 'normal' glass transition. Secondly, these 'fragile' fluids[49] show a marked temperature dependence of the activation energies (E_a) for structural relaxation processes. At low temperatures E_a becomes extremely large and relaxation times may change by one or two orders of magnitude over a few degrees – thus producing a very sharp transition. At high temperatures a much smaller value of E_a means that the same change occurs over a much wider temperature range – thus producing a broad transformation.

2.4.3 Stress–strain behaviour and configurational properties

The characteristic form of a stress–strain curve for a glassy polymer shows an initial linear response typical of any dense solid from which the normal extensional modulus can be derived (see section 2.4.1). At higher strains the response usually becomes non-linear and may even display a distinct yield point and plastic flow regimes. At higher strains still there is nearly always an increase in the modulus which is referred to as strain hardening. This latter property is of great practical importance since the extent of strain hardening is associated with the susceptibility of the material to, for instance, necking and crazing. On a phenomenological level the large strain behaviour of polymers has been rationalized in terms of mechanical models which comprise elements representing the Hookean, viscous and rubber elasticity components of the stress,[50,51] but there is considerable interest in understanding strain hardening in terms of chain topology.

Necking and shear banding can also complicate the interpretation of experimental data since they make true stresses and strains difficult to determine. True stress–strain curves have been obtained for glasses composed of bis(phenol A) copolymerized to produce carbonate and phthalate chains of varying stiffness, and in this case it was noted that there was a correlation between Kuhn length (a measure of chain stiffness) and the large-strain behaviour. Similar results have been obtained with polyisocyanates which have extremely large Kuhn lengths and pronounced strain hardening so that they extend uniformly without necking.[52]

Molecular dynamics simulations can be designed systematically to explore the way in which the true stress–strain behaviour of an amorphous, linear polymer glass depends on its configurational properties. (The boundary conditions employed in our simulations result in a cross-sectional area that is uniform in the direction of the applied tension, consequently the stresses reported are true stresses.) In this case one utilizes the ease with which the system parameters can be controlled to prepare samples of the same glassy polymer with different configurational properties. The aim is to avoid any ambiguities introduced by comparing polymers with different chemical structure.

In principle the configuration of a polymer is completely specified by the sequence of torsion angles, valence angles and bond lengths. Such information is, however, quite intractable and not easy to interpret. Two structural properties that can often be measured experimentally are, however, the fractions of conformers, and in the persistence length of a polymer chain these properties were used as a basis for discussing the computer simulation results.

If we consider the ways in which an amorphous polymer can deform, then elongation can be achieved either by uncoiling the overall chain configuration or by changing the local conformations within the 'tube' formed by neighbouring chains in the entangled structure. This alternative 'local' mechanism corresponds in our model to segments of chains being converted from the *gauche* ('short') form to the long ('*trans*') form; the latter is the lowest energy state and has a planar zig-zag structure. For the polyethylene model used it follows that chains which are either less coiled or contain fewer *gauche* states to begin with will be harder to deform. For the study in question the persistence length was used as a measure of the overall configuration. In an equilibrium polymer melt the overall configuration and the fractions of conformers must be very closely linked, but this restriction does not apply for non-equilibrium glassy states where changes in the preparation procedure can provide some measure of independent control of the two properties.

Flory[27] gives the definition of persistence length a as

$$a = b_0 \sum_{k=0}^{\infty} \langle e_i \cdot e_{i+k} \rangle \qquad [24]$$

where e_i is a unit bond vector, and further shows that

$$a = \tfrac{1}{2} b_0 (1 + C_\infty) \qquad [25]$$

where

$$C_\infty = \lim_{k \to \infty} \frac{\langle r_k^2 \rangle}{k b_0^2} = \frac{\langle (r_i - r_{i+k})^2 \rangle}{k b_0^2} \qquad [26]$$

We have used Eqn [26] to evaluate the averages over five configurations for values of k up to 100. We cannot expect to achieve the asymptotic limit for this rather modest number of monomers. Nevertheless we shall deduce correlation lengths from these data and will assume that to a good approximation the ratios of these values reflect the ratios of the true persistence lengths.

Using the method described in section 2.3 four different sample sets of polymer glass were produced by using different preparation procedures[54]. All of the samples were relaxed at 200 K with $P_0 = 1$ bar for ~ 1 ns. The final densities were all the same to within 1% and we expect all of the

Table 2.1 The percentage of *trans* conformers, %*trans* and correlation length, a_{100}, calculated for each of the four sample sets.

Sample	%*trans*	$a_{100}/\text{Å}$
A	78	5.0
B	82	25.5
C	77	12.0
D	70	5.8

samples to have glass transformations in the range 300–400 K on the simulation time scale. The associated correlation lengths and fractions of *trans* conformers obtained from the final 200 ps of these runs are shown in Table 2.1.

Sample set A was formed by cooling from the melt as described in section 2.4. Set B was produced by growing chains at 200 K using PCG and relaxing them for 1 ns at the same temperature. Set C was obtained by 'flash' heat treatment of set B; this involves raising the temperature instantaneously to 1000 K for 100 ps followed by rapid cooling back to 200 K. This treatment is not sufficiently drastic to alter the overall configuration of the chain, but it does allow conformational transitions to occur and the overall result of this treatment is a net decrease in the *trans* fraction. The more kinked structure shows noticeable retraction of the polymer chain within the 'tube' formed by its near neighbours (see Fig. 2.18).

Sample set D was obtained by phantom chain growth at 200 K with the torsional and 1–5 interaction potentials scaled down by a factor of 50; this has the effect of increasing the *gauche* fraction and produces a highly coiled chain. Van der Waals interactions were then introduced and subsequent relaxation with the full interaction potentials produced a polymer glass with a much reduced fraction of *trans* states.

The stress–strain behaviour of all four sets of samples was obtained by subjecting them to an externally applied uniaxial tension which was increased at a rate of 5 bar ps^{-1}, exactly in the same way as described in section 2.4.1. As a result of the small system size, data were obtained only up to extensions of $\sim 100\%$. In Fig. 2.19 the tension developed within the sample is plotted as a function of the strain; similar plots are shown for the load in Fig. 2.20. The 'load' is determined from the product of the tension and the cross-sectional area and since the samples are very small the loads are also extremely small (of order 10^{-9} N).

The pattern of the stress–strain plots is similar to that discussed in section 2.4.1.1 and shown in Figs. 2.11 and 2.13 – there is an initial elastic response followed by yield and plastic flow. In detail, however, the four

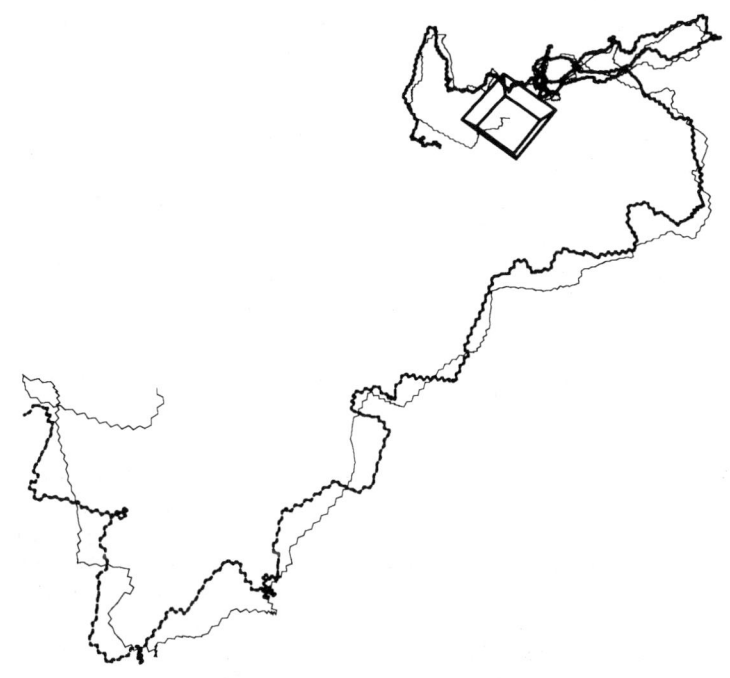

Figure 2.18 A comparison of the chain structure for one sample before (thin line) and after (thick line) heat treatment. Only the coordinates of the continuous primary chain have been shown in both cases. The periodic MD cell which is filled by images of the primary chain is also shown. A higher fraction of *trans* conformers results in the straighter sections visible in the chain represented by the thin line. These change to a more kinked structure after heating with a noticeable retraction of the ends of the chain down the 'tube'. Note in particular how the overall envelope or gross structure is relatively unchanged. From McKechnie et al.[54]

sets of samples show quite different behaviour; the differences begin to be noticeable after about 10% extension. Samples B and C both show enhanced resistance to extension beyond the point at which the A samples yielded ($\sim 20\%$ extension). The high %*trans*-high correlation length samples (set B) show the largest extent of strain hardening and in particular produce significantly more stress than the set C which, within the error, has a similar correlation length but lower *trans* fraction. Conversely, sample set C has practically the same %*trans* as set A so the differences here must be due to their contrasting configurational structure. Set D shows the lowest resistance of all to the applied tension as was expected from its highly coiled structure with a small fraction of *trans* conformations. What these results suggest is that both an increase in the

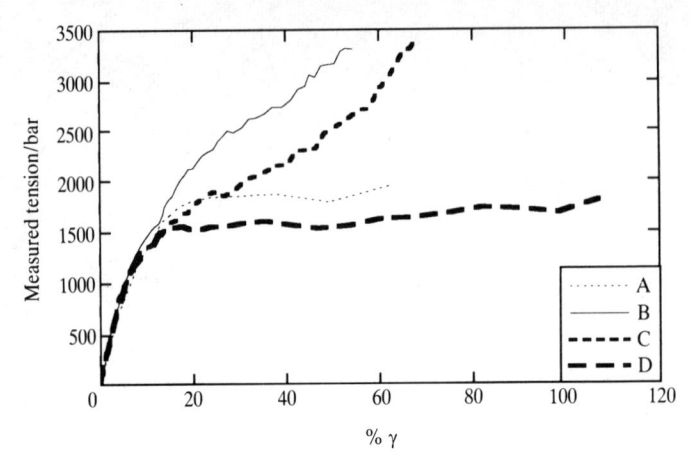

Figure 2.19 The measured tension plotted as a function of the extension for the four sets of samples of the model linear polymer subjected to an applied uniaxial tension increasing at a rate of 5 bar ps^{-1}. The curves represent the results obtained for the four sets of samples A–D of differing configurational and conformational properties as characterized by the persistence lengths and percentage of *trans* conformers shown in the figure and also given in Table 2.1. From McKechnie et al.[54]

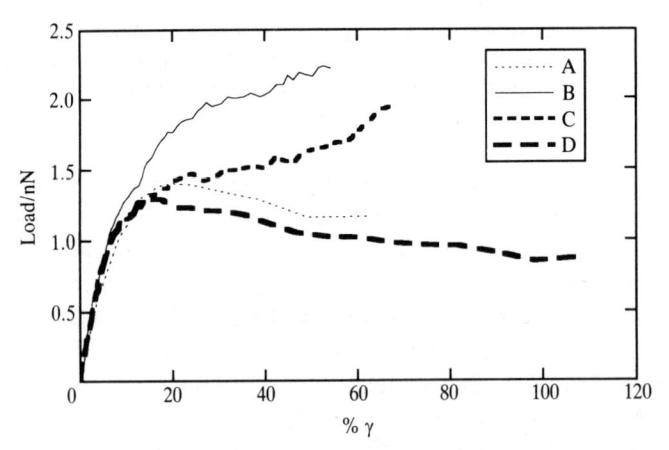

Figure 2.20 The measured load plotted as a function of the extension for the four sets of samples of the model linear polymer subjected to an applied uniaxial tension increasing at a rate of 5 bar ps^{-1}. Notation as for Fig. 2.19. From McKechnie et al.[54]

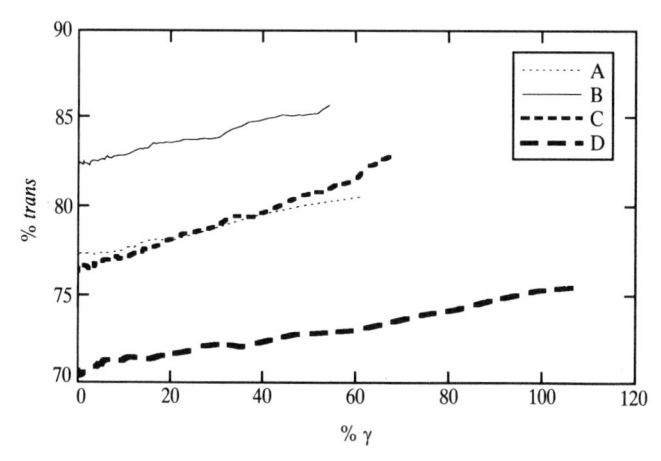

Figure 2.21 The percentage of *trans* conformers plotted as a function of the extension for samples of the model linear polymer subjected to an applied uniaxial tension increasing at a rate of 5 bar ps^{-1}. The curves represent the four sets of samples A–D as in Fig. 2.19. From McKechnie et al.[54]

fraction of *trans* states and an increase in the persistence length can independently contribute to strain hardening.

We can gain further insight into the nature of the changes that occur during deformation of these materials from the strain dependence of the fraction of *trans* conformers. These data are shown in Fig. 2.21. This reveals a common feature of all our uniaxial tension experiments to date – namely that there is a consistently linear dependence of the %*trans* upon the extension. The lack of any discontinuities or breaks in the plots at the yield points confirms the conclusion that this mechanical property has nothing directly to do with the onset of transitions between different conformational states. The gradients of these plots are, however, different and this warrants further examination. In particular we point out the smaller gradients for sets A and D. The small correlation lengths in these samples means that configurations are more highly coiled before straining so that a greater proportion of an extension can be absorbed by uncoiling the chains so there is less increase in the fraction of *trans* states. The large difference between the *trans* fractions of sets B and C that exists before deformation is gradually eroded with increasing extension.

2.5 Conclusions

It is perhaps remarkable that a simple glassy polymer, simulated by molecular dynamics on time scales of order 1 ns, should display very similar mechanical properties to those observed in the laboratory on time scales many orders of magnitude longer; phenomena such as elasticity, yield and plastic flow are all accessible to dynamic atomistic modelling. The simulations further show that these mechanical properties are determined by short-range van der Waals interactions between chains as well as the intrachain structure. In regard to the latter it is necessary to consider not only the overall chain configuration but also conformational properties on length scales down to the level of monomer structure. The transformation from fluid to glassy behaviour is also observable using molecular dynamics simulations although at such ultra-large quench rates the transition is much broader and occurs at a much higher temperature than in the laboratory.

There are still many issues to be resolved. For instance, sample size effects (for a given polymer MW) in relation to, e.g., persistence length have not yet been explored in any detail; such studies are one prerequisite to any successful extension to chemically or structurally more complex polymers. There is much to do in order to establish links between the very high frequency behaviour of polymers to that observed in the laboratory on macroscopic time scales. This will involve examining in more detail the role of chain motions on the whole range of length scales from the size of a monomer to the chain end-to-end separation.

Acknowledgements

We would like to thank E. I. Du Pont de Nemours & Co. Inc. and ICI plc for financial support and the SERC for a generous provision of computing resources.

Appendix 1 The loose-coupling algorithm

The incorporation of such a scheme into the integration of the equations of motion of a system containing constraints via the leapfrog form of the Verlet algorithm is relatively straightforward and is based on that given previously.[22]

We start with the coordinates of the N sites on the chain $\mathbf{r}(t)$ where $(\mathbf{r}_{i+1} - \mathbf{r}_i)^2 = b_0^2$, the corresponding momenta divided by mass at the previous half time step $\mathbf{v}[t - (\Delta t/2)]$, the temperature at the previous step $T(t - \Delta t)$ and the matrix of basis vectors $\mathbf{h}(t) = (\mathbf{abc})$. From Eqn [6] $\mathbf{h}(t)$ determines the volume $V(t)$ and hence the density. We refer to the vector $\mathbf{\textit{v}}$ as a momenta divided by mass in preference to the term 'velocity' as it is important to distinguish the motion due to the forces from that due to the changing shape of the primary cell. Velocity could be interpreted as meaning $\mathbf{\dot{r}}$ which is not the same as $\mathbf{v}$.

The procedure is as follows:

(1) First of all we calculate the scaled positions s(t) of the sites of the chain within the primary cell, i.e. $-1 \leq s(t) \leq 1$, using the following sequence of operations:

$$\mathbf{s}(t) = \mathbf{h}^{-1}(t)\mathbf{r}(t) \quad [A1.1]$$

$$\mathbf{s}'(t) = \mathbf{s}(t) - 2 \times \text{INT}(\mathbf{s}(t)/2) \quad [A1.2]$$

$$\mathbf{s}(t) = \mathbf{s}'(t) - 2 \times \text{INT}(\mathbf{s}'(t)) \quad [A1.3]$$

where by INT we refer to the FORTRAN function which returns the integer part of the expression in the bracket. Note that both transformations given by Eqns [A1.2] and [A1.3] are required as the coordinates of the continuous chain may lie some distance from the primary cell.

(2) Using the scaled coordinates, the site–site separation vectors are calculated using the nearest image convention for those sites deemed to interact via the LJ 12-6 potential, thus

$$\mathbf{s}_{ij}(t) = \mathbf{s}_i(t) - \mathbf{s}_j(t) \quad [A1.4]$$

$$\mathbf{s}'_{ij}(t) = \mathbf{s}_{ij}(t) - 2 \times \text{INT}(\mathbf{s}_{ij}(t)) \quad [A1.5]$$

$$\mathbf{r}_{ij}(t) = \mathbf{h}(t)\mathbf{s}'_{ij}(t) \quad [A1.6]$$

(3) Calculate the force, $f_{ij}(t)$, between sites i and j from the derivative of the LJ potential and the separation vector and also the contribution to the pressure tensor of these non-bonded forces

$$V(t)\mathbf{P}_{ss}(t) = \sum_i \sum_{j>i} \mathbf{r}_{ij}(t) f_{ij}(t) \quad [A1.7]$$

where the double sum is over all interacting pairs and store the total force on each site at this point in a second vector $f^\Phi(t) = f(t)$.

(4) Calculate the forces on the sites due to all other potentials, e.g. torsional and valence angle bending.

(5) From $T(t - \Delta t)$ obtain the temperature scaling factor, λ,

$$\lambda = \left[1 + \frac{\Delta t}{\tau_T}\left(\frac{T_0}{T(t - \Delta t)} - 1\right)\right]^{1/2} \quad [A1.8]$$

where τ_T is an arbitrary relaxation time which controls the rate at which the system tends towards the desired temperature, T_0.

(6) Calculate the new unconstrained positions using the leapfrog algorithm incorporating the temperature scaling factor

$$\mathbf{r}'_i(t + \Delta t) = \mathbf{r}_i(t) + \left[\mathbf{v}_i\left(t - \frac{\Delta t}{2}\right) + \frac{\mathbf{f}_i(t)}{m_i}\Delta t\right]\lambda \Delta t \quad [A1.9]$$

(7) Using an iterative scheme, determine the new constrained positions $(\mathbf{r}(t + \Delta t))$ and from these the constraint forces

$$\mathbf{f}_i^c = \frac{m_i[\mathbf{r}_i(t + \Delta t) - \mathbf{r}_i'(t + \Delta t)]}{\Delta t^2} \qquad [\text{A1.10}]$$

and the new momenta

$$m_i \mathbf{v}_i\left(t + \frac{\Delta t}{2}\right) = \frac{m_i[\mathbf{r}_i(t + \Delta t) - \mathbf{r}_i(t)]}{\Delta t} \qquad [\text{A1.11}]$$

(8) Using the position of sites on the continuous chain, compute the contribution to the pressure tensor from the bonded forces

$$V(t)\mathbf{P}_{\text{bf}}(t) = \sum_{i=1}^{N} \mathbf{r}_i(t)[\mathbf{f}_i(t) - \mathbf{f}_i^{\Phi}(t)] \qquad [\text{A1.12}]$$

where $\mathbf{f}_i$ is now the total force on a site including the forces of constraint, Eqn [A1.10]. The forces due to the LJ potential are subtracted as their contribution has already been calculated, Eqn [A1.7]. The use of the coordinates of the continuous chain for calculating $\mathbf{P}_{\text{bf}}$ is essential to maintain consistency with those used in actually calculating the bonding forces.

(9) From the $\mathbf{v}[t - (\Delta t/2)]$ and the $\mathbf{v}[t + (\Delta t/2)]$ calculate the on-step value of $\mathbf{v}$:

$$\mathbf{v}(t) = \frac{\mathbf{v}\left(t - \frac{\Delta t}{2}\right) + \mathbf{v}\left(t + \frac{\Delta t}{2}\right)}{2} \qquad [\text{A1.13}]$$

(10) Use the $\mathbf{v}(t)$ to calculate the on-step temperature, $T(t)$, using

$$T(t) = \frac{\sum_{i=1}^{N} m_i v_i^2(t)}{k_B N_f} \qquad [\text{A1.14}]$$

where N_f is the total number of degrees of freedom in the system.

(11) From $\mathbf{v}(t)$ again obtain the on-step kinetic contribution to the pressure tensor

$$V(t)\mathbf{P}_{\text{kin}}(t) = \sum_{i=1}^{N} m_i \mathbf{v}_i(t)\mathbf{v}_i(t) \qquad [\text{A1.15}]$$

and thus obtain the total pressure tensor

$$\mathbf{P}(t) = \mathbf{P}_{\text{kin}}(t) + \mathbf{P}_{\text{bf}}(t) + \mathbf{P}_{\text{ss}}(t) \qquad [\text{A1.16}]$$

(12) Update the **h** matrix using Eqn [9]:

$$\mathbf{h}(t + \Delta t) = \mathbf{h}(t) + \left[\frac{\mathbf{P} - \mathbf{P}_0}{M}\right]\Delta t \qquad [\text{A1.17}]$$

(13) Calculate the tensor scaling factor

$$\mathbf{H} = \mathbf{h}(t + \Delta t)\mathbf{h}^{-1}(t) \qquad [\text{A1.18}]$$

(14) Scale the new constrained positions using **H**:

$$\mathbf{r}'_i(t + \Delta t) = \mathbf{H}\mathbf{r}_i(t + \Delta t) \qquad [\text{A1.19}]$$

(15) Return to paragraph (1) and use $\mathbf{r}'(t + \Delta t)$, $\mathbf{v}[t + (\Delta t/2)]$ and $\mathbf{h}(t + \Delta t)$.

The application of step 14 as it stands leads to the destruction of the constraints. At the following step the constraints are reimposed but only after all the forces have been computed from these unconstrained positions. In the past it has been argued[22] that in practice the constraints are only ever satisfied to within some specified tolerance so there is always some error; for the simulations reported here the relative tolerance for the bond length constraint is set to 10^{-6}, i.e. $||\mathbf{r}_{i+1} - \mathbf{r}_i| - b_0| \leq 10^{-6} b_0$ following the imposition of the constraint. We refer to this as Method 1 as distinct from Method 2 in which after step 14 the constraints are reimposed following their destruction by the application of the matrix **H**. A fuller evaluation of these two methods is given in Appendix 2.

For Method 2 the procedure continues:

(15) Using an iterative scheme, reapply the bond length constraints to the set of coordinates $\mathbf{r}'(t + \Delta t)$ to give a new set of constrained coordinates $\mathbf{r}''(t + \Delta t)$.

Note that it is important not to re-calculate a new set of $\mathbf{v}[t + (\Delta t/2)]$.

(16) Return to paragraph (1) and use $\mathbf{r}''(t + \Delta t)$, $\mathbf{v}[t + (\Delta t/2)]$ and $\mathbf{h}(t + \Delta t)$.

To calculate the LJ interactions efficiently a standard Verlet neighbour list technique[53] was used. Briefly, at the first step all the neighbours of each subject particle within a radius of r_c plus a small distance, termed the shell width, are stored sequentially in a large one-dimensional array. For a number of subsequent steps this list is used so pre-eliminating many of the possible interactions. The list is deemed redundant once any particle has diffused a distance greater than or equal to half the shell width. At this point a new list is formed and the process then continues in the same way as before. In using this procedure there is a trade-off between forming a long list, i.e. a large shell width, which is updated less frequently but where the intervening steps take longer to process, and using a small shell width with more frequent updates of a consequently shorter list. To optimize the shell width short test runs of 1000 time steps were carried out on samples of 1000 site chains under conditions of an applied isotropic pressure of 1 bar and at temperatures of 200 K and 500 K in which the shell width was systematically varied.

On the basis of test runs using this scheme[24] the shell width of 1.5 Å was chosen as a reasonable compromise. Using this value the neighbour list is updated roughly every 17 time steps at 500 K where $\Delta t = 2.5$ fs and more or less at the same frequency at 200 K where the time step is twice as large.

Appendix 2 Method 1 vs. Method 2

Here we examine the relative merits of the pressure loose coupling algorithm in the cases where the constraints are left violated following the scaling of coordinates (Method 1) and where the constraints are reimposed prior to the next time step (Method 2). As already shown, the second constraint loop does have an effect on the stability envelope of the algorithm allowing in principal values of M an order of magnitude lower to be used. However, it is questionable as to whether one should want to use values much below $M = M_0$ since there is significant pressure coupling.

To investigate this further the mean absolute relative deviations in the bond length, Δb, defined as

$$\Delta b(k) = \frac{\langle ||\mathbf{r}_i - \mathbf{r}_{i+1}| - b_0| \rangle_k}{b_0} \qquad \text{[A2.1]}$$

have been computed at each iteration, k, of the constraint loop for some short simulations of duration 1 ps ($200\Delta t$) for the system as used before and at the same conditions of $T = 300$ K and $P = 1$ bar. In Fig. 2.22 Δb has been plotted as a function of the iteration number for results obtained using Method 2 with values of $M/M_0 = 0.05$, 0.3 and 1.

As might be expected, lower values of M cause greater deviations in the bond lengths following the scaling of coordinates and cause the second loop to require more iterations to satisfy the desired tolerance of 10^{-6}. Deviations after the 'free flight' phase of the sites, i.e. following their motion due to the forces and momenta alone and before the constraints are applied, are consistently of order 10^{-2}. On average it is found that about 10 passes through the first constraint loop, independent of M, to satisfy the required tolerance of 10^{-6} for all bonds. As Δb is ultimately much less than the required tolerance this suggests that there is a significant spread in the values of the deviation for individual bonds. In fact, as deviations before the kth loop are actually calculated at the kth loop the constraints are actually satisfied to an even greater extent than shown in Fig. 2.22.

Although it seems reasonable to expect that for Method 2 the satisfaction of the constraints following the scaling of coordinates (step 14 in Appendix 1) causes the values for Δb for the first constraint loop to lie

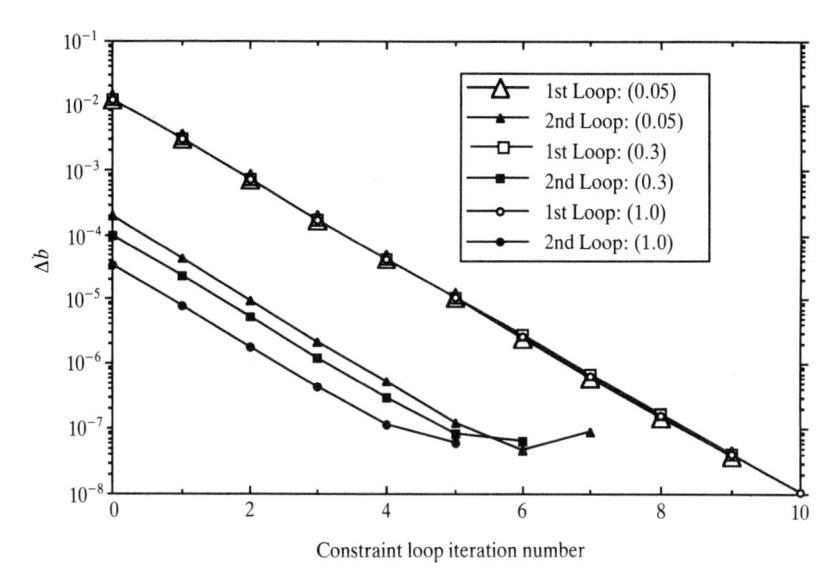

Figure 2.22

Figure 2.22 The absolute relative deviation in the bond length Δb (see Eqn [A2.1]) plotted at each iteration of the constraint loop for the two-constraint loop version of the algorithm (Method 2). Open symbols refer to the first application of the constraints (step 7) and filled symbols to the second following their violation at step 14. The various values of M/M_0 used are given in parentheses in the legend. From Brown and Clarke.[24]

on a common line it is not immediately obvious that this should be the case for Method 1, where the violation of the constraints at step 14 are incorporated into those resulting from the motion due to the forces and momenta at the next time step.

From Fig. 2.23, however, where results for Δb from Method 1 have been compared to the 'common line' obtained from the first loop of Method 2, it seems clear that this is the case. From the results shown in Fig. 2.22, though, this is to a large extent explained by the fact that for the values of M used for Method 1 (shown in Fig. 2.23) the scaling at step 14 typically destroys the constraints by a factor of at least 100 times less than do the forces and momenta. So unless very small values of M are required, which we do not recommend because of the increased coupling, there seems little point in carrying out the extra constraint loop. It will of course slow the calculation down, though by how much will depend to a large extent on the number and complexity of the constraints.

It remains a point of conjecture whether Methods 1 and 2 ultimately give significantly different results. At short times running the two variants from the same configuration produces highly correlated trajectories, but

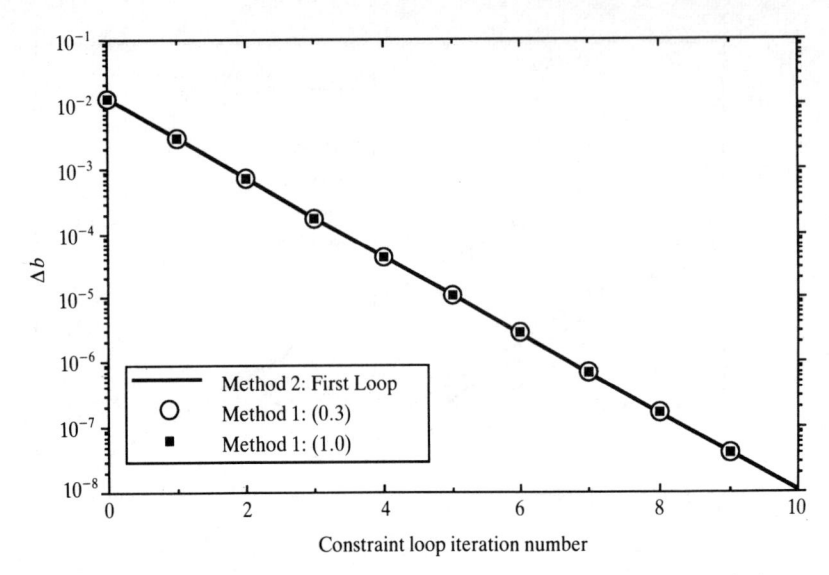

Figure 2.23 As Fig. 2.22 but for Method 1 (one application of the constraints). The solid line is the 'universal' result obtained for the first loop using Method 2. The M/M_0 values are again shown in parentheses. From Brown and Clarke.[24]

these must diverge exponentially and then it becomes a statistical task of analysing whether any observed differences in a quantity are greater than the inherent uncertainties.

Appendix 3 Unbiased Monte Carlo sampling of chain configurations

We describe here a simple procedure for sampling configurations using phantom chain growth in which there are no angle correlations except between neighbouring torsional angles. Averages of a propery, say X, can then be evaluated quite simply using a standard Monte Carlo procedure to estimate the configurational integral as

$$\langle X \rangle = \lim_{k \to \infty} \frac{\sum_{i=1}^{k} X_i \exp(-\beta U_i)}{\sum_{i=1}^{k} \exp(-\beta U_i)} \qquad [\text{A3.1}]$$

where $\beta = 1/k_B T$ and U_i is the total energy of configuration i and is given in this case by

$$U_i = \sum_{j=1}^{n-3} \Phi(\alpha_j) + \sum_{j=1}^{n-4} \Phi(|\mathbf{r}_{j,j+4}|) = \Phi_\alpha + \Phi_{\text{LJ}} \qquad [\text{A3.2}]$$

where $\Phi(\alpha)$ is the torsional potential and $\Phi(|\mathbf{r}_{j,j+4}|)$ is the interaction energy (Lennard–Jones 12-6 potential) between sites separated by three others.

This is a rather naïve procedure, however, as the random choice of dihedral angle leads to a large number of high energy (low probability) configurations. Instead of sampling evenly in α it is expedient to choose angles in the following way. Firstly we form the function $S(\alpha)$ defined as

$$S(\alpha) = \frac{\int_{-\Pi}^{\alpha} \exp(-\beta\Phi(\alpha')) \, d\alpha'}{\int_{-\Pi}^{\Pi} \exp(-\beta\Phi(\alpha)) \, d\alpha} \qquad [A3.3]$$

Clearly $S(-\pi) = 0$ and $S(\pi) = 1$. Now choosing a random number, R_n, such that $0 \leq R_n < 1$ we obtain a dihedral angle as $\alpha = S^{-1}(R_n)$. This is effectively a change of variable in the configurational integral and means that we now obtain averages from the Monte Carlo procedure as

$$\langle X \rangle = \lim_{k \to \infty} \frac{\sum_{i=1}^{k} X_i \exp(-\beta\Phi_{LJ_i})}{\sum_{i=1}^{k} \exp(-\beta\Phi_{LJ_i})} \qquad [A3.4]$$

In computational terms we generate chains of a higher probability more frequently.

References

1. Ferry J D *Viscoelastic Properties of Polymers*, 3rd edn, Wiley, New York, 1980.
2. de Gennes P-G *Scaling Concepts in Polymer Physics*, Cornell, London, 1979.
3. Theodorou D N, Suter U W *Macromolecules* **19**: 139 (1986); *Macromolecules* **19**: 379 (1986).
4. Brown D, Clarke J H R *Macromolecules* **24**: 2075 (1991).
5. Mott P H, Argon A S, Suter U W *Polym. Prepr.* **30**(2): 34 (1989).
6. Brown D, Clarke J H R *J. Chem. Phys.* **92**: 3062 (1990).
7. Roe R J (ed) *Computer Simulation of Polymers*, Prentice-Hall, Englewood Cliffs, NJ, 1991.
8. Brown D and Clarke J H R *J. Chem. Phys.* **84**: 2858 (1986).
9. McKechnie J I, Brown D, Clarke J H R *Macromolecules* **25**: 1562 (1992).
10. Rigby D, Roe R J *J. Chem. Phys.* **87**: 7285 (1987); *J. Chem. Phys.* **89**: 5280 (1988); *Macromolecules* **22**: 2259 (1989).
11. Siepmann J I, Frenkel D *Mol. Phys.* **75**: 59 (1992).
12. Ryckaert J-P, Klein M L *J. Chem. Phys.* **85**: 1613 (1986).
13. van der Ploeg P, Berendsen H J C *J. Chem. Phys.* **76**: 3271 (1982).
14. Ryckaert J-P, Bellemans A *Chem. Phys. Letts.* **30**: 123 (1975).
15. Steele D *J. Chem. Soc. Faraday Trans. II* **81**: 1077 (1985).
16. Maréchal G, PhD Thesis, Université Libre de Bruxelles (1988).
17. Jorgensen W L, Binning R C Jr, Bigot B *J. Am. Chem. Soc.* **103**: 4393 (1981).
18. Richardson M J, Flory P J, Jackson J B *Polymer* **4**: 221 (1964).
19. Parrinello M, Rahman A *Phys. Rev. Letts.* **83**: 4069 (1985).
20. Nosé S *Mol. Phys.* **52**: 255 (1984).
21. Hoover W G *Phys. Rev. A* **31**: 1695 (1985).

22. Berendsen H J C, Postma J P M, van Gunsteren W F, DiNola A, Haak J R *J. Chem. Phys.* **81**: 3684 (1984).
23. Evans D J, Holian B L *J. Chem. Phys.* **83**: 4069 (1985).
24. Brown D, Clarke J H R *Comp. Phys. Comm.* **62**: 360 (1991).
25. Andersen H C *J. Chem. Phys.* **72**: 2384 (1980).
26. Brown D, Clarke J H R *Mol. Phys.* **51**: 1243 (1984).
27. Flory P J in *The Statistics of Chain Molecules*, Hanser Publishers, New York, 1988, p. 57.
28. Ref. 27, p. 57.
29. Flory P J *Polymer Journal* **17**: 1 (1985).
30. Batoulis J, Kremer K *J. Phys. A* **21**: 127 (1988).
31. Brown D Comment on Paper No. 7 ('Comparison of Langevin and MD Simulations', Widmalm and Pastor) of the RSC Faraday Symposium No. 27, 'The Conformations of Flexible Molecules in Fluid Phases', Southampton, 16–18 December 1991.
32. Byron Bird R, Curtiss C F, Armstrong R C, Hassager O *Dynamics of Polymeric Liquids*, Vol. 1, *Fluid Mechanics*, 2nd edn, Wiley, New York, 1987, p. 173.
33. Ward I M *Mechanical Properties of Solid Polymers*, 2nd edn, Wiley, Chichester, 1985, p. 331.
34. Brown N, Ward I M *J. Polymer. Sci. A2* **6**: 607 (1968).
35. Ref. 33, p. 351.
36. Rusch K C, Beck R H Jr *J. Macromol. Sci.-Phys.* **B3**: 365 (1969).
37. Ref. 33, p. 337.
38. Ref. 33, p. 376.
39. Nielsen L E *Trans. Soc. Rheol.* **9.1**: 243 (1965).
40. Newman S, Strella S *J. Appl. Poly. Sci.* **9**: 2297 (1965).
41. Litt M H, Koch P J, Tobolsky A V *J. Macromol. Sci.-Phys.* **B1**: 587 (1967).
42. Whitney W, Andrews R D *J. Poly. Sci. C* **16**: 2981 (1967).
43. Ref. 33; see, e.g., p. 183.
44. Hendra P J, Jobic H P, Holland-Moritz K J *J. Poly. Sci., Poly. Letts. Ed.* **12**: 365 (1975).
45. Lam R, Geil P H *J. Macromol. Sci.-Phys.* **B20**: 37 (1981).
46. Clarke J H R *J. Chem. Soc. Faraday Trans.* **2**: 1371 (1979).
47. Angell C A, Clarke J H R, Woodcock L V *Adv. Chem. Phys.* **48**: 397 (1981).
48. Angell C A, Torrell L M *J. Chem. Phys.* **78**: 937 (1983).
49. Angell C A *J. Phys. Chem. Solids* **49**: 863 (1988).
50. Bueche F, Kinzig, B J, Coven C J *Polymer Letters* 399 (1965).
51. Haward R N, Thackray G *Proc. Roy. Soc.* **302**: 453 (1968).
52. Owadh A A, Parsons I W, Hay J N, Haward R N *Polymer* **19**: 386 (1978).
53. Allen M P, Tildesley D J *Computer Simulation of Liquids*, Clarendon Press, Oxford, 1987.
54. McKechnie J I, Haward R N, Brown D, Clarke J H R *Macromolecules* **26**: 198 (1993).

CHAPTER 3

Monte Carlo studies of collective phenomena in dense polymer systems

K BINDER

Summary

In this chapter the computer simulation of thermodynamical properties of macromolecular materials, including also dynamic phenomena, is reviewed: as examples, concentrated polymer solutions, polymer blends and block copolymer melts are treated. In these problems, phenomena exhibiting non-trivial structure on length scales from 1 Å to 10^3 Å are dealt with, and the time constants for the associated processes range from 10^{-13} s to macroscopic. Therefore meaningful simulation amounts to the construction of a suitable coarse-grained model, which disregards certain aspects of the chemical structure, but still captures the essential features of the collective phenomena that one wishes to describe. It is shown that such studies provide both a sensitive check of approximate theories, such as the crossover scaling theory of semi-dilute solutions, or the random phase approximation in polymer blends or block copolymer melts, etc., and new effects are predicted which can be studied by experiment.

3.1 Introduction: length and time scales

In this section it is explained why the computer simulation of collective phenomena in macromolecular materials is not as straightforward as for simpler problems in condensed matter. In the next section it will be shown that some aspects of these difficulties can be overcome by restricting attention to suitably coarse-grained models. Normally, by computer simulation, one can model a small region of matter in full atomistic detail, e.g. for a simple fluid it often is sufficient to simulate a small box containing of the order of 10^3 atoms which interact with each other with chemically realistic forces.[1] These methods work because simple fluids are homogeneous on a scale of 10 Å. However, for long flexible polymers the situation is very different (Fig. 3.1): a simple chain exhibits structure from

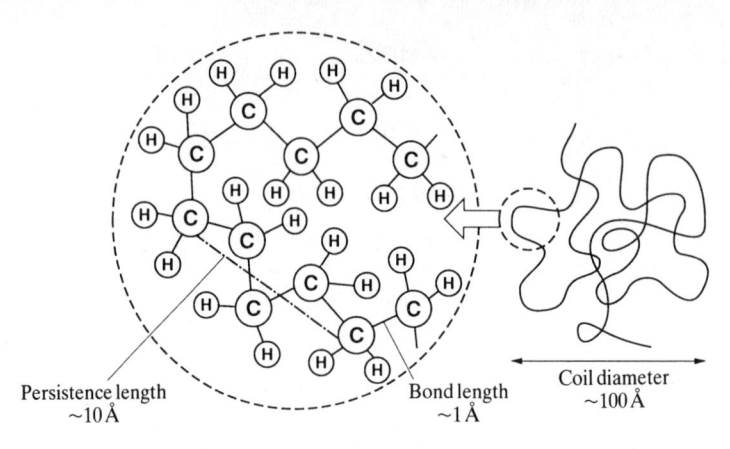

Figure 3.1 Length scales characterizing the structure of a long polymer coil (polyethylene is used as an example).

the scale of a single chemical bond (≈ 1 Å) to the persistence length (≈ 100 Å). Additional length scales occur in a dense polymer solution or melt. Some of these characteristic lengths are of intermediate size. The screening length in a semi-dilute solution[2] describes the range over which excluded volume forces are effective; the tube diameter[3] constrains the motion of a chain in the direction perpendicular to its contour. Collective phenomena related to thermodynamic properties may even lead to much larger lengths, e.g. near the critical point of a polymer mixture the correlation length of concentration fluctuations is predicted[2,4,5] to be much larger than the coil size. The same is true for the time-dependent characteristic wavelength $\lambda_m(t) = 2\pi/q_m(t)$ which describes the scale of concentration inhomogeneities that spontaneously grow after a polymer blend has been suddenly quenched beyond the spinodal curve into the two-phase coexistence region.[4,5] The scale $\lambda_m(t)$ of this 'spinodal decomposition'[4,5] of mixtures can be seen from a peak of the collective scattering function $S(q, t)$; see Fig. 3.2 for an experimental example[6] and corresponding Monte Carlo simulation results.[7]

Another example where collective phenomena define a new length scale occurs in block copolymers,[8–11] (see Fig. 3.3). The enthalpic driving force for this mesophase formation is the same as for macroscopic phase separation in the homopolymer blend, e.g. a repulsive energy $\varepsilon_{AB} > 0$ between monomers of a different kind. Due to the covalent bond between the blocks, the blocks cannot become separated macroscopically, and thus a characteristic wavelength λ^* is stabilized.

As a final example mention is made of ordering phenomena in semi-flexible polymers, e.g. in polyethylene (Fig. 3.4) the energy difference

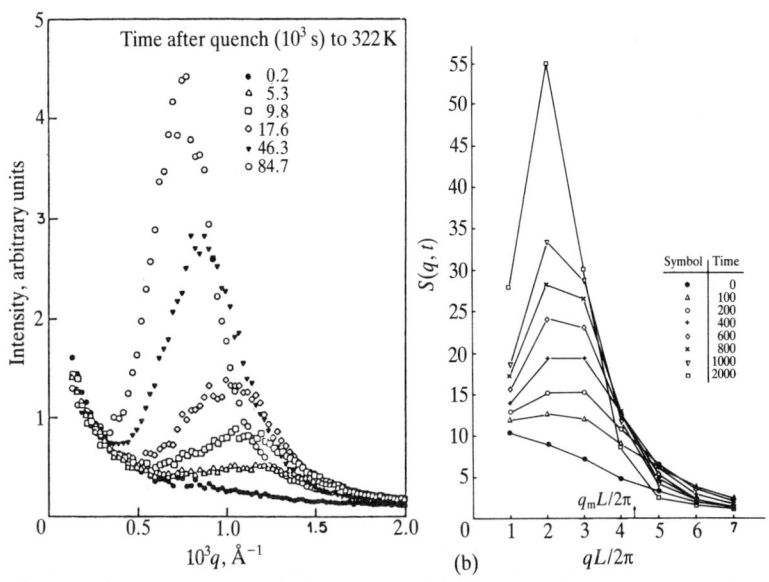

Figure 3.2 (a) Small angle light scattering intensity vs. wavenumber $q \; (= (4\pi/\lambda) \sin \vartheta/2, \; \lambda =$ wavelength, $\vartheta =$ scattering angle) observed after a sudden quench of a mixture of protonated (H) and deuterated (D) polybutadiene to $T = 322$ K (degree of polymerization $N_D = 3550$, $N_H = 3180$, volume fraction ($\phi_D = 0.486$; note that the initial state is at a temperature far above the critical temperature $T_c = 334.5$ K). Note that $q_m(t)$ is found from the maximum of the intensity at time t after the quench (different symbols show different times t, as indicated in the figure). From Bates and Wiltzius.[6]

(b) Collective scattering function $S(q, t)$ plotted vs. wavenumber q for various times t in a quenching 'computer experiment' of a three-dimensional lattice model of a polymer mixture. Polymers are modelled in this example as mutually and self-avoiding walks of $N - 1$ steps on the simple cubic $L \times L \times L$ lattice, for $N = 32$, $L = 40$, and a fraction $\phi_v = 0.6$ of vacant sites. The quench is carried out from 'infinite temperature' (i.e. an athermal mixture) to a temperature $k_B T/\varepsilon = 0.6$, with $\varepsilon = \varepsilon_{AB}$ here being the (repulsive) energy between a nearest-neighbour pair of unlike monomers, which drives the unmixing transition of this model. Due to the periodic boundary conditions applied, only discrete wavevectors $\vec{q}$ can occur, $\vec{q}_{v_x,v_y,v_z} = (2\pi/L)(v_x, v_y, v_z)$ with $\{v_x, v_y, v_z\}$ integer. Thus $S(q, t)$ in this simulation is defined only at discrete points which have been connected arbitrarily by straight lines in the figure to guide the eye. Time is measured in units of attempted moves per monomer. The arrow shows the prediction of the linearized theory of spinodal decomposition[4,5] where initially maximum growth of $S(q, t)$ should occur. From Sariban and Binder.[7]

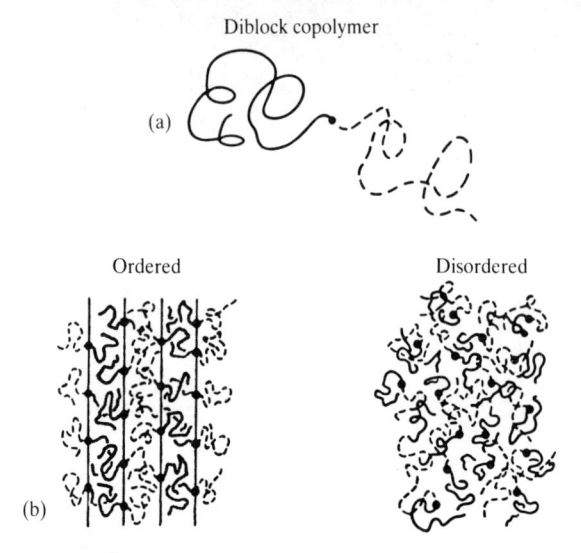

Figure 3.3 (a) Chemical architecture of a diblock copolymer. A diblock copolymer consists of a polymerized sequence of monomers of type A (the A block, indicated as a full line) covalently bound at a junction point (indicated by a dot) to a similar sequence of B monomers (the B block, indicated as a broken line).

(b) The microphase separation transition transforms a compositionally disordered melt (right part) to a spatially periodic, compositionally inhomogeneous phase (left part). For (nearly) symmetric copolymers this ordered phase has a lamellar structure, with the junction points approximately lying in planes (indicated by full straight lines) a distance λ^* apart. From Fredrickson and Binder.[10]

ΔU between the *trans* (t) state and the *gauche* (g_-, g_+) states leads to a tendency for the chains to stretch out as the temperature is lowered, i.e. the persistence length is strongly temperature-dependent.[12–14] This stretching of the chains leads to a tendency to form liquid crystalline-like clusters. While it is difficult to simulate such phenomena with microscopically realistic models,[15] use of coarse-grained models to study them is straightforward;[16–19] see Figs. 3.5 and 3.6.

Thus, in all these cases the large length scales of interest in a simulation where the polymers are treated in full chemical detail necessitate the use of large simulation boxes with linear dimensions of several hundred Å or larger, containing many millions of atoms. Obviously, such an approach would be very difficult.

But the problem is even greater when one considers the second factor that controls simulation feasibility, namely the time scales involved. In

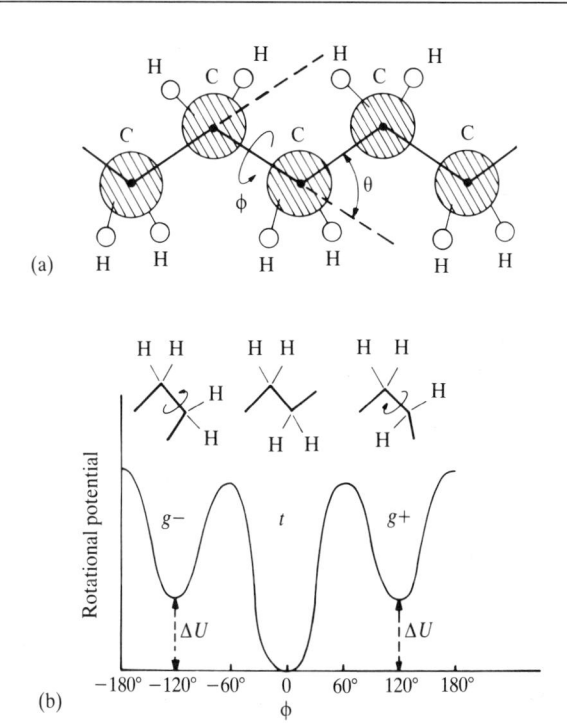

Figure 3.4 (a) Chemical structure of polyethylene $(CH_2)_{N-2}(CH_3)_2$ in the all-*trans* configuration, indicating the definition of bond angle θ and torsional angle ϕ (schematic). (b) Schematic plot of the torsional potential versus torsional angle. Note the energy difference between the *trans* (t) and *gauche* (g_+, g_-) conformations. From Kremer and Binder.[14]

polymeric materials motions occur on many different time scales: fast motions such as the vibrations of the length of a C–C bond (Figs. 3.1 and 3.4) or the bond angle θ (Fig. 3.4) may take the order of $T_{vib} = 10^{-13}$ s or even less; however, here the reorientation jumps in the torsional potential ($t \leftrightarrow g_+, g_-$, Fig. 3.4) are significantly slower, the average time τ_1 between two such jumps being of the order $\tau_1 \approx 10^{-11}$ s for polymer melts at the temperatures of interest. Now a very large number of such slow local motions is necessary to relax the global configuration of a chain (Fig. 3.1) as a whole: if the chains are short enough not to be entangled, the Rouse model [1] applies and then the relaxation time varies proportionally to the square of the 'chain length' (degree of polymerization) N,

$$\tau_N \approx \tau_1 N^2, \quad \text{Rouse model} \qquad [1]$$

omitting prefactors of the order unity in these order-of-magnitude estimates throughout. However, long chains in a melt are mutually entangled

(a)

(b)

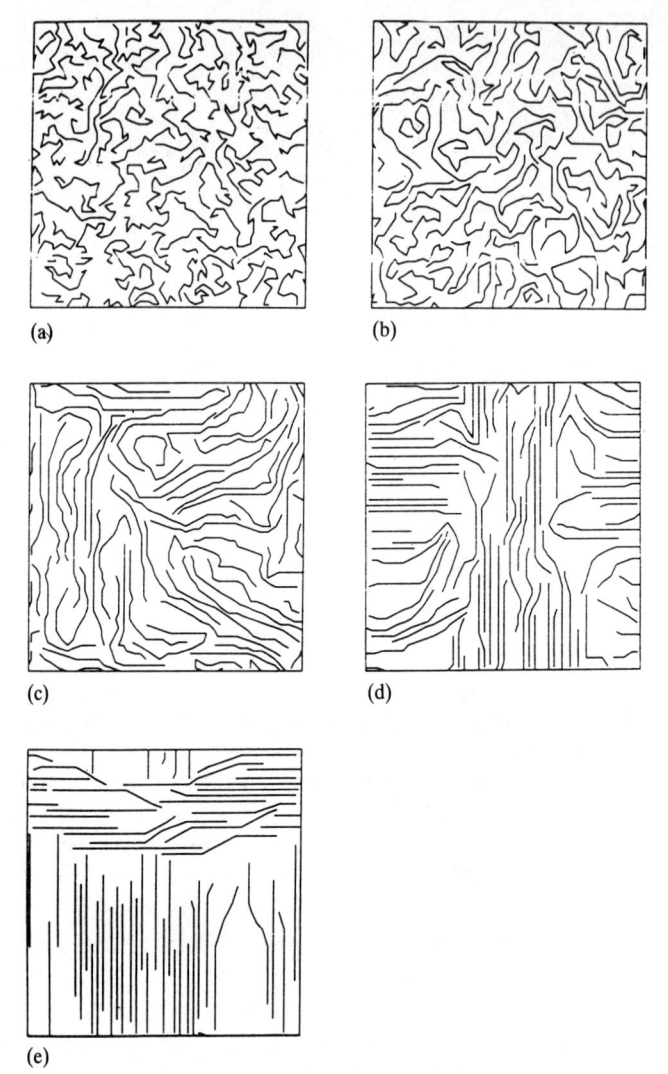

(c)

(d)

(e)

Figure 3.5 'Snapshot pictures' of a simulation of a model for a two-dimensional solution of semi-flexible polymer chains. On a square lattice of 100×100 lattice spacings with periodic boundary conditions, 50 chains containing 20 effective monomers are present. Using the bond fluctuation model,[18] see Fig. 3.6, each effective monomer blocks four sites along a plaquette of the square lattice from further occupation. The bond length of the effective segment connecting two monomers can freely vary from 2 to $\sqrt{13}$ lattice spacings. A potential energy $V(\vartheta) = \varepsilon \sin \vartheta/2$ is introduced, which depends on the angle ϑ between two successive bonds (Fig. 3.6). ε sets the energy scales of this potential. *(Continued opposite)*

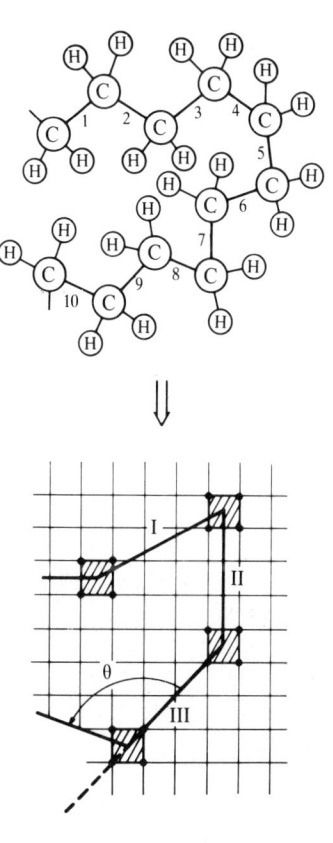

Figure 3.6 Use of the bond fluctuation model on the lattice as a coarse-grained model of a chemically realistic polymer chain (symbolized again by polyethylene). n chemical bonds along the main chain of the polymer are taken together into an 'effective bond' between 'effective monomers'. In the example shown (where $n = 3$) the chemical bonds 1, 2, 3 correspond to the effective bond, the bonds 4, 5, 6 to the effective bond II, etc. From Baschnagel et al.[12]

The snapshots illustrate the gradual onset of orientational order as the temperature T is lowered: $k_B T/\varepsilon = 1.0$ (a), $k_B T/\varepsilon = 0.3$ (b), $k_B T/\varepsilon = 0.15$ (c), $k_B T/\varepsilon = 0.1$ (d) and $k_B T/\varepsilon = 0.05$ (e). Note that at the lowest temperature there is also indication of a polymer–solvent phase separation induced by the ordering. From Lopez-Rodriguez et al.[17]

and then the reptation theory[3] predicts a relaxation time as being proportional to the third power of N,

$$\tau_N \approx \tau_1(N/N_e)N^2 \qquad [2]$$

here N_e is interpreted as the number of monomers between entanglements

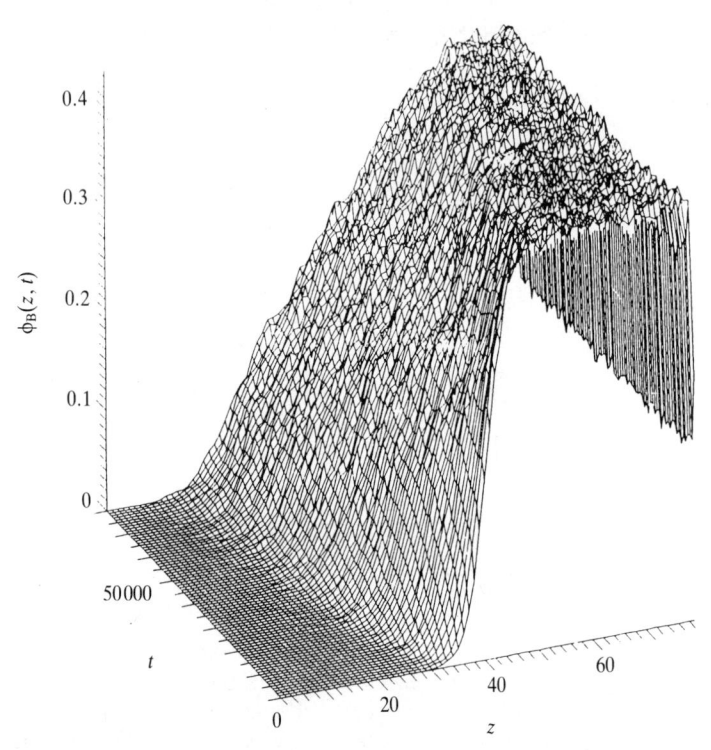

Figure 3.7 Simulation of the time evolution of the concentration profile $\phi_B(z, t)$ of polymer across an A–B interface, using the bond fluctuation model of polymers on the simple cubic lattice, chain lengths $N_A = N_B = 20$, a fraction $\phi_v = 0.58$ of vacant sites, jump rates $\Gamma_A = \Gamma_B = 1$ for both types of monomers, and an attractive energy ε_{AB} between unlike monomers, at a temperature $k_B T/\varepsilon_{AB} = -18/5$. In the initial configuration of a $20 \times 20 \times 80$ lattice with two hard walls of size 20×20 at $z = 0$ and at $z = 80$ and periodic boundary conditions otherwise, all chains with centre of gravity positions $z \geqslant 40$ are B-chains, all chains with centre of gravity at positions $z < 40$ are A-chains. Time t is measured in attempted Monte Carlo steps per monomer (MCS). To gain statistics 48 systems are run in parallel and averaged together. From the broadening of the profile with increasing time one can infer the interdiffusion constant, as done in corresponding experimental work.[22] From Deutsch and Binder.[23]

(typically $N_e \approx 50$).[20] So for $N = 500$, one estimates from Eqn [2] that $\tau_N \approx 10^{-5}$ s, eight orders of magnitude more than the vibration time τ_{vib}! Since in a molecular dynamics simulation the integration time step Δt must be much smaller than the shortest of these times,[1] straightforward application of this technique to chemically realistic models of polymer melts is prohibitively difficult: not only is it hardly feasible due to the lack of stability of the algorithm, it also would require enormous computer time. For example, a recent molecular dynamics simulation of short (unentangled) polyethylene chains[15,21] used only 10 chains with $N = 50$ and needed for one integration step $\Delta t = 10^{-14}$ s a real CPU time of 0.02 s on a CRAY-XMP supercomputer. For $N = 500$ the CPU time for one step would therefore be 0.2 s, and to equilibrate a polyethylene melt for one temperature once one would need $\tau_N/\Delta t = 10^9$ integration steps, i.e. about 10^5 hours of computer time. The situation becomes more difficult for chemically more complicated polymers, and – last but not least – for collective phenomena. This happens because of the 'thermodynamic slowing down' of collective motions – their time constants typically result from multiplying τ_N with the inverse of the 'thermodynamic factor', which represents the thermodynamical driving force which typically is very small for polymers (entropic forces are often of the order of $1/N$). Thus time constants for the interdiffusion of compatible polymers across interfaces[22] are of the order 1 s and more, and hence can be simulated only by a coarse-grained model;[23] Fig. 3.7. The same is true for the simulation (Fig. 3.2b) of spinodal decomposition of polymer blends (Fig. 3.2a) – note that the time unit for the table labelling the different curves is 10^3 s.[6]

Also in homopolymer melts a dramatic slowing down occurs when one approaches the glass transition.[20] Simulating a realistic model of the type of Figs. 3.1 and 3.4, all that one can observe on cooling is[15,21] a freezing in of individual conformation changes such as $t \rightleftharpoons g_+, g_+$ jumps, rather than the slowing down of collective motions. Again it follows that coarse-grained models are the only feasible choice for the study of collective motions in macromolecular materials.

3.2 Some details on 'coarse-graining' models for polymers

From the discussion presented above it should be clear that the only promising way to simulate the very slow motions occurring on very large length scales is to simply omit the very fast motions on very small length scales, thus sacrificing some chemical detail. The idea is that one focuses on the important long wavelength degrees of freedom and their (slower) dynamics. This is illustrated in Fig. 3.6.

Of course, there is no unique prescription how this should be done in practice. One can work in the continuum, for instance, replacing a chain described by potentials $V(l)$, $V(\theta)$ and $V(\phi)$ for bond lengths l, bond angles θ and torsional angles ϕ and an additional interaction between non-bonded monomers[12–14,21] by a simple bead-spring model.[24,25] Although this approach is clearly promising and useful, we shall not discuss it further and proceed to a further level of simplification, namely the introduction of a lattice. The assumption is that for lattice models the necessary information on chain configurations can be handled very efficiently on the computer, and very fast simulation algorithms should be feasible.

This being so, variants of the Verdier–Stockmayer[26] algorithm,[7,14,26–29] (Fig. 3.8) or the 'slithering snake' algorithm,[11,16,30,31] (Fig. 3.9) are used.

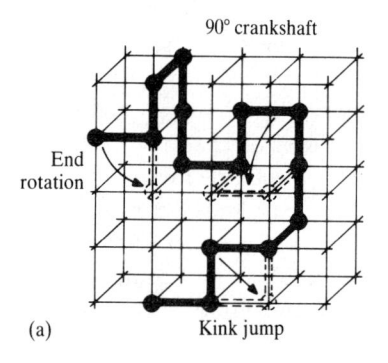

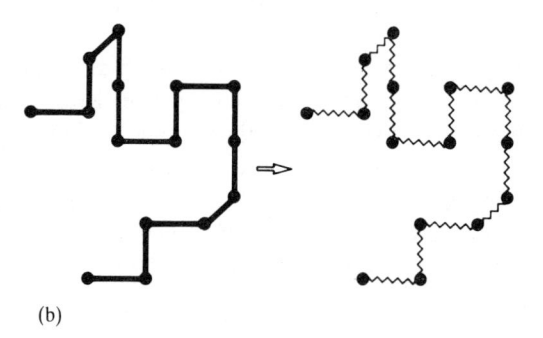

Figure 3.8 (a) Schematic illustration of the standard Verdier–Stockmayer[26] type model for the simplification of polymer dynamics. Chains are modelled as self- and mutually avoiding random walks on a simple cubic lattice. Chain configurations relax by a set of random motions: end rotation, 90° crankshaft rotation and kink-jump motion. (b) For the simulation of a polymer mixture in the grand-canonical ensemble, one considers also attempted moves where an A-chain is replaced by a B-chain of identical configuration, or vice versa.

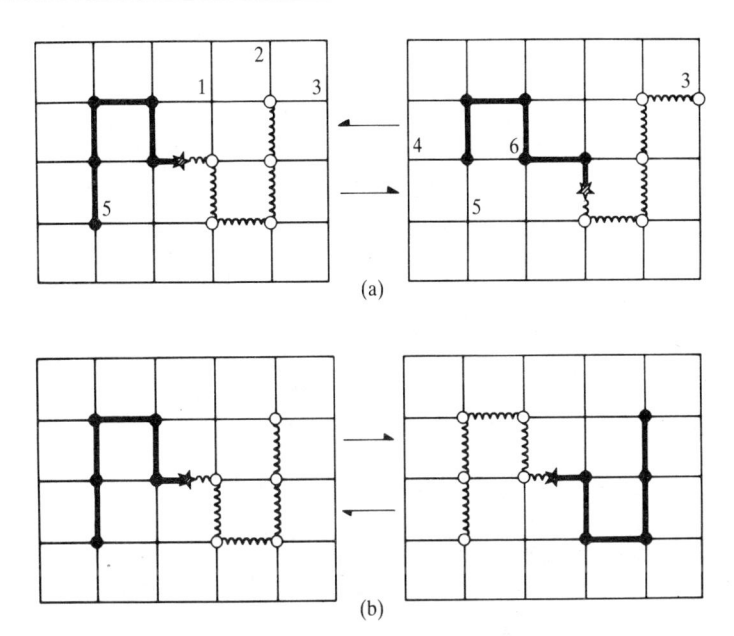

Figure 3.9 Motions used for the relaxation of block copolymer melts. A diblock copolymer is represented by N_A monomers of type A (full dots) and N_B monomers of type B (open circles), the block junction is denoted by a star. In the slithering snake type of move (a) an end monomer (labelled as 5) is removed and at the other end of the chain an end monomer is added (labelled as 3). This motion is accepted only if the randomly chosen site where the end monomer is added is empty (e.g. for the inverse motion shown, sites 4, 5 are acceptable while site 6 would be forbidden). For a block copolymer, the A–B junction has then to be moved along the chain by one lattice unit accordingly. For symmetric block copolymers, case (b) illustrates an interchange of A–B blocks in an otherwise identical configuration. From Fried and Binder.[11]

Results for polymer blends[7,27–29] and block copolymer melts[11,32,33] obtained with such models are reviewed in sections 3.4 and 3.5 below.

This simplest self-avoiding walk model of polymers is clearly very popular but has some technical problems;[14] the further disadvantage is that it is truly universal: differences in chemical structure between different polymers are completely lost. Thus the bond fluctuation model (Figs 3.6 and 3.10) was recently proposed as an alternative coarse-grained polymer model on a lattice.[18,23,34–36] This model has some technical advantages in comparison with the simpler model of Fig. 3.8, and some information on the chemical structure can be kept indirectly by using suitable distributions $P_n(l)$ for the length l of the effective bonds and/or the distribution $P_n(\vartheta)$ for

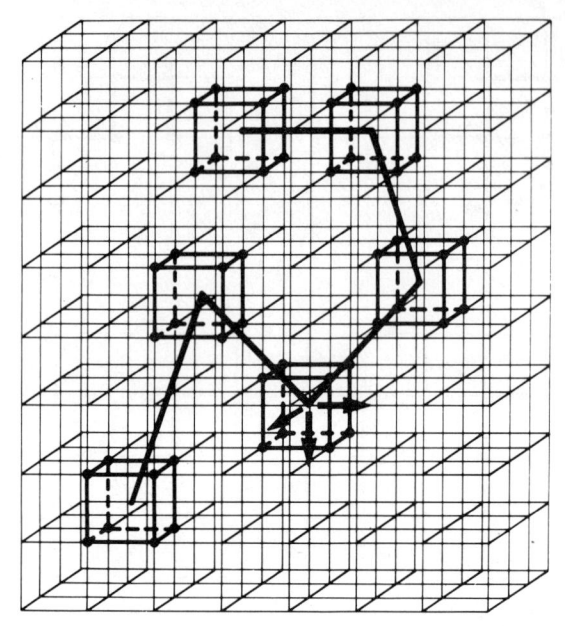

Figure 3.10 Schematic illustration of the bond fluctuation model in three dimensions. An effective monomer blocks a cube containing eight lattice sites for occupation by other monomers. The length l of the bonds connecting two neighbouring cubes along the chain must be taken from the set $l = 2, \sqrt{5}, \sqrt{6}, 3$ and $\sqrt{10}$. Chain configurations relax by random diffusive hops of the effective monomers by one lattice spacing in a randomly chosen direction. From Deutsch and Binder.[23]

the angle ϑ (Fig. 3.6) between subsequent effective bonds (Fig. 3.11).[12] Also in many respects it is closer to the more realistic coarse-grained continuum models.[23]

The main points which make the use of the coarse-grained lattice models so advantageous are: (i) Each 'effective bond' corresponds to several chemical bonds along the main chain of the polymer. The length scale of an effective bond is that of the persistence length (Fig. 3.1) rather than a chemical bond. (ii) The elementary time step for a move (Figs 3.8 or Fig. 3.10), corresponds to the effective monomer reorientation time τ_1: the much faster vibrational motions on very small scales are thus omitted – they only act as a heat bath for the stochastic jumps of the effective monomers. (iii) The lattice structure allows very fast algorithms – both on a CRAY-XMP[34] and on the Mainz parallel computer containing 88 T 800 transputers,[35] a speed of about 10^6 attempted moves per monomer per second is reached for an athermal polymer model. This athermal model

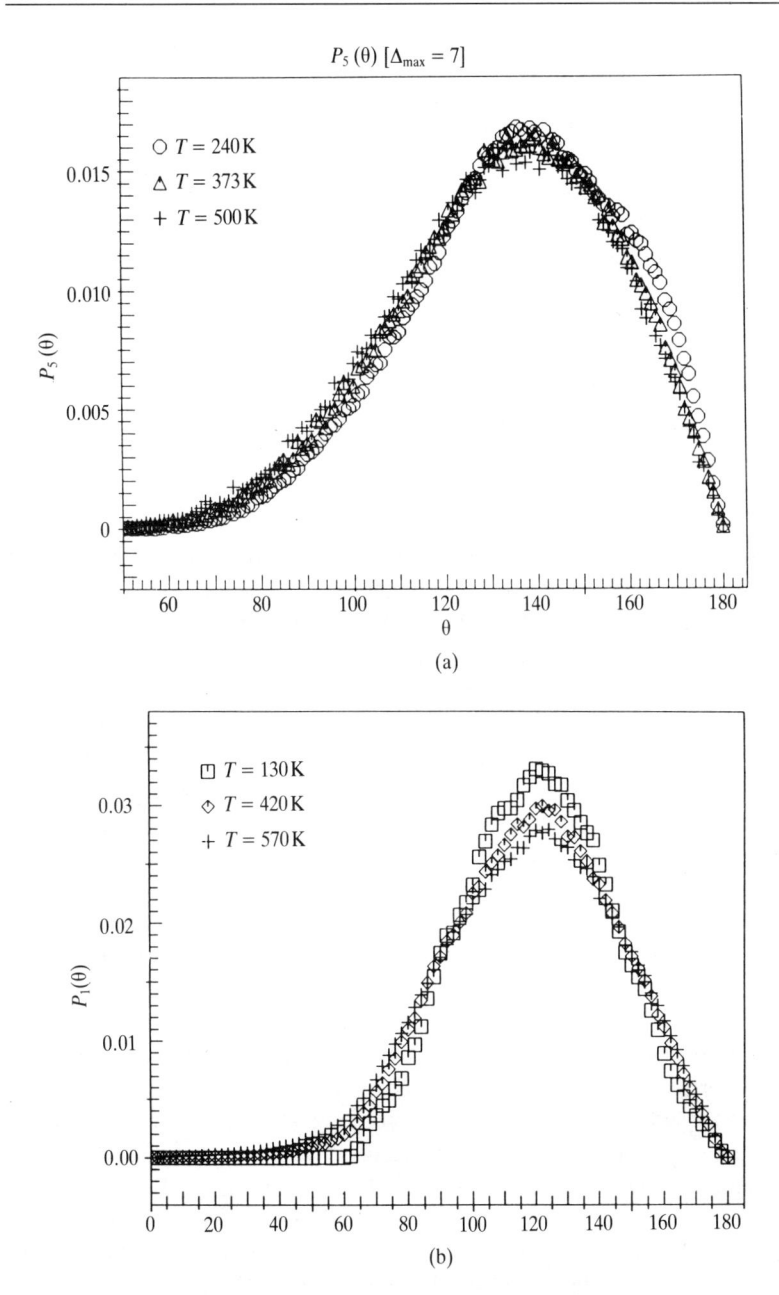

Figure 3.11 (a) Distribution function $P_5(\theta)$ of the angle θ between effective bonds formed from five successive CH_2 units of polyethylene (PE), using the model of Ref. 21, but cutting off the Lennard–Jones interaction for distances exceeding $\Delta_{max} = 7$ units along the chain to avoid chain collapse (this distribution function is
(*Continued overleaf*)

contains only a simplified description of interchain interactions – one deals with self- and mutually avoiding walks that cannot cross each other during their random motions, and thus only the excluded volume among the chains and entanglement constraints are considered. Using interaction potentials $U(l)$ for the bond lengths l[37] or $U(\vartheta)$ for bond angles ϑ, Fig. 3.6,[17,19] or both[38] only reduces the speed a little, but allows a treatment of orientational ordering (Fig. 3.5) or the glass transition,[37,38] Fig. 3.12. There it is shown that the self-diffusion constant $D(T)$ of the chains has a temperature variation consistent with the Vogel–Fulcher[39] law,

$$D(T) = D(T \to \infty) \exp[-A/(T - T_o)] \qquad [3]$$

with an apparent Vogel–Fulcher temperature $T_o \approx 250$ K [note:

$$1/\ln[D(T \to \infty)/D(T)] = (T - T_o)/A$$

and thus T_o is found from the intercept of the straight line in Fig. 3.12 with the abscissa]. It is encouraging that both T_o and A have values which are reasonably compatible with the experimental work: T_o is expected to be below the actual glass transition temperature, which (for large N) is[40] $T_g \approx 420$ K. Thus although the distributions $P_n(l)$ and $P_n(\vartheta)$, obtained from quantum-chemical calculations[12] of the monomer structure and simple-sampling Monte Carlo methods,[14] introduce into the bond fluctuation model (Figs. 3.6, 3.10, 3.12) any information on the chemical structure only in an indirect and simplified way, it still seems possible to predict the physical properties of real macromolecular materials from this approach. In practice, one chooses[38]

$$U(l) = a(l - l_o)^2, \quad U(\vartheta) = b(\cos\vartheta - \cos\vartheta_o)^2$$

and fits the parameters a, b, l_o and ϑ_o to represent the distributions $P(l)$, $P_n(\vartheta)$ such as is shown in Fig. 3.11.

A treatment of binary polymer blends (AB) and block copolymers becomes possible when we also introduce suitable pairwise interaction energies ε_{AA}, ε_{AB}, ε_{BB} for nearby monomers.[7,11,23,27–29,32,33,41] Noting that by the mapping of Fig. 3.6 a volume of $(10^2 \text{ Å})^3$ now corresponds to a lattice containing only a few thousand effective monomers, sufficiently large length scales do become accessible for simulation. Similarly, 1 s CPU

---

computed from a simple sampling Monte Carlo study of a free chain). Three temperatures are shown as indicated in the figure. From Baschnagel et al.[13]

(b) Distribution function $P_1(\theta)$ for the angle between successive monomers (cf. Fig. 3.12) of bisphenol-A–polycarbonate (BPA–PC), at three different temperatures. Both for PE (a) and BPA–PC excluded volume on local scales enforces angles $\theta \gtrsim 60$ degrees. From Baschnagel et al.[12].

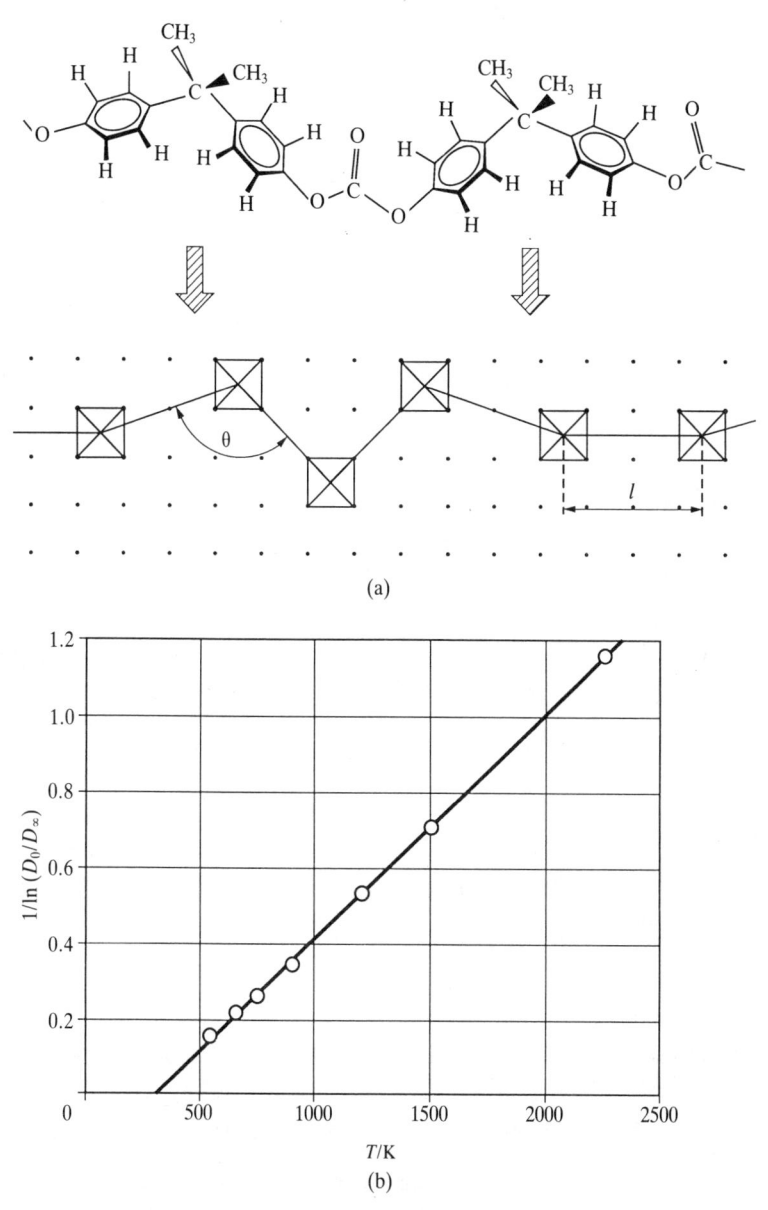

Figure 3.12 (a) Bisphenole-A–polycarbonate contains large monomers with 12 bonds along the chain inside a single monomer and a high degree of internal flexibility. Since the end-to-end distance of a monomer is so large ($l = 10$ Å), one such monomer with 12 chemical bonds is mapped on a chain of three effective bonds, adjusting potentials for bond length l and bond angle θ such that the actual distributions of the chemically realistic chains (Fig. 3.11b) are reproduced as closely as possible.

(Continued overleaf)

time then corresponds roughly to about 10^3 moves per monomer, i.e. a real time of about 10^{-8} s. Thus with about 30 hours CPU time we can access physical time scales of the order of 10^{-3} s, which are reasonable for not too long chains ($N \lesssim 100$ coarse-grained bonds). Thus we expect that collective phenomena in dense polymer systems can be simulated successfully by such methods. The examples reviewed in the following sections show that this expectation is indeed correct. An important aspect also is the choice of a statistical ensemble that is computationally suitable, e.g. for symmetric polymer blends it is convenient to work within the grand-canonical ensemble (Fig. 3.8b) and sample at a given chemical potential difference $\Delta\mu$ between the A- and B-monomers, the numbers v_A and $v_B = v - v_A$ of chains of the type A, B, while the total number v of all chains is kept fixed.

3.3 Crossover in semi-dilute polymer solutions and the dynamics of polymer melts

As is well known[2] polymer coils in good solvents are 'swollen' as long as the solution is dilute, i.e. their gyration radius R_G varies with chain length N as $R_G \sim N^v$ with $v \approx 0.59$ in $d = 3$ dimensions, while in a melt where excluded volume forces are screened out the chains behave like ideal random walks, $R_G \sim N^{1/2}$. When the volume fraction ϕ taken by the monomers of the polymers in solution increases, coils start to overlap, and excluded volume forces can be neglected on length scales exceeding the screening length $\xi(\phi) \propto \phi^{-v/(3v-1)}$ while on scales less than $\xi(\phi)$ the coils still look 'swollen'. In the regime where $\phi \ll 1$ one expects that $\xi(\phi)$ is larger than the persistence length, and then in such 'semi-dilute' solutions the crossover from the swollen coil behaviour to the ideal Gaussian behaviour is expected to be universal.[2,42] This means that the ratio R_G/N^v should depend only on the rescaled chain length $N/N_{\text{blob}}(\phi)$, with the number of monomers inside a 'blob' of size ξ given by $\xi(\phi) \sim N_{\text{blob}}^v(\phi)$, i.e. $N_{\text{blob}}(\phi) \sim \phi^{-1/(3v-1)}$, rather than depending on the two variables N and ϕ separately:

$$R_G(\phi)/R_G(0) = f_G\{N/N_{\text{blob}}(\phi)\} \qquad [4]$$

with f_G being the associated 'scaling function', with $f_G(\mathscr{L} \gg 1) \propto \mathscr{L}^{-(2-1/v)}$.

A similar behaviour might be expected for the dynamics as well: within the limit where in the dilute solution hydrodynamic forces are disregarded and therefore the Rouse model holds,[3] the self-diffusion constant D_N

(b) A melt of such chains with $N = 20$ at a fraction $\phi_v = 0.5$ of vacant lattice sites exhibits a slowing down of the diffusion constant D according to a Vogel–Fulcher law. From Paul et al.[38]

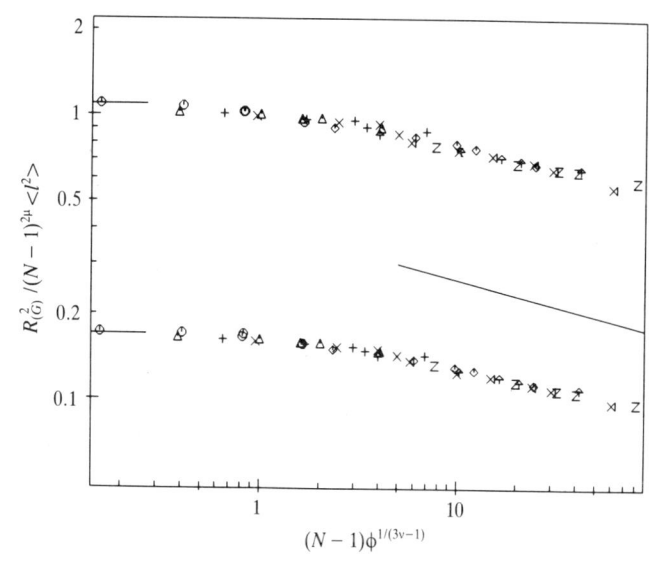

Figure 3.13 Log–log plot of the scaling functions f_G (lower part) and f_R (for the end-to-end distance R, upper part) vs. the scaling variable $(N-1)(\phi l^3)^{1/(3v-1)}$. Note that this variable is essentially the same as N/N_{blob}. Both the squared end-to-end distance R^2 and the gyration radius square R_G^2 are normalized by their asymptotic behaviour in the dilute solution $\langle l^2\rangle(N-1)^{2v}$, where l is the length of a single effective bond, and $N-1$ is the number of effective bonds. The indicated straight line shows the slope of the scaling function, $-(2-1/v) \approx 0.3$. Different symbols indicate different volume fractions: $\phi = 0.025$ (circles), 0.05 (triangles), 0.075 (+), 0.1 ($\times$), 0.2 (diamonds), 0.3 (↑), 0.4 (XI) and 0.5 (Z), respectively. From Paul et al.[35]

of the chains is expected to behave as

$$D_N/D_{Rouse} = ND_N/W\langle l^2\rangle = f_D\{N/N_{blob}(\phi)\} \qquad [5]$$

where now the scaling function $f_D(\mathscr{L})$ must behave as

$$f_D(\mathscr{L} \gg 1) \sim \mathscr{L}^{-1} \qquad [6]$$

in order to yield the expected reptation behaviour.[2,3] In Eqn [5], we have expressed the Rouse diffusion constant D_{Rouse} in terms of an effective monomer jump rate $W \approx \tau_1^{-1}$ and the average mean square bond length $\langle l^2\rangle$.

Using the athermal bond fluctuation model (Fig. 3.10) the first extensive test of this 'blob picture' of crossover scaling has become possible[35,36] (Figs. 3.13 and 3.14). It is seen that this concept provides a useful representation of the data for a wide range of chain lengths (data included

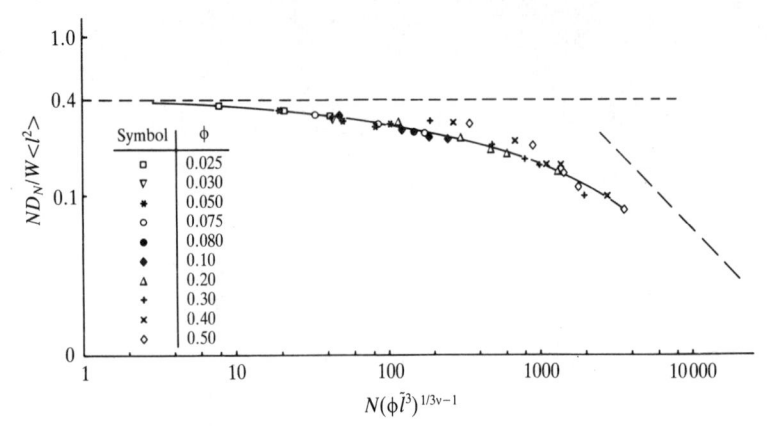

Figure 3.14 Scaling plot of $ND_N/(\langle l^2 \rangle)$ vs. $N(\phi l^3)^{1/(3\nu-1)}$ for the volume fractions as indicated in the figure. Broken straight lines indicate the theoretical limits of the Rouse model (ND_N independent of N) and the reptation model ($ND_N \sim N^{-1}$), respectively. Chain lengths from $N = 20$ to $N = 200$ are included. From Paul et al.[35]

range from $N = 20$ to $N = 200$) and volume fractions (from dilute solutions to melts). Experimentally crossover scaling concepts have been verified with respect to static properties,[42] while for dynamic properties the situation is far less clear:[3,20] in real solutions, the hydrodynamic interactions which are gradually but incompletely screened as the volume fraction increases[43] complicate matters, and the fact that the glass transition temperature $T_g(\phi)$ strongly depends on volume fraction confuses the analysis. In this respect, the Monte Carlo simulation of the coarse-grained model has a decisive advantage in comparison with reality – the model is defined so that hydrodynamic interactions are strictly 'turned off', and in the athermal case (no bond length or bond angle potentials present) any glass transition is absent. Thus one can focus on the problem in hand, the crossover from the Rouse model to the reptation model,[3,35,36] without confusion from other issues. Therefore Fig. 3.14 is a confirmation of crossover scaling from dynamic quantities that has no experimental counterparts yet. At the same time, it is interesting to note that data for very short chains ($N = 20$, and, to a lesser extent, $N = 50$) fall systematically above the crossover scaling function (Fig. 3.14): such short chains remain unentangled even at melt densities, and thus do not exhibit any crossover from Rouse dynamics to reptation dynamics at all.

This point is recognized more clearly when the data are plotted at high volume fractions as D/D_{Rouse} vs. N/N_e, Fig. 3.15:[25,35,44] even for $\phi = 0.5$ where the radii follow Gaussian statistics down to very short chains and hence one has reached a melt density, one still has $N_e \approx 30$, and since

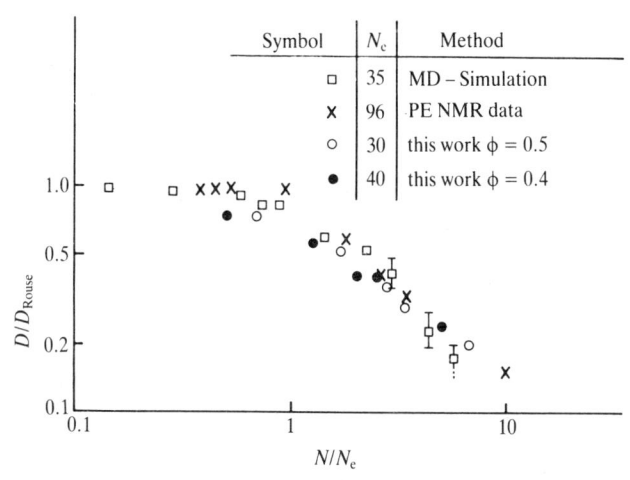

Symbol	N_e	Method
□	35	MD – Simulation
×	96	PE NMR data
○	30	this work $\phi = 0.5$
●	40	this work $\phi = 0.4$

Figure 3.15 Log–log plot of the normalized self-diffusion constant D of polymer melts vs. normalized chain length. D is normalized by the diffusion constant of the Rouse model limit, D_{Rouse}, which is reached for short chain lengths. N is normalized by the entanglement chain length N_e, which in the simulation is extracted from the observation of crossover points in the time dependence of mean square displacements (Fig. 3.16). Experimental data from Pearson et al.[44] for polyethylene (PE) are included as well as molecular dynamics (MD) data from a bead-spring model of Kremer and Grest.[25] Open and full circles denote Monte Carlo data for the bond fluctuation model at monomer volume fractions $\theta = 0.5$ and $\theta = 0.4$, respectively. Estimates for N_e are included in the legend. From Paul et al.[35]

reptation behaviour is seen only for $N \gg N_e$, it is never seen for $N = 20$ and $N = 50$. It is very gratifying, however, that in the normalization of Fig. 3.15 the Monte Carlo data practically superimpose with MD results[25] as well as with experimental data:[44] all material parameters are absorbed in the coordinate scales, and this reduced representation allows a comparison between simulation of our model and the experiment. This comparison (Fig. 3.15) also shows that the results are not sensitive to the simulation technique – both Monte Carlo (MC) and MD methods yield essentially equivalent results. This is only possible because the motions of interest are slow and diffusive and such motions are well modelled by MC. It is often believed that the MD method is superior to the MC method due to its realistic treatment of fast motions such as vibrations; however, for the simulation of very slow motions the MD method need not be advantageous at all; MC methods are competitive for simulations of self-diffusion (Figs. 3.14 and 3.16) and even superior for still slower processes such as spinodal decomposition of blends (Fig. 3.2).

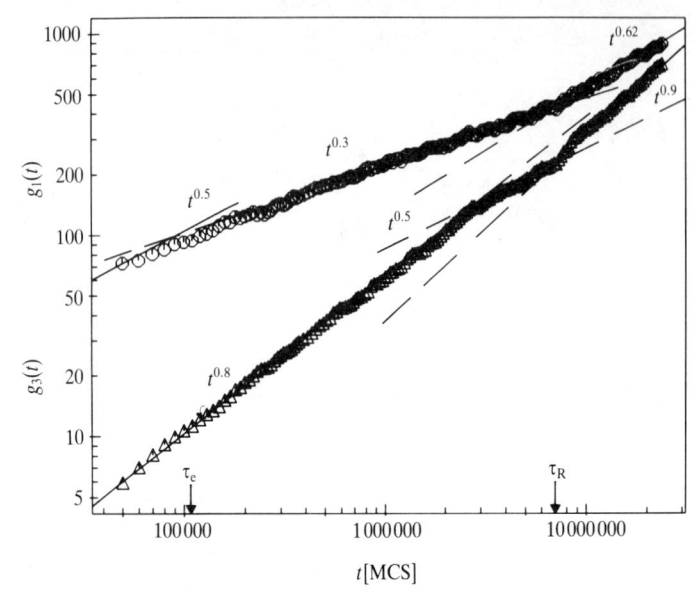

Figure 3.16 Log–log plot of mean-square displacements $g_1(t)$ of inner monomers and of the centre of gravity $g_3(t)$ versus time t for a model melt represented by the athermal bond fluctuation model at $\phi = 0.5$ and $N = 200$. Straight lines indicate effective exponents characteristic for the different regimes of behaviour predicted by reptation theory. Intersection points of these straight line fits define various relaxation times τ_e, τ_R as indicated by arrows. From Paul et al.[36]

A great advantage of the simulations in comparison with other experiments is that one can focus on any quantity that is useful to analyse, e.g. the mean-square displacement of inner monomers of a chain can be followed (Fig. 3.16) which would be hard to measure experimentally. The observation of several crossovers is possible: $g_1(t) \equiv \langle [\vec{r}_i(t) - \vec{r}_i(0)]^2 \rangle$ behaves as $g_1(t) \sim t^{1/2}$ in the Rouse regime, followed by the reptation law $g_1(t) \sim t^{1/4}$ and then a regime $g_1(t) \sim t^{1/2}$ where the chain diffuses along the contour of the tube as a whole. Due to the fact that even a chain length $N = 200$ is not yet deep in the reptation regime, slightly different values for the exponents in these power laws are observed. Similarly a slow intermediate behaviour $g_3(t) \sim t^{1/2}$ is also seen for the centre of gravity displacement $g_3(t) \equiv \langle [\vec{r}_{cG}(t) - \vec{r}_{cG}(0)]^2 \rangle$. These crossovers occur in a mutually consistent fashion (Fig. 3.16) and provide compelling evidence that the reptation idea is correct; other theories[45] leading also to the law $D_N \propto N^{-2}$ could not explain our data as well. It is the first time that both successive crossovers (from $g_1(t) \sim t^{1/2}$ to $t^{1/4}$ and then to $t^{1/2}$ again) have been seen in simulations (Fig. 3.16): these data have needed about 10 weeks CPU time at the Multitransputer facility at the University of Mainz. Earlier work studying chain dynamics with lattice models at varying polymer concentrations[46] obtained results qualitatively consistent with those reported in Paul et al.[35,36] but with distinctly less statistical precision; hence those in Kolinski et al.[46] did not allow a definitive analysis to be carried out in terms of crossover scaling, as presented here. One important result which follows from Fig. 3.14 is that in semi-dilute solutions N_e and N_{blob} should be simply proportional,

$$N_e \approx 7N_{blob} \approx 750(\phi l^3)^{1/(3\nu-1)} \qquad [7]$$

3.4 Unmixing in polymer blends

For about 50 years the Flory–Huggins lattice model[47–51] has been the standard basis for discussing the statistical thermodynamics of polymer miscibility (Fig. 3.17). Although the approximate theory built on this model is broadly used to interpret experimental data, the validity of these approximations has never been seriously tested: possible defects of the theory are hidden in the phenomenological Flory–Huggins χ-parameter which is allowed to depend on both volume fractions ϕ_A, ϕ_B (Fig. 3.17) and on temperature.[50] The numerous approximations involved in this

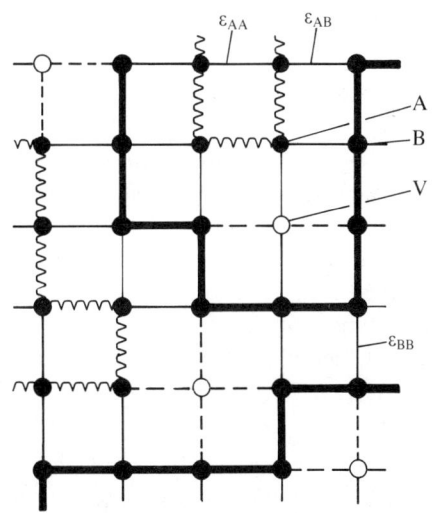

Figure 3.17 The Flory–Huggins lattice model[47–50] describes the two types of polymer chains (A, B) as mutually and self-avoiding walks of $N_A (N_B)$ steps on a lattice. Each lattice site is thus either taken by an A-monomer, a B-monomer, or vacant (V). The volume fractions of these monomers are ϕ_A, ϕ_B and $\phi_V = 1 - \phi_A - \phi_B$. Interaction energies ε_{AA}, ε_{AB} and ε_{BB} occur between nearest neighbour AA, AB and BB pairs, respectively.

theory (neglect of the condition that the chains must not intersect; neglect of site occupancy correlations; approximate count of the number of neighbours of distinct (AB) monomers, etc.) are of doubtful accuracy, however. By Monte Carlo simulations, applying the techniques outlined in Fig. 3.8, the model (Fig. 3.17) may be readily tested and no approximations or adjustable parameters whatsoever are introduced. In this way we can distinguish possible shortcomings both of the model (Fig. 3.17 does not allow for any asymmetry in size and shape of the two types of monomers, A, B, for instance[50]) and of the approximations of the theory, which is a mean field theory: the free energy of mixing is written as a sum of entropy of the mixing terms and of an enthalpy term,

$$\frac{\Delta F}{k_B T} = \frac{\phi_A \ln \phi_A}{N_A} + \frac{\phi_B \ln \phi_B}{N_B} + \phi_v \ln \phi_v + \chi \phi_A \phi_B \quad [8]$$

where χ is related to the energy parameters ε_{AA}, ε_{AB} and ε_{BB} (Fig. 3.17) as follows:

$$\chi = z\varepsilon/k_B T, \qquad \varepsilon = \varepsilon_{AB} - (\varepsilon_{AA} + \varepsilon_{BB})/2 \quad [9]$$

z being the coordination number of the lattice. For a symmetric mixture $(N_A = N_B = N)$ Eqns [8], [9] predict that unmixing occurs if T is less than the Flory–Huggins critical temperature T_c^{FH}, with

$$k_B T_c^{FH}/z\varepsilon = (\chi_{crit}^{FH})^{-1} = N(1 - \phi_v)/2 \quad [10]$$

Figure 3.18 shows the first Monte Carlo test[28] of this prediction: it is seen that T_c^{FH} exceeds the actual critical temperature almost by a factor of two, even much more in the case of high vacancy content. For low polymer concentrations $(1 - \phi_v \lesssim 0.4)$ it also no longer is true that T_c depends on ε_{AA}, ε_{AB} and ε_{BB} only via the single parameter ε, since then the three curves in the upper part of Fig. 3.18 should superimpose.

Very recently there have been various attempts to remedy the neglect of correlations in the Flory–Huggins theory, either by more sophisticated treatments of the lattice models[52] or by integral equation theories for off-lattice models of blends.[53,54] A spectacular prediction of Schweizer and Curro[53] is a dramatic renormalization of the effective χ parameter from its 'bare' value (Eqn [9]) to a renormalized value that is reduced by a factor $\sqrt{N}$, and hence one would have $k_B T_c^{SC}/z\varepsilon \propto \sqrt{N}$ rather than $k_B T_c/z\varepsilon \propto N$ (Eqn [10]), for sufficiently large N. This result would imply that $T_c^{FH}/T_c \to \sqrt{N}$ as $N \to \infty$, and the increase of the ratio T_c^{FH}/T_c in Fig. 3.18 was taken[54] as evidence for the predicted novel $T_c \propto \sqrt{N}$-behaviour. On the other hand, in their original papers[28,29] Sariban and Binder suggested that the effective χ parameter was reduced from its 'bare' value (Eqn [9]) by a finite factor, which was physically explained by an overestimation in the number of interchain contacts as directly established

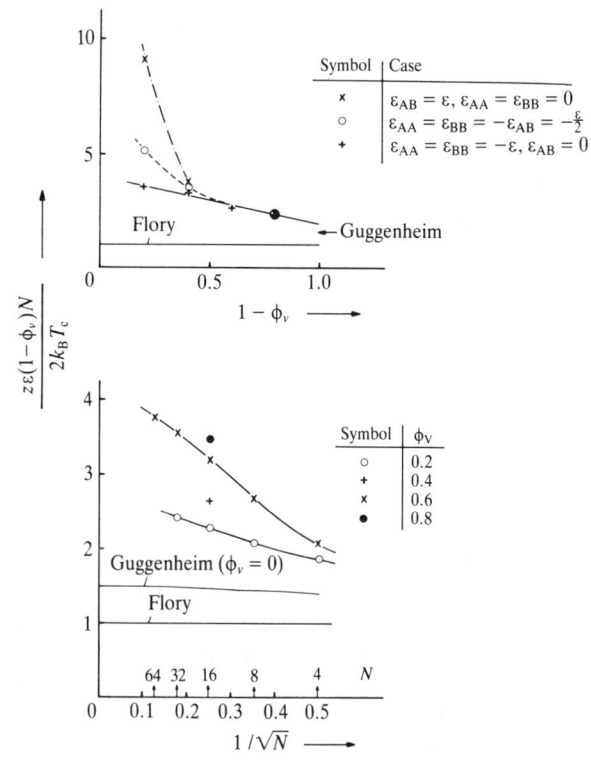

Figure 3.18 Ratio between the Flory–Huggins critical temperature T_c^{FH} (Eqn [10]) and the actual T_c observed in the Monte Carlo simulation of the model of Fig. 3.17. Upper part plots T_c^{FH}/T_c vs. the volume fraction of monomers $\phi_A + \phi_B = 1 - \phi_v$ for $N = 16$, for three choices of energy parameters ε_{AA}, ε_{AB} and ε_{BB} but the same value as ε, as indicated in the figure. Lower part shows the N-dependence of the ratio T_c^{FH}/T_c for several choices of ϕ_v. Note that the Guggenheim[51] approximation is available for $\phi_v = 0$ only. From Sariban and Binder.[28]

from the simulation itself (Fig. 3.19). This explanation differs from the 'correlation hole' effect of Schweizer and Curro[53,54] but it is possible that the chain lengths employed in Ref. 28 were too short to see this correlation hole effect. However, a recent study of symmetric polymer mixtures[41] in the framework of the bond fluctuation model (Fig. 3.10) employed chains from $N = 16$ to $N = 256$ and obtained very clear evidence for a linear variation of T_c with N (Fig. 3.20). Again the prefactor in the relation $T_c \propto N$ is reduced in comparison with the Flory–Huggins prediction, and in addition there is a correction for small N due to the free ends of the chains. Thus the conclusion in Deutsch and Binder[41] was that the integral equation theories[53,54] cannot account correctly for the behaviour observed in the simulations.

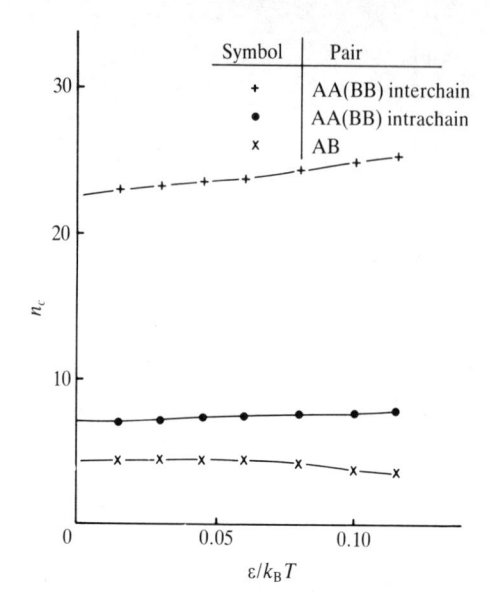

Figure 3.19 Number of nearest-neighbour contacts n_c between various types of monomer pairs plotted versus inverse temperature, for the model of Figs. 3.8 and 3.17 with $N = 32$, $\phi_v = 0.6$, $\varepsilon_{AB} = 0$, $\varepsilon_{AA} = \varepsilon_{BB} = -\varepsilon$. The composition of the mixture is here chosen as $\phi_A/(\phi_A + \phi_B) = 0.9$, $\phi_B/(\phi_A + \phi_B) = 0.1$. Dots represent nearest-neighbour contacts inside a chain, which do not contribute to phase separation but rather to a contraction of the chain radii (see Fig. 3.22). Thus the total number of interchain contacts is about $n_c \approx 27$, while Flory–Huggins theory would predict $n_c^{FH} = zN(1 - \phi_v) \approx 77$. From Sariban and Binder.[28]

Now for a fluid mixture neither the coordination number z nor the number of contacts $[n_c^{FH} = zN(1 - \phi_v)$ in Flory–Huggins theory[28]] is very meaningful, and thus one might think that the renormalization of χ by a factor n_c/n_c^{FH} is not so important. It must be realized, however, that this reduction is really due to correlation effects which are strongly concentration dependent. This is clearly seen if one follows the experimental procedure and extracts an effective χ parameter from a fit of the peak intensity for the collective structure factor (describing coherent scattering of light or neutrons from concentration fluctuations) to the Flory–Huggins theory (or its extension via the random phase approximation [RPA]²),

$$[S_{coll}(\tilde{q} = 0)]^{-1} = \frac{1}{4N} \frac{1 - \phi_v}{\phi_A \phi_B} - \frac{1}{2} \chi_{eff} \qquad [11]$$

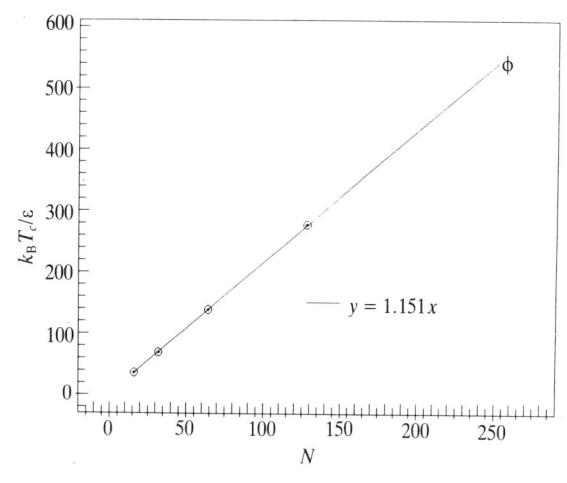

Figure 3.20 Plot of $k_B T_c/\varepsilon'$ versus N, for a choice of energy parameters $\varepsilon_{AB} = -\varepsilon_{AA} = -\varepsilon_{BB} = \varepsilon'$, and the range of these interactions was chosen such that the effective coordination number in the bond fluctuation model at $\phi_v = 0.5$ is $z \approx 14$. For $N \leqslant 128$ the errors are smaller than the size of the symbols, while for $N = 256$ an error bar is given. The data can be very well represented by a law $k_B T_c/\varepsilon' \approx 2.15N + 1.35$, while Flory–Huggins theory[49] would yield $k_B T_c/\varepsilon' \approx 7N$. From Deutsch and Binder.[41]

Figure 3.21(a) shows that a characteristic U-shaped concentration dependence of χ_{eff} for $T = T_c$ results, similar to corresponding experiments on symmetric polymer mixtures[55] (Fig. 3.21b). We conclude that such a U-shaped concentration dependence of χ_{eff} does not signify any physical effect of the local enthalpy but rather reflects an artefact of an inaccurate description of the free energy (Eqn [8]).

A further effect which invalidates both Flory–Huggins theory,[47–51] de Gennes RPA[2] and the Schweizer–Curro integral equation theory,[53,54] which all assume Gaussian chain configurations unaffected by the interactions ε_{AA}, ε_{AB}, ε_{BB} between the monomers, is the fact that the chains contract significantly when one approaches T_c (Fig. 3.22a). As expected, the contraction for the minority component is stronger – in this way the minority chains can avoid making too many energetically unfavourable contacts with surrounding chains. If the minority chains are very dilute this can even lead to dramatic collapse of the chains.[56] Qualitatively such effects have very recently been explained by a self-consistent calculation of the effective potential acting on a chain in a blend.[57] It is interesting to note that a time-dependent contraction of chains has also been observed in Monte Carlo simulations of the early stages of spinodal

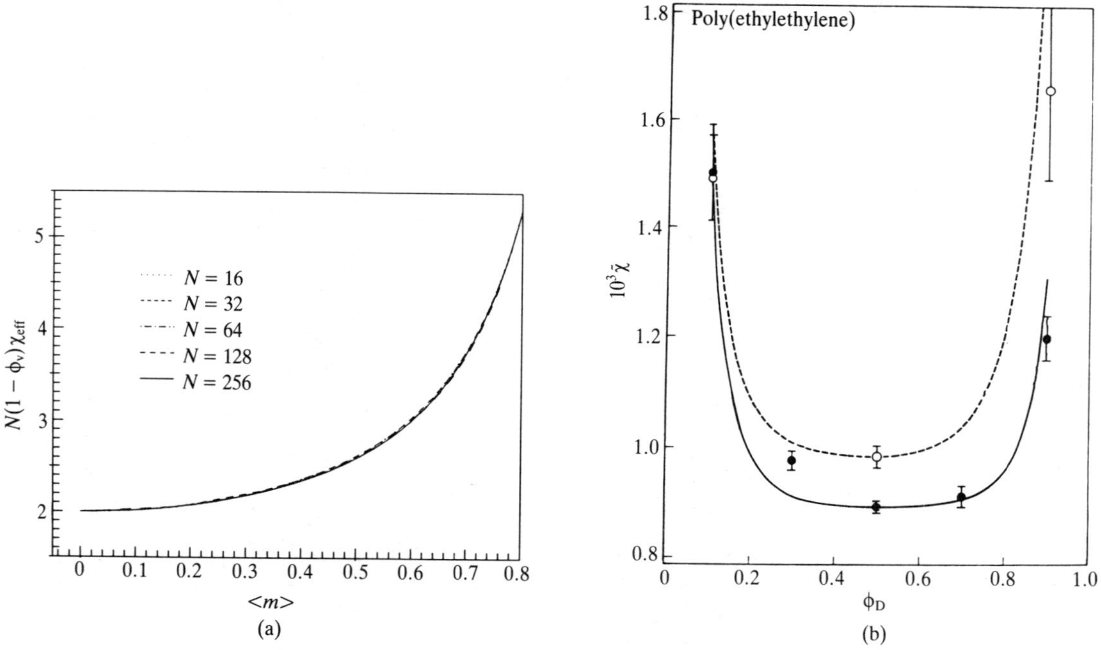

Figure 3.21 (a) Plot of χ_{eff} vs. the order parameter $\langle m \rangle = 2\phi_B/(\phi_A + \phi_B) - 1$, as obtained from a fit of Eqn [11] to the data for $S_{\text{coll}}(\vec{k}=0)$ observed in the simulation of the bond fluctuation model, for $T = T_c(N)$ and various N as shown in the figure. Since the model is precisely symmetric with respect to the sign of $\langle m \rangle$, only positive values of $\langle m \rangle$ (i.e. $\phi_B/(\phi_A + \phi_B) \geq 1/2$) are shown. From Deutsch and Binder.[41]

(b) Effective Flory–Huggins parameter $\tilde{\chi}$ plotted as a function of ϕ_D, the volume fraction of deuterated (D) polyethylene in a mixture with protonated (P) polyethylene. Open circles refer to $N_P^w = 818$, full dots to $N_P^w = 1710$, while $N_D^w = 1330$. From Bates et al.[55]

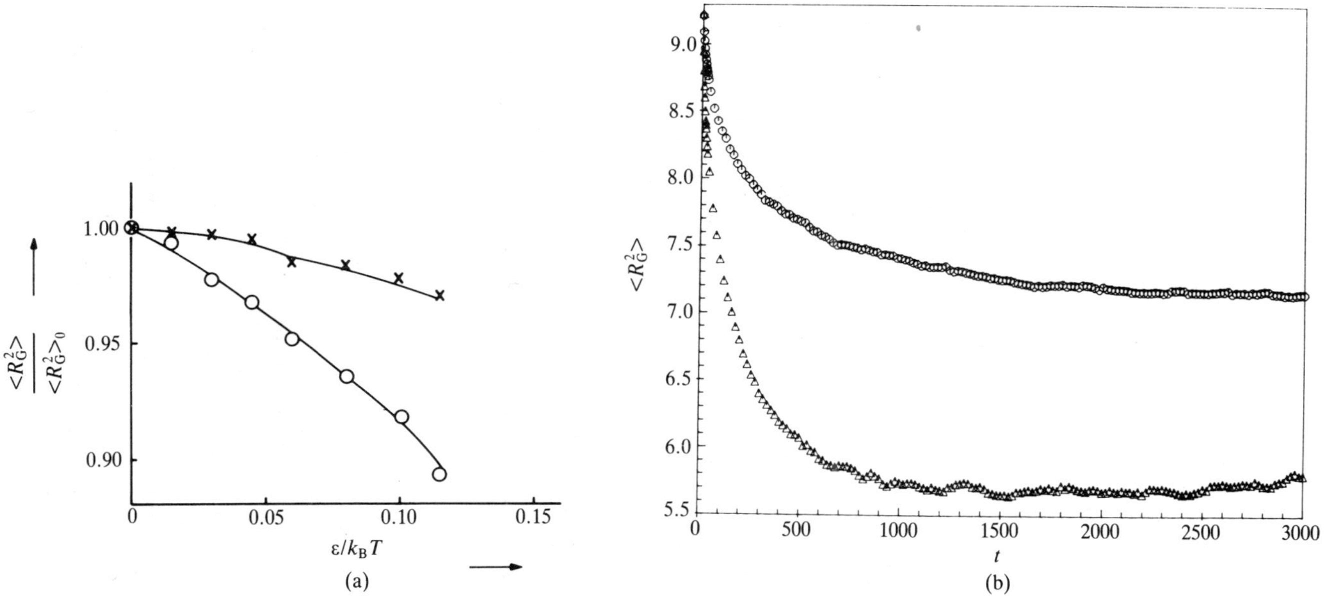

Figure 3.22 (a) Mean-square gyration radius of A-chains (crosses) and B-chains (dots) plotted vs. inverse temperature, for the model of Figs 3.8, 3.17 with $N = 32$, $\phi_v = 0.6$, $\varepsilon_{AB} = 0$, $\varepsilon_{AA} = \varepsilon_{BB} = -\varepsilon$, and a composition $\phi_A/(\phi_A + \phi_B) = 0.9$, $\phi_B/(\phi_A + \phi_B) = 0.1$. The radii are normalized by their values in the non-interacting case ($\varepsilon/k_B T = 0$). The last points ($\varepsilon/k_B T \approx 0.115$) correspond to the coexistence curve of the polymer blend. From Sariban and Binder.[28]
(b) Mean-square gyration radius plotted vs. time for $\phi_A = 0.32$, $\phi_B = 0.08$, $N = 32$, $\phi_v = 0.6$, following a quench from $\varepsilon/k_B T = 0$ to $\varepsilon/k_B T = 0.6$ ($\varepsilon_{AA} = \varepsilon_{BB} = -\varepsilon$, $\varepsilon_{AB} = 0$). Octagons refer to the majority component; triangles to the minority component. From Sariban and Binder.[7]

decomposition,[7] when a blend is suddenly quenched from the one-phase region deeply into the two-phase coexistence region (Fig. 3.22b).

Another feature where the Flory–Huggins theory fails is the critical behaviour near the critical point of the blend.[18,27,58] Since the analysis of critical behaviour from systems of finite size requires application of the finite size scaling theory,[59–61] which cannot be given here due to lack of space, and since these results have been reviewed elsewhere,[62] we summarize only the main points here:

1. The coexistence curve does not have a parabolic shape:
$$(\phi_A^{coex} - \phi_a^{crit}) \propto (1 - T/T_c)^\beta$$
(with mean-field critical exponent $\beta = 1/2$), but is distinctly flatter [$\beta \approx 0.32$ as in the Ising model[59,63,64]].
2. The inverse critical scattering intensity $S_{coll}^{-1}(\vec{q} = 0)$ does not vanish linearly as T_c is approached for $\phi_A = \phi_A^{crit}$ from above $[S_{coll}^{-1}(\vec{q} = 0) \propto (T - T_c)^\gamma$ with mean-field exponent $\gamma = 1$], but according to a curved function [$\gamma \approx 1.24$ as in the Ising model[64]]. These predictions (Fig. 3.23a) are consistent with recent experiments (Fig. 3.23b).[65,66]

As a final topic on the simulation of blends, we discuss the 'quenching experiments' performed from an athermal initial condition ($\varepsilon/k_B T = 0$), where all A, B-chains are well mixed, to a temperature T well below T_c, so that the system is inside the coexistence curve and phase separation sets in via the mechanism where random concentration fluctuations become spontaneously amplified ('spinodal decomposition'[4–7]). As has been shown in Fig. 3.2, this phenomenon can be studied by analysing the collective structure factor $S(\vec{q}, t)$ describing concentration fluctuations:

$$S(\vec{q}, t) = \left\langle \left\{ \sum_j \exp(i\vec{q} \cdot \vec{R}_j)[\phi_B^j(t) - \phi_A^j(t) - \langle \phi_B^j - \phi_A^j \rangle_T] \right\}^2 \right\rangle_T \bigg/ L^3 \quad [12]$$

where $\phi_A^j(t)[\phi_B^j(t)]$ is the local concentration which is unity if an $A[B]$ monomer occupies site j at time t after the quench, $\vec{R}_j$ is the coordinate of that site and L is the linear dimension of the lattice. Note that the simulation of spinodal decomposition is very demanding in computer resources, since $S(\vec{q}, t)$ is not 'self-averaging';[67] thus one has to repeat the 'quenching computer experiment' M times and average over these M samples ($M = 30$ was used for Fig. 3.2b).

These simulations not only compare favourably with the experiment (Fig. 3.2a) but again can be used to check the associated theory. The linearized theory of spinodal decomposition, adapted to polymers,[4,5] predicts an initial growth rate of

$$\frac{1}{q^2} \frac{d}{dt} S(\vec{q}, t) \bigg|_{t=0} \propto \left\{ \frac{1}{2} \frac{T_c}{T} - \frac{1}{x} \left[1 - \frac{1}{x}(1 - e^{-x}) \right] \right\}, \quad x = q^2 R_G^2 \quad [13]$$

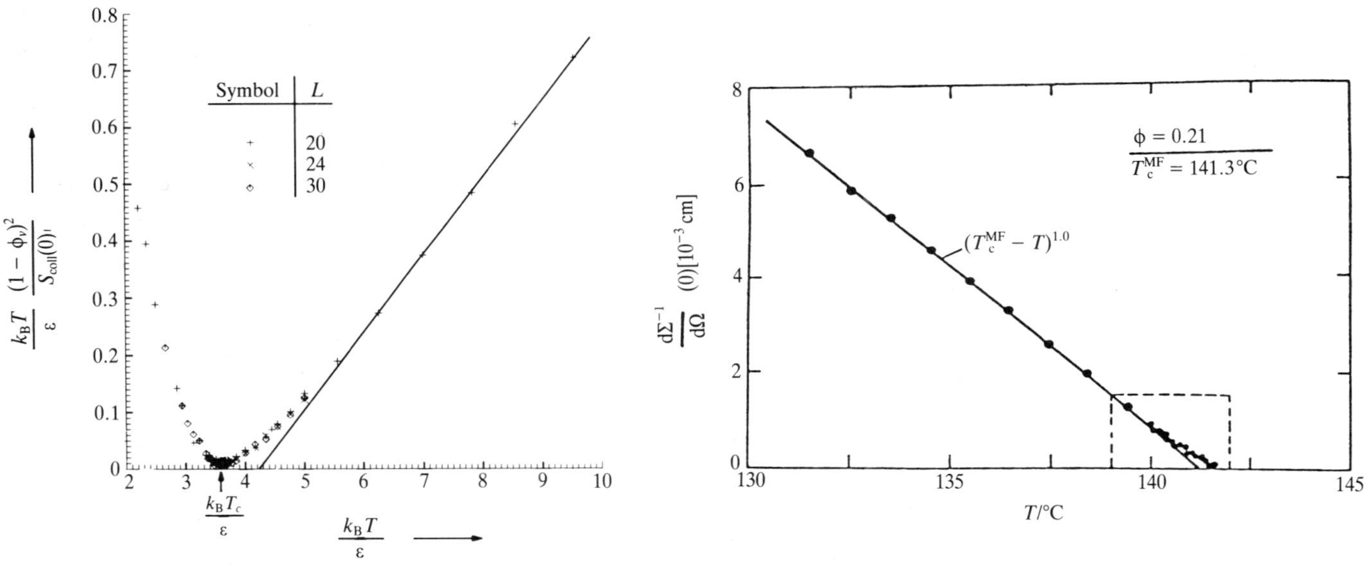

Figure 3.23 (a) Normalized inverse collective scattering function $(k_B T/\varepsilon)(1-\phi_v)^2 S_{coll}^{-1}(\vec{k}=0)$ plotted vs. temperature, for the model of Figs 3.8, 3.17 and parameters $N=8$, $\phi_v=0.6$, $\varepsilon_{AB}=0$, $\varepsilon_{AA}=\varepsilon_{BB}=-\varepsilon$; three different linear dimensions L of the simple cubic lattice are included, as shown in the figure (unit of length is the lattice spacing). Note that the linear extrapolation of the data from high temperatures deviates from the actual critical temperature, indicated by an arrow. From Sariban and Binder.[28]

(b) Inverse neutron small angle scattering intensity plotted vs. temperature for a mixture of polystyrene ($M_{PS}^w = 232\,000$) and polymethylvinylether ($M_{PMVE}^w = 89\,000$) at critical volume fraction of PS $\phi = \phi_{crit} = 0.21$. Note that this system shows a lower critical consolute point rather than the upper critical consolute point implied by the model of Fig. 3.17, cf. also Eqns [8]–[10], and hence one approaches T_c from the one phase region by raising rather than lowering the temperature. From Schwahn et al.[65]

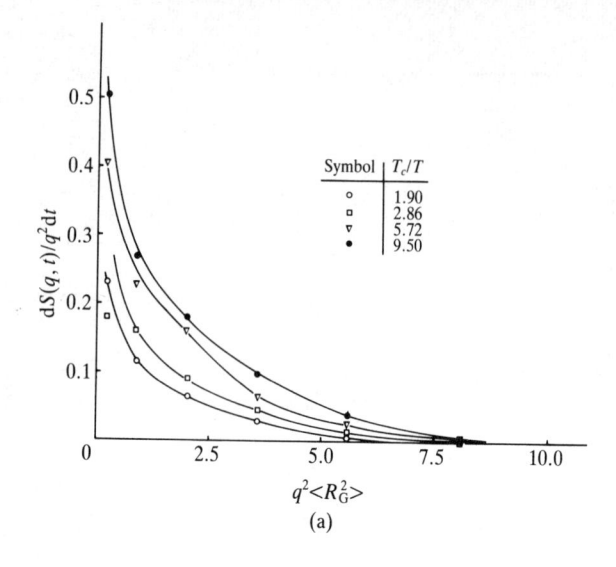

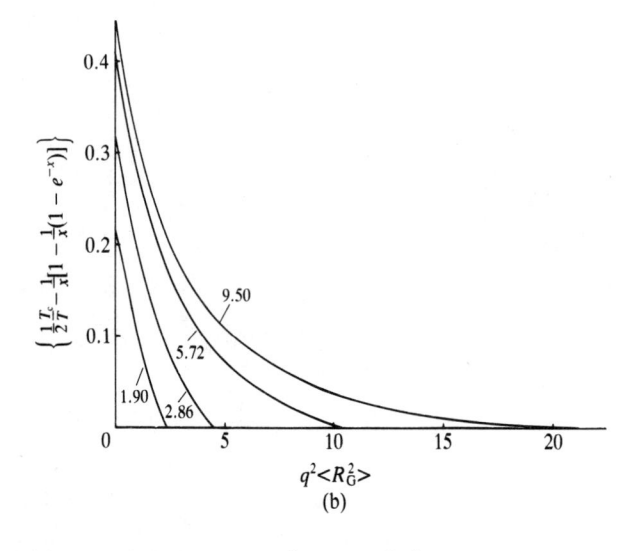

Figure 3.24 (a) Plot of $dS(\vec{q}, t)/(q^2\, dt)$ vs $R_G^2 q^2$, as obtained from Monte Carlo simulation of the model of Fig. 3.17 for $\phi_A = \phi_B = 0.2$, $N = 32$, $\varepsilon_{AA} = \varepsilon_{BB} = 0$, $\varepsilon = \varepsilon_{AB}$. Four different temperatures are shown, as indicated in the figure. To gain statistics, a sampling of 20 to 40 Monte Carlo runs was performed. (b) Plot of the theoretical growth rate of Ref. (4), as given by Eqn [13], for the same choice of reduced temperatures T_c/T as in (a). Note that an undetermined factor setting the time scale in the ordinate of this plot is suppressed. From Sariban and Binder.[7]

which is non-zero for $q < q_c$, the critical wavevector. Figure 3.24 compares the actual growth rate observed in the simulation for various choices of T_c/T with Eqn [13]. While the general trend in theory and simulation is similar (the initial growth rate is generally larger for increasing quench depth $T_c/T - 1$, and becomes a strongly curved function of x for deep quenches) there are also important differences: in particular, a sharply defined position q_c where the initial rate vanishes cannot be identified easily. Rather this position seems to be smeared out over a broad range of q values. Such a smearing of q_c is expected due to the statistical fluctuations and non-linear effects,[68] which are not included in the simple linearized theory given in Binder.[4] Also the position of the initial maximal growth rate $q_m(t = 0)$, marked by an arrow in Fig. 3.2(b), is clearly somewhat overestimated. The fact that non-linear effects set in from the start is obvious from Fig. 3.2(b) since the peak intensity grows somewhat more slowly than linearly with time t, rather than exponentially as predicted in Binder,[4] and also the shift of the peak position to smaller values of q sets in almost immediately. Much longer chains would be needed to verify the predicted[4,5,68] validity of the linearized theory of early-stage spinodal decomposition for polymers from simulations, consistent with experiments.[6]

3.5 Local ordering and chain stretching in block copolymer melts

Here we consider the problem of lamellar mesophase formation in melts of symmetric diblock copolymers (Fig. 3.3) and use a lattice model as explained in Fig. 3.9. This 'microphase separation' is similar to the initial stages of macroscopic phase separation as just discussed: the A-blocks want to separate from the B-blocks, but due to their junction they can do this on a microscopic scale only. Consequently, one expects a peak in the collective structure factor $S_{\text{coll}}(\vec{q})$ at a characteristic wavenumber $q^* = 2\pi/\lambda^*$, λ^* being the wavelength describing lamellar ordering. Extending de Gennes'[2] RPA to this case, Leibler[8] predicted for the symmetric diblocks ($N_A = N_B = N/2$)

$$NS_{\text{coll}}^{-1}(\vec{q}) = F(x) - 2\chi N \equiv \frac{x^4}{2}\left(\frac{x^2}{4} - e^{-x^2/2} - \frac{1}{4}e^{-x^2} - \frac{3}{4}\right)^{-1} - 2\chi N,$$
$$x = qR_g \qquad [14]$$

Equation [14] predicts a maximum at all temperatures $T > T_c$ at $x^* = q^*R_G = 1.95$, with T_c being given by the condition $\chi N = \chi_c N = 10.495$.[8]

This theory has been tested by carrying out simulations for a model with $\phi_v = 0.2$, $\varepsilon_{AA} = \varepsilon_{BB} = 0$, $\varepsilon_{AB} \equiv k_B T\varepsilon > 0$, and N in the range from $N = 16$ to $N = 60$.[11] Figure 3.25(a) shows typical results. Our data,

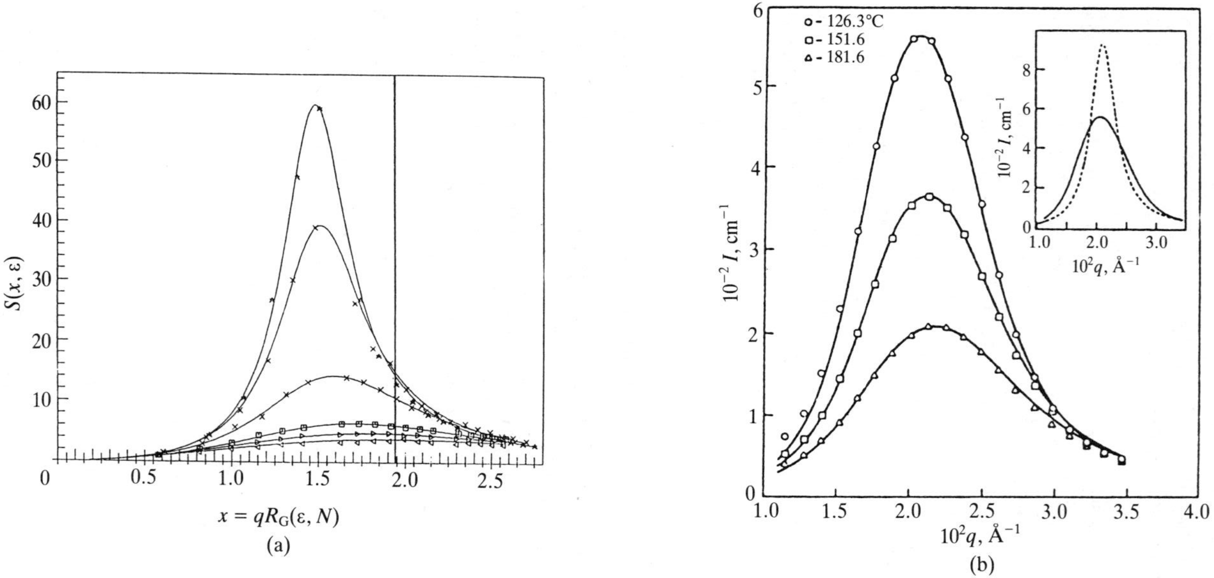

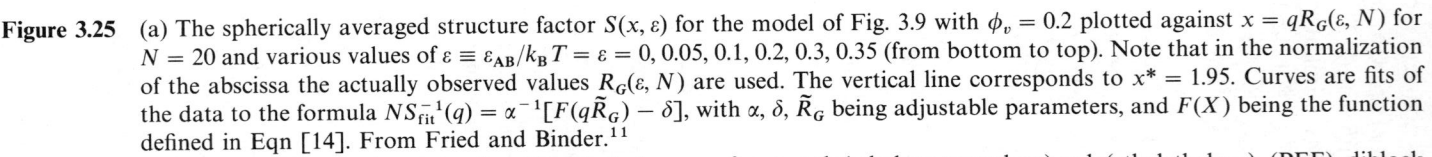

Figure 3.25 (a) The spherically averaged structure factor $S(x, \varepsilon)$ for the model of Fig. 3.9 with $\phi_v = 0.2$ plotted against $x = qR_G(\varepsilon, N)$ for $N = 20$ and various values of $\varepsilon \equiv \varepsilon_{AB}/k_BT = \varepsilon = 0, 0.05, 0.1, 0.2, 0.3, 0.35$ (from bottom to top). Note that in the normalization of the abscissa the actually observed values $R_G(\varepsilon, N)$ are used. The vertical line corresponds to $x^* = 1.95$. Curves are fits of the data to the formula $NS_{fit}^{-1}(q) = \alpha^{-1}[F(q\tilde{R}_G) - \delta]$, with $\alpha, \delta, \tilde{R}_G$ being adjustable parameters, and $F(X)$ being the function defined in Eqn [14]. From Fried and Binder.[11]

(b) Neutron scattering intensity $I(q)$ plotted vs. q, for a poly(ethylene-propylene)-poly(ethylethylene) (PEE) diblock copolymer containing 55% PEP, for a molecular weight $M_W = 57\,500$, $M_W/M_N = 1.05$. The microphase separation transition occurs at $T_c = 125°C$. Three temperatures $T > T_c$ are shown. From Bates et al.[69]

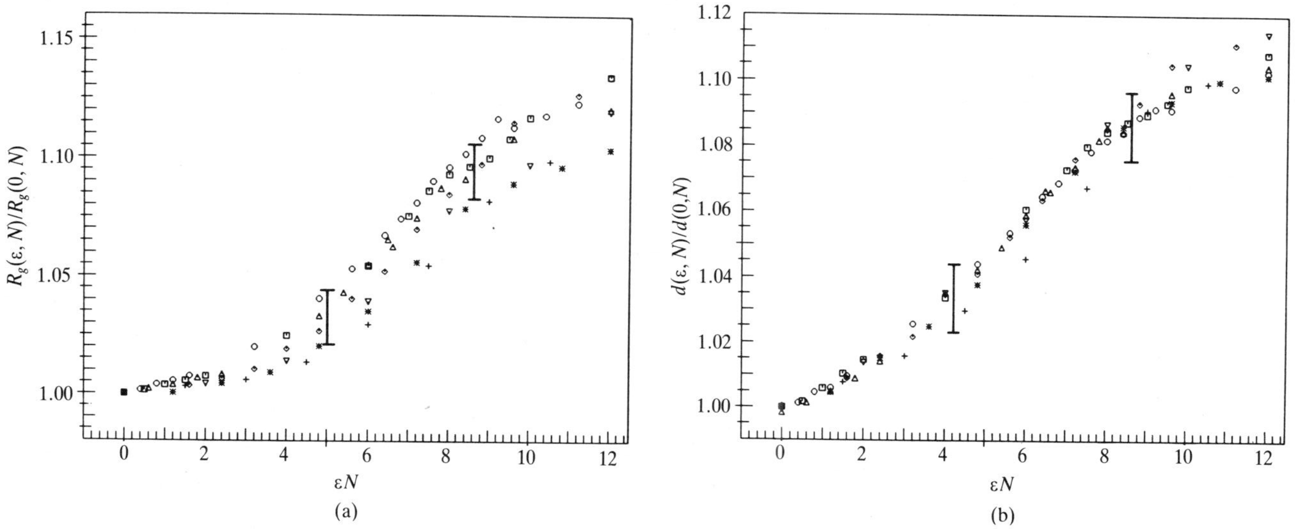

Figure 3.26 (a) Plot of the ratio $R_G(\varepsilon, N)/R_G(0, N)$ versus εN. Different symbols refer to different values of N: $N = 16$ (circles), 20 (squares), 24 (triangles), 32 (diamonds), 40 (inverted triangles), 48 (stars) and 60 (crosses).
(b) Same as (a) but for the ratio $d(\varepsilon, N)/d(0, N)$ where $d(\varepsilon, N) \equiv R_{AB}(\varepsilon, N)/R_G(\varepsilon, N)$. Error bars indicate maximal expected error due to insufficient statistics and finite lattice size effects. From Fried and Binder.[11]

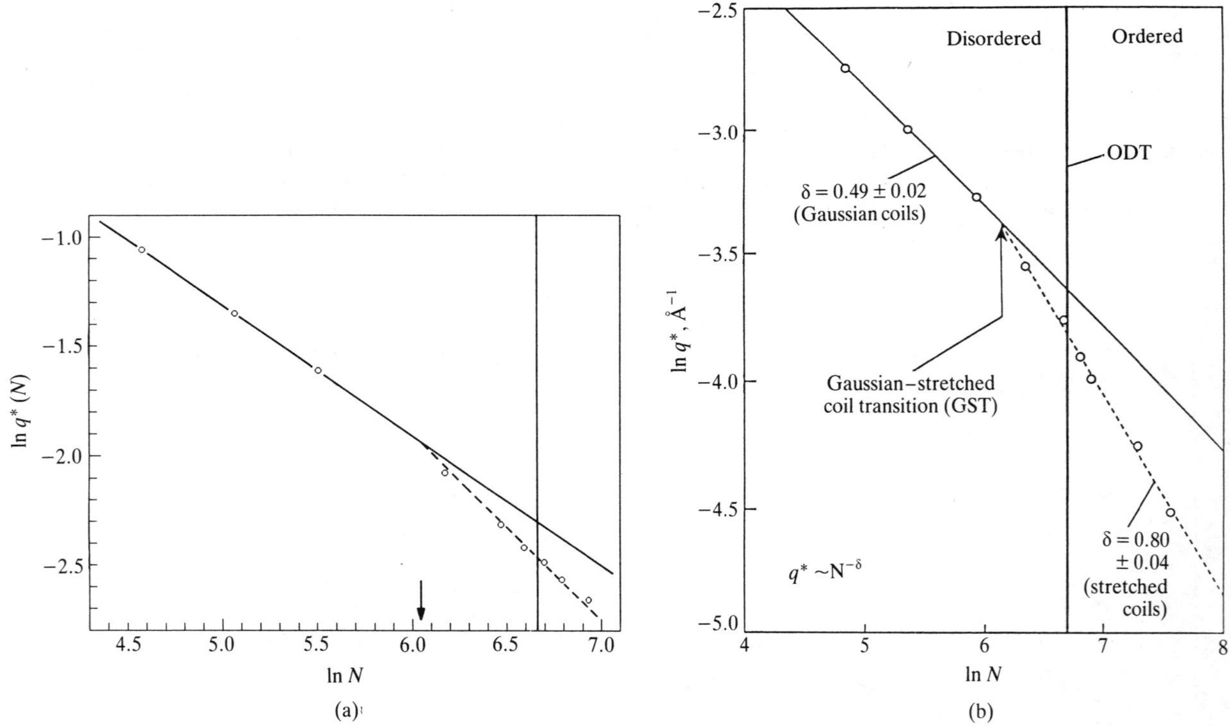

Figure 3.27 (a) Log–log plot of q^* vs. εN fixing ε such that the critical chain length at the microphase separation transition (shown by a vertical line) is $N_{MS} = 803$. Arrow shows the positions of a 'transition' from Gaussian to stretched coils (this is not a sharp transition, however). From Fried and Binder.[33]
(b) Dependence of the small angle neutron scattering peak position q^* on the degree of polymerization N. Data were taken at $T = 23°C$ for poly(ethylene-propylene)–poly(ethylethylene) diblock copolymers containing 55% by volume PEP. Order–disorder transition (ODT) was located from measurements of dynamical response. From Almdal et al.[71]

moreover, look qualitatively very similar to corresponding experiments,[69] Fig. 3.25b: but unlike the experiment where only $S_{coll}(\vec{q})$ can be measured and R_G is not known for the block copolymer but rather is treated as a fitting parameter, the simulation readily yields both $S_{coll}(\vec{q})$ and $R_G(\varepsilon, N)$ simultaneously. Thus we have been able to show[11,32,33] that the length scales q^{*-1} and R_G satisfy the Leibler[8] relation $q^*R_G = 1.95$ in the athermal case $\varepsilon = 0$ only, while with increasing interaction strength $\varepsilon \equiv \varepsilon_{AB}/k_B T$ q^* significantly decreases, until a value of about $q^*R_G(\varepsilon, N) \equiv 1.45$ near the order–disorder transition. At the same time, both $R_G(\varepsilon, N)$ and the distance $R_{AB}(\varepsilon, N)$ between the centre of gravity of the A-block and the B-block distinctly increase with increasing interaction strength (Figs. 3.26a, b). This signifies a breakdown of the RPA again, which implies that the chains remain Gaussian in the presence of the interaction ε and that R_g, R_{AB} are not affected by ε at all. What actually occurs, however, is a gradual stretching of the chains in the melt, since $d(\varepsilon, N) \equiv R_{AB}(\varepsilon, N)/R_G(\varepsilon, N)$ also increases (Fig. 3.26b) we do not have a uniform expansion of the chains. Within the accuracy of the simulations,[11] these effects are simply controlled by the parameter εN and hence do apply to arbitrarily long chains – Leibler's theory[8] hence is predicted to fail even in the limit $N \to \infty$, contrary to the conclusions of various other theories.[9,70] A further point in favour of the belief that the results of this simulation are relevant to the experiment is the fact that there is a close analogy of the behaviour of q^* vs. N at fixed ε between simulation (Fig. 3.27a) and corresponding experiment,[71] (Fig. 3.27b). There is now a strong theoretical activity in order to account for these effects[72–75] but the theoretical situation at present is still unclear. Since some of the recent experimental[71] and theoretical [74,75] work was manifestly motivated by the simulation results,[11,32] the case of block copolymers is a good example for the close interaction of simulations with both analytical theory and experiment.

3.6 Final remarks

The Monte Carlo simulation of coarse-grained models of polymers is a very active field, and many different problems can be treated: only a small selection has been given here (other examples include studies of the structure of grafted polymer layers,[76–78] polymers confined between plates,[79] surface enrichment of polymer blends,[80] etc.). Both thermodynamic properties of macromolecular materials and relaxation processes occurring in them can be studied; bulk properties as well as interfacial properties are accessible. Although the coarse-grained models which have been discussed here disregard the details of the chemical structure and focus on long wavelength degrees of freedom, they are not only useful for checking the validity of various theories, but also can be compared to experiments and give valuable hints for the interpretation of experimental data and also predict new effects which can be researched. As an example of this interplay between simulation and both analytical theory and experiment we have described static and dynamic properties of dilute and concentrated polymer solutions, the thermodynamics of polymer mixing as well as the kinetics of unmixing, and concentration fluctuations in diblock copolymer melts. We have shown that standard theories which so far have been accepted for a long time, such as the Flory–Huggins theory of polymer mixtures, the linearized theory of spinodal decomposition, the RPA description of collective scattering functions in blends and block copolymer melts, etc. suffer from severe shortcomings. Moreover, the simulations in many cases have given deep insight into where these theories need improvement.

Due to the lack of chemical structure information the comparison between simulation and experimental data either is necessarily of a qualitative nature (Figs 3.2, 3.21, 3.23, 3.25 and 3.27) or material parameters can be absorbed in constants normalizing coordinate scales (Fig. 3.15). The theoretical prediction of such material parameters, correlated with the chemical structure, may also be possible within the framework of the present methods: the chemical structure may be used indirectly as an input via suitable effective potentials appearing in the coarse-grained model, e.g. a potential $U(l)$ for bond lengths and $U(\vartheta)$ for bond angles in the bond fluctuation model. First results constructing these potentials from microscopically calculated distribution functions (Fig. 3.11) are very promising; they yield the glass transition temperature of BPA–PC in reasonable agreement with experiments undertaken (Fig. 3.12). But clearly this approach still needs to be explored in much greater detail, to assess its strengths as well as its weaknesses. Thus we feel that the study of collective phenomena in dense polymer systems is far from being fully exploited.

Acknowledgements

The author has benefited from a fruitful collaboration with A Sariban, W Paul, H P Deutsch, H Fried, K Kremer, D W Heermann, H-P Wittmann, I Carmesin, B Minchau, B Dünweg, J Baschnagel, A Lopez-Rodriquez and K Qin. This research was supported in part by grants from the Deutsche Forschungsgemeinschaft (DFG), No. SFB41/S38, SFB262/D1, Bi 314/3, the Max Planck Institut für Polymerforschung Mainz, and by BMFT No. 03M4028 and the BAYER AG.

References

1. Cicotti, G, Hoover W G (eds) *Molecular Dynamics Simulations of Statistical-Mechanical Systems*, North-Holland, Amsterdam, 1986.
2. de Gennes P G *Scaling Concepts in Polymer Physics*, Cornell University Press, Ithaca, New York, 1979.
3. Doi M, Edwards S F *Theory of Polymer Dynamics*, Clarendon Press, Oxford, 1986.
4. Binder K *J. Chem. Phys.* **79**: 6387 (1983).
5. Binder K *Colloid & Polymeric Sci.* **265**: 273 (1987).
6. Bates F S, Wiltzius P *J. Chem. Phys.* **91**: 3258 (1989).
7. Sariban A and Binder K *Macromolecules* **24**: 578 (1991).
8. Leibler L *Macromolecules* **13**: 1602 (1980).
9. Bates F S, Fredrickson G H *Ann. Rev. Phys. Chem.* **41**: 525 (1990).
10. Fredrickson G H, Binder K *J. Chem. Phys.* **91**: 7265 (1989).
11. Fried H, Binder K *J. Chem. Phys.* **94**: 8349 (1991).
12. Baschnagel J, Binder K, Paul W, Laso M, Suter U, Batoulis I, Jilge W, Bürger T *J. Chem. Phys.* **95**: 6014 (1991).
13. Baschnagel J, Qin K, Paul W, Binder K *Macromolecules* **25**: 3117 (1992).
14. Kremer K, Binder K *Computer Phys. Repts.* **7**: 259 (1988).
15. Rigby D, Roe R-J *J. Chem. Phys.* **89**: 5280 (1988).
16. Baumgärtner A *J. Chem. Phys.* **81**: 484 (1984).
17. Lopez-Rodriguez A, Wittmann H P, Binder K *Macromolecules* **23**: 4327 (1990).
18. Carmesin I, Kremer K *Macromolecules* **21**: 2819 (1988).
19. Wittmer J, Paul W, Binder K, *Macromolecules* **25**: 7211 (1992).
20. Ferry J D *Viscoelastic Properties of Polymers*, 3rd edn, Wiley, New York, 1980.
21. Rigby D, Roe R-J *J. Chem. Phys.* **87**: 7285 (1987).
22. Jordan A W, Ball R C, Donald A M, Fetters L J, Jones R A L, Klein J *Macromolecules* **21**: 235 (1988).
23. Deutsch H P, Binder K *J. Chem. Phys.* **94**: 2294 (1991).
24. Kremer K, Grest G S *Phys. Rev.* **A33**: 3628 (1986).
25. Kremer K, Grest G S *J. Chem. Phys.* **92**: 5057 (1990).
26. Verdier P H, Stockmeyer W H *J. Chem. Phys.* **36**: 227 (1962).
27. Sariban A, Binder K *J. Chem. Phys.* **86**: 5859 (1987).
28. Sariban A, Binder K *Macromolecules* **21**: 711 (1988).
29. Sariban A, Binder K *Colloid & Polymer Sci.* **267**: 469 (1989).
30. Kron A K *Polymer Sci. USSR* **7**: 1361 (1965).
31. Wall F T, Mandel F *J. Chem. Phys.* **63**: 4592 (1975).
32. Minchau B, Dünweg B, Binder K *Polymer Commun.* **31**: 348 (1990).
33. Fried H, Binder K *Europhys. Lett.* **16**: 237 (1991).
34. Wittmann H P, Kremer K *Computer Phys. Commun.* **61**: 1309 (1990).
35. Paul W, Binder K, Heermann D W, Kremer K *J. Phys. (Paris)* **II1**: 37 (1991).
36. Paul W, Binder K, Heermann D W, Kremer K *J. Chem. Phys.* **95**: 7726 (1991).
37. Wittmann H-P, Kremer K, Binder K *J. Chem. Phys.* **96**: 6291 (1992).
38. Paul W, Binder K, Heermann D W, Kremer K *Macromolecules* **24**: 6332 (1991).

39. Vogel H *Phys. Z.* **22**: 645 (1921); Fulcher G S *J. Am. Ceram. Soc.* **8**, 339 (1925).
40. Macho E, Allegria A, Colmenero J *Polym. Eng. and Sci.* **27**: 810 (1987).
41. Deutsch H P, Binder K, *Europhys. Lett.* **17**: 697 (1992).
42. Des Cloiseaux J, Jannink G *Polymers in Solution: Their Modelling and Structure*, Oxford University Press, Oxford, 1990.
43. Richter D, Binder K, Ewen B, Stühn B *J. Phys. Chem.* **88**: 6618 (1984).
44. Pearson D S, Verstrate G, von Meerwall E, Schilling F C *Macromolecules* **20**: 1133 (1987).
45. Schweizer K S *J. Chem. Phys.* **91**: 5802, 5822 (1990).
46. Kolinski A, Skolnick J, Yaris R *J. Chem. Phys.* **86**: 1567, 7164 (1987).
47. Huggins M J *J. Chem. Phys.* **9**: 440 (1941).
48. Flory P J *J. Chem. Phys.* **10**: 51 (1942).
49. Flory P J *Principles of Polymer Chemistry*, Cornell University Press, Ithaca, 1953.
50. Koningsveld R, Kleintjens L A, Nies E *Croat. Chim. Acta* **60**: 53 (1987).
51. Guggenheim E A *Proc. Roy. Soc.* **A183**: 203, 231 (1945).
52. Dudowicz J, Freed K F *Macromolecules* **24**: 5076 (1991).
53. Schweizer K S, Curro J G *J. Chem. Phys.* **91**: 5059 (1989); **94**: 3986 (1991).
54. Schweizer K S, Curro J G *Chem. Phys.* **149**: 105 (1990).
55. Bates F S, Muthukumar M, Wignall G D, Fetters L J *J. Chem. Phys.* **89**: 535 (1988).
56. Sariban A, Binder K *Makromolekulare Chem.* **189**: 2357 (1988).
57. Vilgis T A, Borsali R *Macromolecules* **23**: 3172 (1990).
58. Deutsch H-P, Binder K *Macromolecules* **25**: 6214 (1992).
59. Fisher M E in *Critical Phenomena*, Green M S (ed), Academic Press, New York, 1971, p. 1.
60. Binder K *Ferroelectrics* **73**: 43 (1987).
61. Privman V (ed) *Finite Size Scaling and Numerical Simulation*, World Scientific, Singapore, 1990.
62. Binder K *Colloid & Polymer Sci.* **266**: 871 (1988).
63. Fisher M E *Rev. Mod. Phys.* **46**: 597 (1974).
64. Le Guillou J C, Zinn-Justin J *Phys. Rev. B* **21**: 3976 (1980).
65. Schwahn D, Mortensen K, Yee-Madeira H *Phys. Rev. Lett.* **58**: 1544 (1987).
66. Bates F S, Rosedale J H, Stepanek P, Lodge T P, Wiltzius P, Fredrickson G, Hjelm R P Jr *Phys. Rev. Lett.* **65**: 1893 (1990).
67. Milchev A, Binder K, Heermann D W *Z. Physik B* **63**: 521 (1986).
68. Binder K in *Materials Science and Technology, Vol. 5: Phase Transformations in Materials*, Haasen P (ed), VCH Verlagsges., Weinheim, 1991, p. 405.
69. Bates F S, Rosedale J H, Fredrickson G H, Glinka C J *Phys. Rev. Lett.* **61**: 2229 (1988).
70. Fredrickson G H, Helfand E *J. Chem. Phys.* **87**: 697 (1987).
71. Almdal K, Rosedale J H, Bates F S, Wignall G D, Fredrickson G H *Phys. Rev. Lett.* **65**: 1112 (1990).
72. Melenkovicz J, Muthukumar M *Macromolecules* **24**: 4199 (1991).
73. Olvera de la Cruz M *Phys. Rev. Lett.* **67**: 85 (1991).
74. Barrat J L, Fredrickson G H *J. Chem. Phys.* **95**: 1281 (1991).
75. Tang H, Freed K F, *J. Chem. Phys.* **96**: 8621 (1992).
76. Murat M, Grest G S *Macromolecules* **22**: 4054 (1989).

77. Chakrabarti A, Toral R *Macromolecules* **23**: 2016 (1990).
78. Lai P-Y, Binder K, *J. Chem. Phys.* **95**: 9288 (1991); **97**: 586 (1992).
79. Kumar S, Vacatello M, Yoon D Y *Macromolecules* **23**: 2189 (1990); *J. Chem. Phys.* **89**: 5209 (1988); ten Brinke G, Ausserre D, Hadziioannou G *J. Chem. Phys.* **89**: 4374 (1988); Theodorou D N *Macromolecules* **21**: 1400 (1988).
80. Wang J S, Binder K *J. Chem. Phys.* **95**: 8537 (1991).

CHAPTER 4

Molecular modelling of crystalline polymers

E A COLBOURN AND J KENDRICK

4.1 Introduction

Because of the large numbers of conformations of polymer chains which have similar energies, polymers pose a unique challenge to simulation. A protein chain, unless denatured, takes up a well-defined conformation which can be determined crystallographically; polymers in general show no such neatly defined behaviour. Polymers can have varying degrees of crystallinity which depend not only on the chemical composition of the chains, but also on the process history of the material. As discussed elsewhere in this book,[1] simulation of polymer crystallization is an active research field in its own right.

The degree of crystallinity would be of minor interest were it not for the marked effect that it has on polymer properties. Crystalline polymers are generally dimensionally stable. The mechanical properties (evidenced in elastic strain moduli and Young's moduli, for example) depend strongly on the degree of crystallinity. A polymer with little crystallinity is in general quite tough, but can be relatively easily deformed. As the amount of crystallinity increases, the material becomes more rigid, but also more brittle. The effect of crystallinity on mechanical properties is perhaps exemplified most strongly in the self-reinforcing liquid crystalline polymers, where the rod-like nature of the individual chains enables them to pack effectively in a crystalline or pseudocrystalline fashion. It is also worth noting that crystalline polymers can be drawn to give fibres and the very act of drawing the polymer tends to increase the crystallinity because of the increased orientation of the polymer chains.

Barrier properties also depend critically on the degree of crystallinity. Perhaps unsurprisingly, diffusion occurs much more easily in amorphous polymers than in crystalline ones. In amorphous polymers there is sufficient free volume, and sufficient chain motion, for diffusion to occur. Crystalline polymers in general do not permit access to diffusing molecules, certainly not of any appreciable size. They therefore provide better barrier properties. This is related to the observed resistance to swelling shown by crystalline polymers.

All of these properties are desirable ones for polymers, so the control of the degree of crystallinity is an important part of polymer preparation and processing. In some cases, such control is exerted by synthesizing a polymer of defined tacticity; in other cases, thermal annealing, a draw process, or some other mechanical process, is part of the production of the final material which aims to control its degree of orientation and possibly of crystallinity.

It must be emphasized that crystalline polymers do not form perfect crystals, in the way that small molecules do. They are also not analogous to protein crystals, where each chain is identical, with the spaces between the protein chains filled with water molecules. A single polymer chain might be found in several different crystallites, traversing the amorphous regions between them. Nonetheless, crystalline regions can be found in the polymer; regions where the monomer units along the chain pack together in a regular manner. Chain ends and sidegroups can be difficult to accommodate in such crystalline regions so tend to reduce the proportion of crystallinity in a polymer.

Although the packing of chains within the unit cell can be determined by X-ray diffraction, this alone is not sufficient for a full understanding of crystalline polymers. It must be supplemented with information about the higher level of arrangement of the polymer chains in the system, which controls morphology and mechanical properties.

This chapter describes the application of molecular modelling to the simulation of perfect polymer crystallites. X-ray diffraction techniques have been able to supply information as to the space group of the crystal, but only in general with the assumption of model polymer chain configurations. Molecular modelling takes this process one step further, with the ability to predict the chain configuration from first principles. To do this, an accurate description of the inter- and intramolecular forces is required; this is discussed later in this chapter. Various methodologies are available for describing the packing of chains in the polymer crystal, from the cluster approach to the use of fully periodic boundary conditions. Again, these are discussed in detail below. Finally, we treat specific applications of these techniques to some semi-crystalline polymers.

4.2 Polymer morphology

Polymers can crystallize either from the melt or from a supersaturated solution. In either case, the initial crystallization forms lamellae of a defined thickness. In general the lamellar thickness decreases as the temperature at which crystallization occurs increases. The polymer chains are required to 'fold over' and to re-enter the lamella. Superimposed on this is a larger-scale structure of spherulites, caused by the branching of the lamellae and their growth in all directions. The closer the crystallization temperature to the melting temperature of the polymer, the lower the nucleation rate, and hence the larger the spherulites will be.

Even the most crystalline polymers, grown under the most careful conditions, will not show a degree of crystallinity above about 90%. In general, the amount of crystallinity is much lower. The molecular simulation of crystalline polymers usually makes the initial assumption that the polymers are perfectly regular. Thus, all considerations of lamellar and spherulitic structure are neglected and the polymer is taken to be a collection of single, infinitely long chains packed into a unit cell which creates a perfect, boundless system. From this starting point it is possible to introduce defects such as kinks, folds and surfaces, and to study their energetics relative to this idealized bulk.

The advantage of such an approach is that it allows sufficient computational simplification of the problem to permit the determination of the optimum chain configuration, packing and unit cell dimensions. The unit cell dimensions from this model system can then be compared directly with X-ray diffraction studies of the crystallites. This allows the force field to be tested and verified, and provides additional information on chain conformation which may not be available from X-ray work alone. Once the validity of the force field is established, then further work on more realistic models can be undertaken with confidence that inter- and intrachain interactions are being considered with reasonable precision.

In this chapter we will not concentrate on the formation of spherulites, or indeed even of lamellae. Such topics are discussed elsewhere in this book. Here we concentrate on the packing of the chains within the unit cell determined by X-ray diffraction, and the particular requirements that this makes of computer simulation. This chain packing depends on the shape of the chains and on the more subtle details of the interaction between them; for example, that provided by electrostatics.

4.3 Molecular mechanics force fields

Molecular mechanics gives a classical description of the molecule which can provide a reliable estimate of energies and conformations.[2] Chain geometry, and the interactions between chains, can be described, at least in principle, by molecular mechanics. Developed originally for describing large biological macromolecules, the method has been extended to treat a variety of carbon and silicon based polymeric systems.

Molecular mechanics assumes that the forces which an atom in a molecule experiences can be calculated from a potential energy surface, which can be written as a summation of analytical expressions representing physical concepts like bond stretching, angle bending and torsional rotation. In addition to these interactions between bonded atoms, non-bonded interactions must be taken into consideration. A typical expression for the potential energy of a molecule is of the type:

$$E_{total} = E_s + E_b + E_t + E_{vdw} + E_{elec} + E_{nb}$$

where E_s represents bond stretching, E_b describes bond angle bending, and E_t describes torsional rotations. E_{elec} covers the electrostatic interactions, and E_{nb} describes the so-called non-bonded interactions which are used for all intermolecular forces, and those intramolecular interactions between atoms which are separated by more than two bonds.

In addition to the simple terms covered explicitly above, a number of other interactions might need to be included. These involve any hydrogen-bond interaction term and cross terms connecting the simple interactions; for example, the coupling between bond stretching and angle bending may need to be added to reproduce experimental information on molecular vibrations as determined by infrared spectroscopy. This has been shown to be necessary for the accurate representation of even in simple systems, such as poly(ethene), by Goddard and coworkers.[3]

The exact form of the functions chosen to describe the fundamental interactions (bond stretch, angle bending, torsional rotation) is not clearly defined, but in most simulations a well-established force field is assumed. Among these are the Cosmic,[4] Amber[5] and CharMm[6] force fields, the Biosym force field[7] (including the Class II case[8]) and the Dreiding force field[9] described by Goddard and coworkers. All of these have subtly different forms and, of course, different parameterizations.

4.3.1 Bond stretching terms

In the simplest case, it is usual to describe bond stretching within the harmonic approximation, assuming that the bond corresponds to a spring which obeys Hooke's Law:

$$E_s = \tfrac{1}{2} k_s (r - r_0)^2$$

where k_s is force constant, and r_0 is the equilibrium bond length. This is a reasonable approximation near the equilibrium position of the bond, although of course it does not describe bond dissociation, so is limited in its range of validity. For this reason, for example in the MM2 force field, additional cubic terms can be added;[10] MM3 goes to a further degree of sophistication by including quartic terms.[11] Still other force fields rely on a Morse potential. Although this is in principle more accurate, it leads to problems when the starting geometry is far from equilibrium. In this case, the cruder harmonic approximation will lead to more rapid convergence of the geometry optimization.

4.3.2 Bond bending terms

The angle bending is usually described by a harmonic approximation, with further refinements being included in some force fields. Again, simple functional expressions for the energy ensure that the optimization of the molecular geometry is not complicated by the presence of spurious minima in the overall potential energy surface, or that the potential energy is so flat that no progress can be made towards the minimum. However, as more structural information about small molecules is becoming available, the force fields are also being continually refined. In particular the MM3 force field of Allinger[11] includes cubic and sextic terms.

A common expression for the bond bending energy is the so-called 'theta expansion':

$$E_b = \tfrac{1}{2}k_b(\theta - \theta_0)^2$$

which is used, with higher terms, in the MM2 and MM3 force fields. An alternative expression which is used in the Dreiding force field,[9] for example, is a quadratic function based on a cosine expansion:

$$E_b = \tfrac{1}{2}k_b(\cos\theta - \cos\theta_0)^2$$

4.3.3 Torsional terms

Torsional potentials are more complicated, and are frequently represented as a Fourier expansion of the torsional angle, ϕ. This takes the general functional form:

$$E_t = \sum \frac{V_n}{2}(1 \pm \cos n(\phi - \phi_0))$$

The Fourier coefficients are difficult to determine as they reflect the barrier heights for rotation, and experimental information is normally available only for the minima. *Ab initio* calculations are often used to provide the necessary potential energy curves to determine these parameters, but care must be taken to use sufficiently large basis sets and, in some cases, to include electron correlation effects.

4.3.4 Van der Waals and electrostatic interactions

The van der Waals and electrostatic terms can also be described by simple functional forms, but the choice for them is not as obvious as it is for the simple bond stretches and angle bends, and different forms tend to be assumed in different force fields.

The most common form for the van der Waals interaction is the Lennard–Jones 12-6 function:

$$E_{LJ} = Ar^{-12} - Br^{-6}$$

The first term, involving the distance to the inverse 12th power, represents the repulsive interactions between two overlapping charge clouds. This can be replaced by an exponential form, which has a better foundation in theory, and results in the exponential-6 function:

$$E_{exp6} = A\,e^{-Cr} - Br^{-6}$$

In both of these expressions the second term, involving the inverse sixth power of the distance, represents the attractive, van der Waals interaction arising from the correlated motion of the electrons on the different atoms. As will be shown, it is possible to determine values for A, B and C by fitting the parameters so that the calculated packing of small organic molecules agrees well with experimental results for X-ray diffraction studies. However, there are not always sufficient data to do this. In such cases it is possible to use simple combination rules to take information on the homonuclear interactions and to calculate the heteronuclear interaction parameters as either geometric or arithmetic means of the relevant homonuclear ones.

The electrostatic interaction for small molecules has been represented as a sum over bond dipoles.[10,11] Whilst this approach often provides a highly transferable set of parameters, for large molecules such an expression is not feasible, and the majority of polymer simulations use an expression involving the electrostatic interactions of point charges centred on the atoms. Here the interaction can be calculated simply from Coulomb's law

$$E_{elec} = \frac{q_i q_j}{Dr_{ij}}$$

Unless periodic boundary conditions are being used, both the electrostatic and van der Waals interactions are assumed to 'cut off' at some distance. The use of the same cutoff for both the van der Waals and electrostatic interactions is questionable, since at large separation the former falls off as the inverse 6th power while the latter decays as the reciprocal of the distance (i.e. considerably more slowly). Nonetheless, it is common practice to assume the same cutoff for both types of interactions. To improve this approximation, some practitioners[12] suggest using a distance-dependent dielectric constant, which represents to a limited extent the screening that is observed between point charges at large separation. Such ideas, developed for study of biological systems, may not be as applicable to polymeric ones, and the assumption needs further investigation. To avoid discontinuities in the energy during geometry optimization or molecular dynamics simulation, it is also common to apply a smoothing function, so that at a distance close to the cutoff the potential decays smoothly to zero.

The expressions given above still only approximate the potential energy surface. For some systems where either inter- or intramolecular hydrogen bonding is important, it is necessary to include additional terms in the electrostatic interaction, to describe their angular dependence. The values for these terms depend very much on the way that the electrostatic interactions are represented. If the atomic charges are relatively small, as is often the case if they have been obtained from a Mulliken population analysis of the molecular wavefunction, then the hydrogen bond term itself must be larger to give realistic hydrogen bond energies. If, however, the charges are relatively large, which frequently happens when they are determined by fitting to the electrostatic potential,[13] then the specific hydrogen bond term can be smaller.

If spectroscopic information is available about the vibrational frequencies of a molecule then the terms described above may not be sufficiently flexible to produce a really good fit to the experimental information. To overcome this problem, some force fields include additional terms representing the coupling between the different modes of motion in the molecule. Thus, for instance, MM2 and MM3 include a stretch-bend term to account for the change in bond angle during stretching of the bonds.

Because the functional form of the force field terms is quite simple, with relatively few parameters, it is necessary to introduce different 'atom types' to reflect the different chemical environments in which the atoms find themselves. In the case of nearly all force fields, for example, there are different types of carbon atoms for the different possible hybridizations – sp3, sp2, sp and 'resonant' carbons, as well as different types of nitrogen, oxygen and most other common atoms. This increases the flexibility and reliability of the force field parameterization.

4.4 Determination of force field parameters

Of course, any force field is only as good as the parameters which are used to describe the molecular interactions. It is necessary, but not sufficient, to obtain a physically realistic functional form for each of the interactions; parameter determination remains of prime importance for an accurate representation of the system. Although a complex functional form for the potentials may be necessary to reproduce available experimental information accurately, the additional parameters in the functions may lose their physical significance during the fitting process. In addition, attempts to transfer the parameters or the functional form to other systems where experimental information is not available may not be successful. It follows that for force fields trying to treat a wide range of molecules and elements, the force field functions should be as simple as possible. However, for systems where many experimental data are available, the force field can be made more complex to represent the information more accurately.

It is also desirable that the parameters be reasonably transferable, so that a set of parameters in one molecule will be equally useful in another molecule containing the same moiety. However, it must be realized that parameters are inter-related and are not necessarily transferable to a different force field with different functional form. As will be seen below, the inter-relation between parameters is especially evident in the non-bonded van der Waals and electrostatic interactions.

The determination of parameters required in a force-field description of polymers remains an evolving technology, since the parameters developed for proteins are often inadequate to describe the different linkages in polymers. This is especially true for aromatic polymers, and for links with a high degree of resonance character, such as esters. Much work has been expended in finding parameters for the more popular polymer linkages, as we discuss in more detail later.

4.4.1 Determination of bonded parameters

Bonded interactions are typically determined from small molecular species containing the linkages of interest. The parameters may be fitted to experimental information on barriers to rotation, equilibrium geometries and heats of formation. More recent has been a trend to develop force fields which fit the results of *ab initio* calculations.[14] There are several advantages in performing quantum mechanical calculations to determine the force field parameters. The most important of these is the ability of such calculations to probe regions of the molecular potential energy surface which are not readily accessible to experiment. Experimental investigations are often limited to studying the equilibrium structure of a molecule. They can also be limited when investigating condensed phases as it becomes difficult to separate the inter- and intramolecular interactions. However, using quantum mechanical calculations it is also possible to probe modes of motion independently. Thus, for example, it is possible to stretch one bond, while keeping all the other bonds, bond angles and torsion angles fixed.

There are, however, several drawbacks to the quantum mechanical approach. One is that the most accurate methods can only be applied to small molecules. This often means that the calculations have to be performed on model systems and the results transferred to the entity of interest, hoping that the model does indeed represent the real system. Another drawback is that it is not clear how best to perform the quantum mechanical calculations to extract the maximum amount of information. A method has been developed recently which promises to provide the answer.[12,14] Here, the *ab initio* calculation on a molecule is used to provide knowledge of the energy and its first and second derivatives with respect to the nuclear coordinates of the atoms in the molecule, at several geometries. The use of first and second derivatives is very important. For a molecule containing N atoms, these matrices provide, approximately, $3N$ and $3N(3N + 1)/2$ additional pieces of information at each point on the potential surface, to which the proposed force field can be fitted. The computational cost of calculating the derivatives is probably only five or six times the cost of calculating the energy alone. The force field parameters can then be determined by fitting to the potential energy surface and its derivatives. If a sufficiently accurate *ab initio* calculation is done, this approach can provide very good parameters for the bond, angle and torsional parameters, but this approach is not suitable for studying the non-bonded interactions. Unfortunately the choice of non-bonded parameter can be quite important in determining the bonded force field parameters. Procedures for overcoming these problems have been put forward by Dinur and Hagler.[15]

Non-bonded van der Waals interactions are usually determined from X-ray diffraction data on single crystals and from heats of sublimation.[2] They may also be obtained from the dispersion interactions between atoms and molecules, where the dispersion interactions can be related to polarizabilities.

4.4.2 Determination of atomic charges

The calculation of suitable atomic charges for the representation of the electrostatic interaction has been the subject of considerable recent investigation,[16] particularly in the simulation of biological compounds, where the electrostatic interaction is very significant in determining the interaction with polar solvents such as water.

Strictly speaking it is not the charges themselves which are important, but the electrostatic potential around the molecule. This potential represents the strength of interaction felt by a probe positive charge with the molecule. It is not readily amenable to experimental measurement, but it is possible to calculate the potential at any point in space from the quantum mechanical, molecular wavefunction.[13]

At a long distance from the molecule, the electrostatic potential is well represented by a multipole expansion in which the leading terms are the molecular dipole and quadrupole moments. Unfortunately, at short range this approximation breaks down and it is necessary to calculate the potential explicitly. However, because the non-bonded interactions prevent molecules from approaching each other too closely, it is not necessary to know the electrostatic potential within the van der Waals envelope of the molecule. Many researchers, therefore, calculate the potential on successive surfaces around the molecule which are essentially magnified van der Waals envelopes.[17] Three or four of these surfaces, ranging from a scaling of one to two times the van der Waals radius of the molecule, usually suffice to build up a picture of the potential away from the molecule. It is also possible in these surface calculations to incorporate the size of the probe molecule, exploring the molecular surface by shifting the surfaces away from the molecule by a distance representative of the probe radius.

Once the electrostatic potential in the region around the molecule is known, it is possible to fit the atomic charges to reproduce the potential as well as possible, in a least-squares sense. During the fitting procedure it is also a simple task to constrain the charges to reproduce the molecular charge and the molecular dipole if necessary. In practice the unconstrained charges usually predict the molecular charge and dipole extremely well.

The use of atomic charges alone to perform this fitting is clearly an approximation. The fit is not exact and could be improved by including atom-centred dipoles and higher multipoles. If sufficient multipoles are included on each atom then the representation of the quantum mechanical electrostatic potential becomes essentially exact. An alternative approach is to analyse the wavefunction directly with a multipole expansion. Such an approach has been put forward by Stone and Alderton[18] and used successfully in the calculation of crystal structures of organic molecules.[19] This distributed multipole approach is much simpler computationally, as the analysis of the molecular wavefunction expanded in Gaussian functions is relatively simple. However, there are not very many molecular modelling packages which allow anything other than atom-centred charges, so these distributed multipole methods are often not a viable option for obtaining charge information for use in molecular mechanics packages.

Although the calculation of the electrostatic potential appears to offer an excellent procedure for calculating the strength of interaction between two molecules, there are still several approximations to consider. In the first place the procedure assumes that the molecules are static and do not change shape. In practice the molecular structure can alter as molecules interact with each other and it is necessary to consider the changes in the electrostatic potential that this brings about. Some work in this area has been done by Stouch and Williams.[20] In some cases[21] it has been noted that charges can change by as much as 0.7 esu during the rotation of a bond; in practice this may not be very significant as the potential itself may not be changing very much. A second consideration is that atoms and molecules are polarizable; their electron distribution distorts in the presence of an electric field. In the solid state this effect can be quite significant since the additional electrostatic interaction which arises from increased charge separation leads to increased stability. A final consideration is that, despite rapid increases in computational power, the size of molecular system for which it is possible to do accurate *ab initio* calculations is still extremely limiting. This restricts the above procedures to relatively small organic systems and inorganic systems involving elements in the first three rows of the periodic table. Recent research[22] has shown that some of the semi-empirical computer programs, the MNDO method in particular, are capable of calculating electrostatic potential surfaces which are in very close agreement with the *ab initio* results and this does allow larger systems to be considered. However, the application of the method to all elements of the periodic table is not yet feasible, mainly because the requisite semi-empirical parameters are not available.

One approach which has met with considerable success, and is able to treat very large systems, is a method proposed by Gasteiger and Marsili,[23] based on partial equalization of orbital electronegativity, for non-conjugated organic molecules. This was later extended by Gasteiger and Saller[24] to conjugated systems. The method uses atomic ionization potentials and electron affinities to calculate suitable atomic charges, which reproduce known experimental dipole moments of organic molecules well. A similar methodology has also been put forward and used extensively by Abraham *et al.*,[25] who use a one, two and three bond additive scheme, incorporating electronegativity differences between neighbouring atoms and polarizability effects. This sort of approach has recently been extended by Goddard and others[26] in the so-called charge equilibration methods. Such approaches allow charge to flow from atom to atom as the geometry is altered and may provide the answer to some of the criticisms mentioned above for the use of atomic charges which are independent of molecular geometry. However, little experience is available in the literature for these latter methods and they do suffer from the drawback that the electronic structure optimization (which the determination of atomic charge has now become) is not fully integrated into the geometry optimization of the molecule.

4.4.3 Application to aromatic polyester force fields

Determination of force field parameters for polymer systems has formed an important research thrust in recent years. Here, we summarize selected

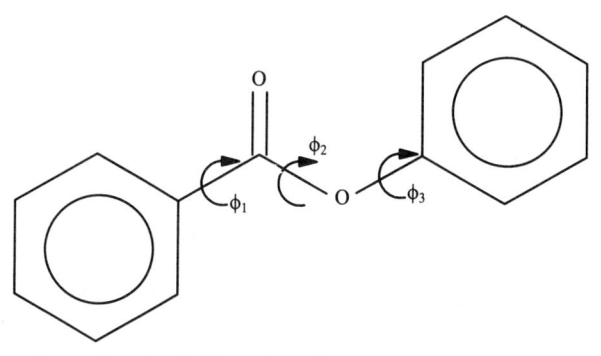

Figure 4.1 Definition of torsion angles in an aromatic ester.

examples which show the use of the methodology; our survey is not intended to give a comprehensive listing of the many studies on parameter evaluation which have been published.

Main-chain liquid crystalline polyesters usually consist of rigid ester-linked aromatic units with flexible hydrocarbon spacers. The aromatic polyester linkage is one which is not well treated in conventional 'biological' force fields, but is of special interest in materials problems. In their work Coulter and Windle[27] concentrated on an accurate description of the torsional potential for rotation of the ester group attached to an aromatic system, which is expected to have a high barrier due to conjugation between the aromatic ring and the adjacent carbonyl group. The torsional motion of the ester unit is defined by three angles, ϕ_1, ϕ_2 and ϕ_3 as shown in Fig. 4.1. Coulter and Windle searched the Cambridge Crystallographic Database for all molecular crystals containing the Ph–CO$_2$–Ph moiety, subject to certain restrictions. From the 12 structures obtained, they generated an 'idealized' structure and noted that there were large variations in the dihedral angles of the ester. The angle ϕ_1 was found to occupy a value of $0 \pm 10°$ but with an absolute minimum at $\pm 6°$. The dihedral angle ϕ_2 had a similar range of values around $0°$, but with a smaller deviation, indicating steeper walls. The dihedral angle ϕ_3 was shown to have a high inverse correlation with ϕ_1 and was seen to have a much broader range of angles, although some clustering around $67°$ was noted.

They then went on to summarize the quantum mechanical calculations which have been performed at the *ab initio* and semi-empirical levels, including their own work, which could be used to assign barrier heights to the different dihedral rotations. These often involved calculations on model systems, such as methyl benzoate, benzoic acid and phenyl acetate. In some of these calculations, a rigid rotor approximation was applied, which assumes that no geometry change takes place during the rotation. This will lead to an overestimate of the barrier to rotation.

Finally they went on to fit the MM2 force field parameters to the information which they had gathered. They used an 'iterative intuitive' fitting process to determine the parameters, which produced good agreement with experiment for the bond angles around the ester carbonyl carbon and reproduced the quantum mechanical estimates for the barriers to rotation. Insight gained about the nature and origin of the parameters was then used to estimate how the barriers to rotation would change on substitution of the phenyl ring.

4.4.4 Application to linear aromatic polymer force fields

The linear aromatic polymers, formed by linking aromatic units such as benzene or naphthalene rings together with more flexible groups such as ester, ether, sulphide, methylene, sulphone, amide, or ketone, exhibit a very rich variety of chemical and physical properties, which can be tuned by judicious choice of linking group, copolymers and polymer blends to meet the exacting requirements demanded of materials in the modern world. A large number of these polymers can possess a high degree of crystallinity. This is probably due in the main to the aromatic component of the polymer chain, since the rings are able to pack together to provide additional stability to the crystalline structure. Some of the polymers, such as the poly(aryl ether sulphones) and poly(aryl ether imides), form predominantly amorphous polymer which is relatively easy to process. Others, such as the polyetherketones, are semi-crystalline polymers, which are more resistant to solvents and show improved thermal stability. The naming convention of the polyetherketones, where each benzene ring is linked either by a ketone or an ether group, is quite straightforward. Thus PEK refers to a polymer where the linking groups, between phenylene groups, are alternately ether and ketone; while in PEEK, there are always two adjacent ether linking groups between each ketone linking group. The chemical structures of PEK and PEEK are shown in Fig. 4.2.

The crystal structures of poly(phenylene oxide), PE;[28] poly(phenylene ether ketone), PEK;[29] and poly(phenylene ether ether ketone), PEEK have been extensively studied experimentally.[30] A considerable amount of theoretical work has also been performed by Abraham and Haworth[31,32] to understand the nature of the interactions in the polymer crystal, including development of the requisite force field parameters.

Abraham and Haworth[31] first considered the non-bonded interactions between the aryl units in the polymer since, in the crystal structure as determined by X-ray diffraction, they show edge-to-face packing of the benzene rings. The approach they adopted to find the non-bonded

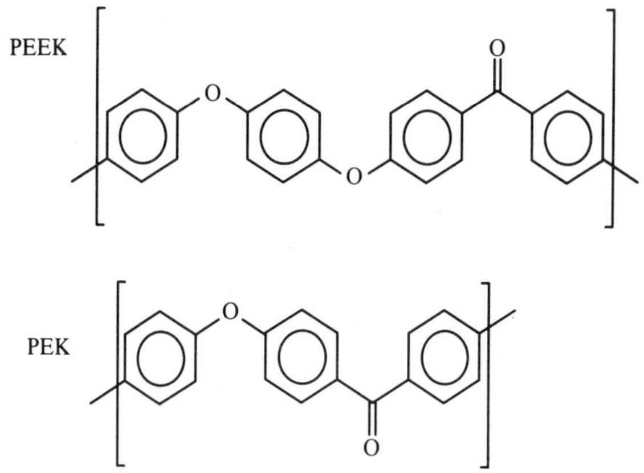

Figure 4.2 The chemical repeat unit for PEK and PEEK.

interactions demonstrates the difficulties which can be encountered in determining a suitable set of transferable parameters and highlights the interdependencies which can occur between the various non-bonded interactions. They chose a charge of −0.13 on the carbons and +0.13 on the hydrogens, which is consistent with the measured quadrupole moment of benzene and with an infra-red study of the gas-phase band intensities. Based on this, the CHARGE2[25] model which was developed for calculating the charges of molecules was modified to reproduce these charges. It is of interest to note that Powell et al.[33] have shown that it is possible to reproduce the crystal structure of benzene satisfactorily, without the inclusion of an electrostatic term at all, but rather through modification of the non-bonded potential function. This provides an illustration of the interdependent nature of the non-bonded potentials.

Having established the electrostatic interaction parameters, Abraham and Haworth went on to determine a suitable set of non-bonded van der Waals type interactions based on an atom-centred Morse function. The parameters of the function were determined to reproduce crystal geometries, and, where available, sublimation energies of some representative small organic molecules. They used the same repulsive part of the Morse function for all interactions: C—H, C—C, C—O, O—H, O—O and H—H, arguing that the repulsive non-bonded interaction will be dominated by the H—H interaction.

This approximation reduces the number of unknown variable parameters to just six: E_{min}, R_{min} for the homonuclear interactions; C—C, O—O and H—H. The other interaction parameters were calculated using the arithmetic and geometric means respectively of the parameters for the relevant contributing atoms. The energy of a unit cell of the crystal was calculated using a cluster approach, where a reference unit cell was considered with a surrounding lattice extending up to three crystallographic axes in each direction. A 6 Å cutoff was used for the non-bonded van der Waals interactions and the molecules were assumed to remain rigid during the calculations.

The optimum non-bonded parameters were chosen in the first place to reproduce the experimental neutron diffraction data on crystalline benzene, as these are the most reliable data for the atomic coordinates which include hydrogen atom positions. Then the transferability of the hydrocarbon potentials was tested by calculating the optimum crystal structure of other hydrocarbons and comparing the calculated structure and enthalpy of sublimation with the experimental values. The interactions involving oxygen were determined by fitting the experimental crystal structure data on a few oxygen-containing aromatic molecules.

In a second paper, Abraham and Haworth[32] went on to apply the force field they had developed to the calculation of the crystal structures of PE, PEK and PEEK. They assumed that the ether and ketone linkages were equivalent. This assumption is borne out by experimental X-ray diffraction work on these systems. They therefore assigned the same torsional potential to both linkages, by averaging the potential curves from STO-3G calculations obtained at MNDO optimized geometries for benzophenone and diphenyl ether.[34] Apart from the torsion angle, the other geometric variables were kept fixed at idealized values.

With this model they were able to calculate successfully optimized crystal structures for PE, PEK and PEEK which agreed well with experimental information. However, for PEEK it was found necessary to exclude the torsion potential to obtain a satisfactory twist angle for the aryl groups, compared with experiment. This step was justified on the grounds of NMR data which indicate that the introduction of the neighbouring ether linkages in a chain reduces the conjugation in the chain substantially. The torsion potential they have developed appears not to be transferable between all the systems of interest because of the delocalization possible in some of these aromatic systems.

The work described by Abraham and Haworth exemplifies the approach adopted by many other groups to the task of modelling the structure of semi-crystalline materials. The non-bonded and bonded interactions need to be defined, and at the moment they can best be obtained by testing the potential against experimental crystal structures of small molecules. The crystal structure is most important in determining the non-bonded and electrostatic interactions; however, the potentials are not unique. It is often possible to subsume the electrostatic component into the van der Waals, non-bonded interaction. At the same time the potentials are tested for transferability since this increases their chances of being realistic in polymer calculations.

For bonded interactions, the situation is more complicated. Very often it assumed that the bond and bond angles remain fixed at some idealized value. This is quite a dangerous assumption since the c-axis length, for instance in the poly(aryl ether ketones), will be very sensitive to the value of the bond angle at the linkage. The torsion angle potentials are determined where possible from experimental information, but usually there are insufficient data to give a full potential curve, and quantum mechanical calculations are used to provide supplementary information.

Besides the approaches mentioned above for the study of semi-crystalline polymers, the experimental study of the crystal structure of oligomers gives much useful information and allows more checks on the validity of the potential functions. It is imperative to have good experimental information on the packing of important polymer linkages to develop force fields which are sufficiently accurate and reliable to be used in a wider range of simulations.

4.5 Simulations of crystal structure

4.5.1 Single chain calculations

Polymer chains can adopt an extended conformation, at least for local segments, and in general this can be described as a helix. Thus, local conformation within the chain, and chain flexibility, are important in determining polymer crystallinity. To a large extent, conformation and flexibility are determined solely by the torsional and non-bonded interactions of the chain. Indeed, the Rotational Isomeric State (RIS) method developed by Flory[35] uses only information about the torsional potentials in a single chain to obtain significant properties such as radius of gyration, persistence length and average end-to-end distance. Single chain calculations provide vital information for understanding the crystalline phase of the polymer, as well. Unlike the unit cells of small organic molecules, the long polymer chain results in a unit cell which has an axis (usually the c-axis) dominated by helix formation along that axis. The helix itself is described using the nomenclature m_n, where n turns of the helix are made by m repeat units of the polymer. Thus the structure of a single chain underpins our interpretation of the crystal structure.

The earliest applications of molecular mechanics to polymeric systems involved use of single chain calculations in the interpretation of fibre X-ray diffraction studies.[36] It is usually possible to deduce the unit cell dimensions and the space group from these experiments, but the molecular conformation is more difficult to determine. Tadakoro and coworkers[37–40] have had remarkable success in performing single chain calculations to determine several optimum molecular configurations consistent with the fibre period of the polymer. These few conformations could then be used to predict a fibre pattern which could be compared with the experiment. With this approach, Tadakoro and coworkers have determined the crystallographic unit cell, and the molecular conformation, of several polymers including polyisobutylene, polyesters and polydiketene.

4.5.2 Chain packing

The starting point for understanding the interactions between polymer chains in crystals lies in the packing of chains (helices) together. This is dominated by geometrical and conformational factors, although electrostatics plays a significant role as well.

The calculation of chain packing in a crystal was pioneered in large part by Corradini and coworkers. In an early study, Corradini and Avatabile[41] calculated the energetics of isotactic poly-acetaldehyde in 4_1 helical conformation for six tetragonal space groups. The calculations assumed a fixed chain geometry, and confirmed the experimental result that the most stable packing corresponded to the $I4_1/a$ space group. This approach remains popular, as more recent studies on poly(*cis*-1,4-butadiene),[42] poly(*trans*-1,4-butadiene),[43] isotactic 1,2-poly(1,3)butadiene, poly(methylvinyl ether)[44] and *trans*-1,4-polyisoprene[45] show.

Traditionally, crystal structures have been approached by taking a regular chain conformation, then assuming that it does not change further when it is packed into crystals. It is presupposed that intramolecular forces are unaffected by intermolecular forces, although clearly this cannot be the case in the real physical system. This methodology proves inadequate in a number of cases. One such was highlighted by Kusanagi *et al.*[40] who investigated the alpha form of poly(ethylene oxybenzoate). Poly(ethylene oxybenzoate) can take up two crystal forms. In the beta form the chain is essentially extended, with a *trans* conformation about the ethyl linkage. It is relatively easy to see how these chains pack in crystals. The alpha form involves zig-zag chains, where the conformation about the ethyl linkage is gauche. In the case of the alpha form, packing in the crystal affects the chain geometry significantly, and assuming that the chain conformation is unaltered by interactions with the neighbouring chains does not give a good fit with the observed X-ray diffraction data. A refinement process is necessary to give an improved fit with a different chain conformation.

This can be done either by use of a 'cluster' approach, or through application of true periodic boundary conditions (PBCs).

4.5.3 Cluster calculations

A detailed application of the molecular mechanics method to solid, crystalline polymers requires the effect of neighbouring chains to be taken into account. One way of doing this is to create a 'cluster' of polymer chains so that the chain at the centre of the cluster is representative of the bulk material. This approach requires the cluster to be large enough that the interaction between polymer chains on the surface of the cluster and the one at the centre is negligible. To ensure this, the most important terms in the force field are the non-bonded and electrostatic interactions, since the other terms are only influential over distances of up to about three bond lengths, at most 5 Å. Computationally, the non-bonded interactions are often assumed to cut off after 8 to 10 Å. This results in a realistic minimum size of the simulation involving a box of at least 512 Å^3 filled with polymer chains.

The cluster approach was adopted by King et al.[46] in studying chain packing in crystalline PEEK, poly(ether ether ketone); Fig. 4.2. The space group of PEEK is Pbcn; the polymer chains lie in an extended zig-zag conformation. The oxygen atoms lie in planes parallel to [100], and the aromatic rings are inclined to this plane at angles of $+37°$ alternating down the chain. As discussed previously, in PEEK, the ether and ketone linkages have a similar geometry; all but the most detailed experimental investigations suggest that there are only two aromatic rings in the c-direction of the unit cell, whereas an examination of the chemical repeat unit indicates that there must be at least three. This suggests that there is no strong preference for the ketone groups to lie either 'in register' (Fig. 4.3a) or 'out of register' (Fig. 4.3b). King et al. performed calculations of the lattice energy of six different structures, each consisting of a large cluster of 59 chains, one of which was a central 'probe' chain. The size of cluster was based on trial calculations which suggested that the cutoff needed to be taken at about 15 Å, and no geometry optimization was allowed. The calculations indicated that there was no significant energetic preference for 'in register' conformations. At first sight, this appears rather surprising; the ketone linkage is expected to be significantly bulkier than the ether group. However, calculations of the electron density and electrostatic potential for the two linkages show their remarkable similarity.[47]

Abraham and Haworth[32] have also studied the chain packing in crystalline PEEK and PEK, with respect to the alignment of the ether and ketone linkages, and suggested that increased packing order involves both horizontal ether–ketone alignment and vertical ether–ether and ketone–ketone alignment.

PEKK has been shown experimentally to comprise two different crystal structures[29] – one which is like PE, PEEK and PEK, and another

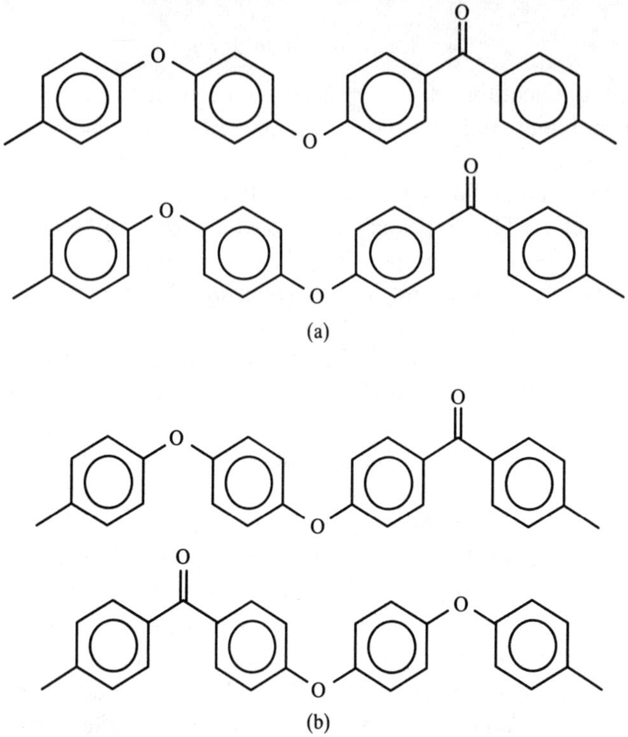

Figure 4.3 (a) In register packing in PEEK; (b) out of register packing in PEEK.

dominant form which has the structure illustrated in Fig. 4.4. We have investigated chain packing in PEKK for both structures. Although energy minima could be found for both, that for the PEEK-like geometry was somewhat lower in energy.[47]

In the method used by King *et al.* and employed more widely by Tadakoro and colleagues, the single chain was not allowed to relax once it was placed in a crystal environment. As discussed previously, this is not always a reliable assumption, since intermolecular forces can change the intramolecular geometry.

Sorensen and coworkers[48] have therefore proposed a scheme by which chain conformation can be affected by the crystal environment, with both intramolecular and intermolecular forces taken into account simultaneously. In their method, a trial helix is generated, and placed on a trial packing grid. Of course, the particular choice of trial helix and grid affects the final solution, since a number of packing arrangements is likely to lead to local minima.

In their work, the trial helix is described by several parameters,

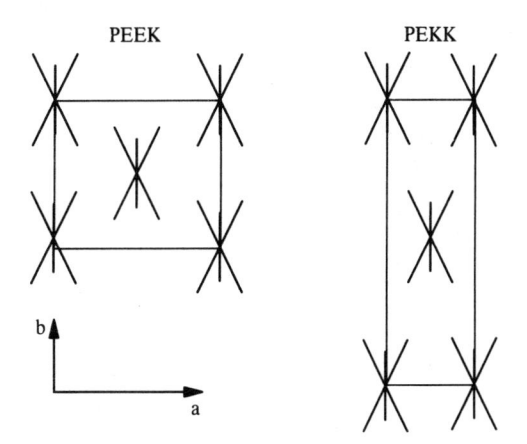

Figure 4.4 Schematic view of the unit cell structures of PEEK and PEKK as viewed along the chain axis.

including the cartesian coordinates of the atoms in the monomeric repeat unit (RU), the helix advance angle and the helix translation distance. The helix advance angle is the angle between one RU and the next; by the time the helix translation distance is traversed, the helix advance angle has passed through 360°. Unlike the work of King et al., where the probe was of finite length, the helices of Sorensen et al. are infinite in length. Two different cutoffs are assumed – one in the z-direction, along the helical axis, and the other in the x–y-direction, between different chains. The reference Helical Repeat Unit is selected in the middle of the centre chain, and the energies, including intramolecular and intermolecular effects, are calculated. It is possible to do an isolated chain calculation, or to assume rigid chains within a crystal, or to take flexible chains in a crystal. It is also possible to incorporate defects. The method has been applied to aliphatic polyesters,[50] polyethylene and poly(oxymethylene).[49] Poly(oxymethylene) was of special interest since it has two different crystal structures.

The method developed by Sorensen and coworkers allows a number of properties of crystalline polymers to be calculated. The elastic constants at zero strain, and under compression, tension and shear can all be determined, in the latter cases by deforming the unit cell and holding it deformed while doing the energy minimization. Vibrational dispersion curves, refractive indices and heat capacities at constant volume or pressure can be obtained as well.

4.5.4 Simulation of pseudo-crystalline polymers

The assumption that polymer chains pack to form perfectly periodic systems has already been shown to be only partially true, on a limited size scale within a crystallite. For those situations where intermolecular forces are particularly high, even within a crystallite perfect translational periodicity is often not maintained since the large intermolecular forces considerably influence the local chain conformations, leading to defects and non-ideal or pseudo-crystalline structures. Rutledge and Suter[51] have developed a static, molecular mechanics method specifically for simulating such systems. Their method begins with the determination of the conformational behaviour of a single chain and goes on to consider explicit nearest neighbour and next nearest neighbour interchain interactions. Periodicity is invoked only to estimate the long range compaction forces. The atomic coordinates of their representative chain are generated by application of Flory generator matrices[35] to a conformational repeat unit (CRU). The CRU is usually taken to be one chemical repeat unit. The chain itself is a general, possibly non-rational, helix where the unique conformational parameters are taken from the CRU. The chains themselves are packed onto a two-dimensional lattice of 'perfect' lattice points defined by a set of two-dimensional lattice vectors, with the chain axis running perpendicular to the set of lattice points. The chains are essentially copies of the parent chain moved onto the two-dimensional net of lattice points with the possibility of inversion or translation along the c-axis. The simulation considers two zones: the explicit zone consisting of the reference chain and its nearest neighbour chains or in some cases its next nearest neighbour chains; and the implicit zone which treats the polymer as a perfect crystalline polymer.

Rutledge and coworkers[52,53] have applied their methods to studying poly(p-phenyleneterephthalamide), abbreviated PPTA, as discussed in more detail in section 4.8.2.

4.5.5 Periodic boundary conditions

An alternative to the cluster approach uses periodic boundary conditions to give a useful representation of a periodic system. By taking a unit cell of the polymer and replicating it in all three dimensions, all parts of the unit cell are treated equally; no surface effects occur since there is no surface. The non-bonded interactions are frequently still calculated using a cutoff distance, but methods have been proposed which remove this limitation, including the use of Ewald summations.[54] For crystalline material, the unit cell usually contains one or two polymer chains running along the c-axis. The packing of the chains is determined by the directions and lengths of the a- and b-axes, and the relative positions of the chain(s) in the cell. The energy of the unit cell can be calculated by considering the bonded and non-bonded interactions of the atoms in the cell with other atoms in the cell and with the images of atoms in surrounding cells.

The use of a non-bonded cutoff makes this type of calculation very similar in practice to the cluster calculations mentioned previously. However, cluster calculations assume that all atoms in the calculation move independently. Here we will define periodic boundary condition calculations to require that only atoms in the unit cell can move independently. The position of atoms outside the reference unit cell is determined by the translational symmetry of the space group to which the polymer crystal belongs.

This translational symmetry is very important and can be used to simplify dramatically the computational cost of the calculation and the accuracy to which it can be performed. Work done on the simulation of the structure of ionic materials developed very sophisticated ways of calculating the electrostatic interaction in systems with periodic boundary conditions. Pair-wise summation of the real space interactions between point charges is known to be very slowly convergent and methods for accelerating the convergence have been developed which require performing part of the summation in real space and part in reciprocal space. Similarly the non-bonded interactions in periodic boundary condition simulations of polymers are very slowly convergent, and Karasawa and Goddard[55] have shown how the acceleration techniques originally proposed by Ewald for purely ionic systems can be modified to calculate accurately both the non-bonded electrostatic interactions and the non-bonded van der Waals interactions which decay as r^{-6}.

In addition to using the techniques for accelerating the convergence of the non-bonded interactions developed previously for ionic systems, the use of periodic boundary conditions has allowed the application of methods developed by Born and Huang[56] to calculate the elastic constant tensor of the material as well as the response of the material to external electric and magnetic fields. This methodology allows the calculation of the modulus of the material in a fairly straightforward manner. Such developments of the periodic boundary condition methodology have been published by Karasawa and Goddard[55] and applied to crystalline poly-(ethene),[3] as discussed later in this chapter. These calculations avoid the use of cutoffs and are therefore very accurate.

4.5.6 Molecular dynamics simulation

So far in this chapter we have concentrated on the simulation of crystalline polymers using static or minimization methods. These concentrate on determining the optimum molecular configuration of the polymer chain with respect to itself and its interactions with other chains. Such calculations have the advantage of being simple to interpret, there being only a single molecular configuration to consider. However, they assume that these configurations are representative of the polymer. In reality, at any given temperature, the polymer chain is in motion, sampling its configurational space according to the energy of the configuration and therefore its Boltzmann probability. The molecular dynamics (MD) method, discussed in detail in Chapter 2 of this book, solves Newton's equations of motion for the atoms in the polymer chains according to the forces on each of the atoms, and follows the evolution of the polymer motion in time. For the case of periodic boundary condition simulations, methods have been developed which allow the effects of constant temperature, and either constant pressure or constant volume, to be included. For the case of cluster-type simulations, it is possible to treat constant temperature effects, but the concept of volume and pressure in such calculations are not very well defined.

From the MD simulation it is feasible to study the effect of temperature or pressure on the molecular motion of the polymer chain and on the density of the polymer. It is also possible to extract information on mechanical properties and to simulate phase changes within the crystal.

4.6 Computer simulation of poly(ethene) crystals

4.6.1 Perfect crystals

The simplest common polymer is polyethylene, or poly(ethene), which is made up of repeating $-CH_2-$ units. Synthesis of the polymer often takes place via a free radical mechanism, and experimentally it has been found that, depending on synthesis conditions, varying degrees of branching can be introduced into the polymer through a 'backbiting' mechanism. This produces 'low density' polyethylene, which is highly amorphous because the side-chain branches cannot be accommodated into the crystal. However, high density polyethylene can be synthesized by controlling the degree of branching rigorously, and in this case very high degrees of crystallinity, in excess of 90%, can be obtained.

Single crystal polyethylene has been studied by Keller and O'Connor,[57] who determined that it had a flat tabular crystal habit, with molecules perpendicular to the layers which themselves were about 100 Å thick. To produce this morphology, molecules need to be sharply folded, and simple molecular models showed that this was possible, with three to five carbon atoms involved in each fold. The fold lies in the [110] plane, which represents the growing face of the lozenge-shaped crystal.

Polyethylene has a simple structure, with the hydrocarbon chain forming a planar zig-zag in which all the C–C bonds can lie trans to each other. Interchain interactions are expected to be weak. The crystal structure was determined as long ago as 1928 by Müller;[58] electron density maps were reported in 1939 by Bunn.[59] The unit cell was determined to be orthorhombic, with a Pnam space group and four CH_2 moieties in the unit cell. At the temperatures used in Bunn's work, the lattice parameters were determined to be $a = 7.4$ Å, $b = 4.93$ Å and $c = 2.534$ Å. Studies have been undertaken at other temperatures to determine the variation of the lattice parameters with temperature; as Avitabile et al.,[60] show, using neutron diffraction that, at 4 K, the lattice parameters are $a = 7.121$ Å, $b = 4.851$ Å and $c = 2.548$ Å. The largest change is seen to be in the a parameter.

A monoclinic crystal structure has also been reported;[61] this develops when the system is under high pressure, and some effort has been devoted to understanding the orthorhombic-to-monoclinic transition. Initial work by Yemni and McCullough[62] indicated that the monoclinic packing would be slightly more stable than orthorhombic, in contradiction to the experimental results. The difference between the two forms was relatively small, 0.15 kJ/mole per CH_2 unit. These findings were confirmed by a subsequent investigation by Tai et al.,[63] who found that, regardless of the potential parameters they used over a reasonable range, the monoclinic form was very slightly more stable. The study of Yemni and McCullough considered only intermolecular forces, and they therefore attributed the discrepancy to the lack of inclusion of intramolecular energy, particularly of chain folds. They believed that inclusion of entropy would not change their results, since the energy wells associated with the orthorhombic and monoclinic forms were essentially of the same shape and width. However, Kobayashi and Tadakoro,[64] in their calculation of the vibrational free energy term, found it to be about 0.5–0.3 kJ/mole per CH_2 unit, enough to make the orthorhombic form more stable than the monoclinic form. In their work, intermolecular forces were taken into acccount for two hydrogen atoms separated by a distance of less than 3 Å, with both Lennard–Jones and Buckingham-type potentials being investigated; intramolecular forces were also included.

In many ways, polyethylene serves as a 'model' system for studies on more complex polymers. This is reflected in the work which has been done in determining suitable potentials for studying the system. In the simplest case, polyethylene can be represented as simple 'beads' on a chain, with each $-CH_2-$ unit forming one bead. Thus, the hydrogen atoms are not treated explicitly. Potentials of this form have been used for example by Brown and Clarke.[65]

At a higher level of complexity, the non-bonded interactions between the hydrogen atoms, and between hydrogen and carbon atoms, can be considered explicitly. Yemni and McCullough,[62] for example, used explicit H... H interactions in their study of the relative stabilities of monoclinic and orthorhombic forms, as did the early work of Kobayashi and

Tadokoro.[64] Later work of Kobayashi and Tadokoro[66] showed the importance of including C... H interactions as well as H... H, in order to get good agreement with vibrational spectra. A notable improvement was seen in the agreement with experiment when the carbon–hydrogen interactions were included.

Williams and Cox[67] have done a careful determination of intermolecular forces, obtaining van der Waals parameters for H and C by examining X-ray structures of a series of hydrocarbons. In their work the intramolecular structure of each molecule was fixed, and the C and H exponential-6 parameters were optimized to give accurate unit cell parameters and sublimation energies.

Using the Dreiding force field, Yang and Hsu[68] report elastic constants for polyethylene, together with calculated lattice parameters of 7.21 Å, 5.15 Å and 2.58 Å. These, it must be noted, are calculated effectively at a temperature of 0 K; when compared with the values of Avitabile et al.,[60] the largest error (of about 6%) is seen to be in the b-direction.

Work on refining the potentials for polyethylene continues. Sorensen et al.,[49] showed the effect of including intramolecular as well as intermolecular functions, using the potentials of Schactschneider and Snyder,[69] with some re-parameterization, to represent the intramolecular interactions. There has been extensive recent work by Karasawa et al.[3] in developing a very accurate force field for polyethylene. They considered three different possible cases:

1. no cross terms in the force field;
2. with bond–angle and bond–bond cross terms included;
3. with angle–angle cross terms included.

In their 'Hessian-biased' force field, experimental vibrational frequencies for n-butane were input together with a second derivative matrix (Hessian matrix) calculated by ab initio quantum mechanics. The charges on the hydrogen atoms were determined carefully by examining the charge distribution for methane; periodic boundary conditions were used for treating the system, including the use of Ewald summations for treating the electrostatics. This work highlighted the importance of including the cross terms: the angle–angle cross term was seen to have a significant effect on the Young's modulus in the c-direction. It also showed the importance of using a reasonably large cutoff for the non-bonded interactions, and it was their view that the value of 4 Å used by previous workers was too small. Evidence for this was taken to be the discrepancies in the calculated elastic constants.

Interestingly, Karasawa et al. saw a new (and mechanically unstable) orthorhombic crystal structure (with $a = 7.727$ Å, $b = 4.483$ Å and $c = 2.547$ Å) when they took relatively large deformations in the

a-direction. These compare with their calculated values for the lowest-energy orthorhombic form of $a = 7.202$ Å, $b = 4.795$ Å and $c = 2.546$ Å, and with the experimental values (at 4 K) of $a = 7.121$ Å, $b = 4.851$ Å, $c = 2.548$ Å. The increase in the a lattice parameter is compensated by a decrease in the b parameter, thereby maintaining a similar density.

4.6.2 Defects in polyethylene crystals

The above discussion has focused on the computer simulation of non-defective polyethylene crystals formed by packing of chains, all of which are in a *trans* planar zig-zag conformation. However, clearly, real crystals will have imperfections. Indeed, these imperfections can have an important role to play, for example in lamellar thickening. Crystal thickening requires chain transport; it is not clear what the mechanism of this is, but almost certainly it requires defect migration.

One of the early studies on defects in polyethylene crystals was undertaken by Boyd,[70] who investigated 'kinks' by using model systems which consisted of crystalline arrays of the long-chain hydrocarbons $C_{14}H_{30}$, $C_{18}H_{38}$ and $C_{22}H_{46}$ packed into an orthorhombic structure, including various numbers of co-ordination shells. The minimum energy position was found for one chain, with all other chains held fixed, in an iterative procedure which treated the chains sequentially. The particular defect studied was one in which the defect TGTG'T was transformed to TG'TGT, in a 'flip flop' motion. The energy of the defect in the crystal was calculated to be about 27.6 kJ/mole; the barrier to re-orientation was about 42 kJ/mole. There was a smaller barrier to 'unkinking' to give the all *trans* form, of only 8 kJ/mole.

Although interesting, kink re-orientation or even kink translation of this type does not result in translation or rotation of a whole chain; rather they are localized phenomena. They therefore cannot account for chain transport leading to lamellar thickening, although they are likely to contribute to disorder near the crystal melting point. To do this, it is necessary to introduce defects which introduce an effective shortening of the chain. One such is the so-called Reneker defect,[71] a localized, 180° buckled twist of planar zig-zag stems relative to each other. When a 180° twist is introduced, together with an extra CH_2 unit in the chain, Reneker et al.[72] show that the chain can still be accommodated within the lattice. There is an adjustment of the dihedral angles around the defect, affecting five or six neighbouring bonds, with the dihedral angles changed from the 180° of the all-*trans* configuration to angles varying up to 15° from this. Their investigation of possible defect migration showed that there was a relatively low energy barrier (about 17 kJ/mole), which provides evidence that the motion of a defect carrying an extra CH_2 unit can account for chain diffusion.

Without the introduction of an extra CH_2 unit in the chain, twists of the chain can lead, in principle, to a translational mismatch in the crystal; as Mansfield and Boyd point out,[73] a rotation of a polyethylene chain through 180° is equivalent to a translation by one CH_2 unit, so that rigid rotation of a whole chain leads to a translational mismatch.

Mansfield and Boyd addressed an alternative but somewhat different defect, involving a twist of 180° which was delocalized over about 12 CH_2 units. The twist was incorporated into the chain through a slight modification of the torsional angles, by about 15° each, from the perfect trans conformation. In common with the Reneker defect, the twist of 180° introduces a translational lattice mismatch. However, there is a negligible shortening of the chain, and the translational mismatch is localized on one side of the twisted region. The crystal forces as a result of this mismatch produce a tension (or a compression) of the entire chain, and gradually force the chain into lattice register far away from the twisted region. The twist propagated through the crystal smoothly, and they saw no evidence for a hopping type of motion. This they refer to as the Rotation–Translation–Twist–Tension model. The defects are expected to migrate rapidly, and their motion has been interpreted in terms of solitons.[74,75]

Other defects in polyethylene have been studied extensively by Reneker and his collaborators.[72,76] In addition to the twist of 180°, accompanied by an extra CH_2 unit, they have looked at a defect which has no net twist, with two extra CH_2 groups, which can dissociate into two defects with opposite senses of twist. They have also considered a 360° twist with no additional CH_2 units, and a wide range of higher energy, more complex defects.

Increasingly, molecular dynamics methods are being used to simulate defect processes, and defect migration, in crystals. Pioneering studies by Noid et al.[77] have addressed the issue of pulling a single chain out of a polyethylene crystal which can be related to the lamellar thickening process.[78] Comparisons have been made with the experimental IR spectra, using the MUltiple SIgnal Classification (MUSIC) method for estimating frequencies from short dynamics simulations. Simulations have been undertaken of the crystal-to-melt transition in polyethylene,[79] including the effect of chain folding.[80] Simulations are also reported of the correlation of very localized defects within polyethylene crystals,[81] and of the transition from crystals to 'condis' (conformationally disordered) crystals, which will be discussed in more detail later. In addition, MD has been able to provide information on the twist motion in polyethylene proposed by Mansfield and Boyd; studies by Sumpter et al.[82] showed that the defect motion was energetically downhill to the nearby crystal surface, that it migrated very rapidly, and that no special soliton-like motion was observed.

4.7 Simulation of condis crystals

The study of defects in crystals leads naturally to an investigation of conformational disorder within crystals. A distinct and different type of mesophase with dynamic conformational disorder has been shown to exist, and has been christened the 'condis crystal'. Noid et al.[83] have used molecular dynamics to study the conformational disorder realistically (i.e. explicitly accounting for the dynamic nature of the processes). They describe a model for the polyethylene crystal which has seven dynamic chains surrounded by a shell of twelve fixed ones. Therefore, in the terminology which we introduced earlier, this is a cluster calculation, and is similar to the model used by Reneker and Mazur.[76] The transition from a crystal to a conformationally disordered crystal was observed, but no specific defect type or movement was especially obvious. The fibre period was seen to reduce a little, and the heat capacity fell by a small amount as the condis state was approached. Both of these seem to fit with observed behaviour.

A larger study of the same system was reported subsequently.[84] This confirmed that rotational isomeric defects were created at random; specific defects had a very short lifetime and there was little or no correlated motion for defects. There is a continuous source of defects which can be used to explain lamellar thickening, due to a free enthalpy gradient on crystal surfaces and deformation by external forces; the MD studies are also consistent with a decrease in lamellar end-to-end distance as temperature increases.

The simulation of condis crystals has so far been undertaken only for polyethylene, the simplest of polymers. However, experimentally condis mesophases can be expected in trans-1,4-polybutadiene, polytetrafluoroethylene, poly-para-xylene, polyoxybenzoate among other polymers,[85] so this is likely to remain an active and growing field for simulation.

4.8 Modelling studies of polyamide systems

The polyamide polymers form an important industrial family of materials which possess a wide range of material properties. The aliphatic polymers such as nylon 6, nylon 6,6 and those with higher aliphatic content are extremely important to the fibres industry and are widely used in the manufacture of carpets and synthetic fibres for clothing. The aromatic nylons such as KEVLAR are very high performance polymers with exceptional properties, including high thermal stability and good tensile strength. The unusual properties of the polyamide systems are in general at least partially attributable to the interchain hydrogen bonding which can take place between the chains. In the aliphatic polymers this can be adjusted almost at will by altering the aliphatic content of the monomer units. These strong intermolecular interactions lead to polymers that tend to be semi-crystalline, but which, because of their hydrogen-bonding tendency, also have a strong affinity for water. We discuss them in some detail here because they provide a substantial contrast with polyethylene, discussed earlier, with its weak interchain interactions.

4.8.1 Molecular modelling of nylon 6,6

Nylon 6,6 is a semi-crystalline polymer which is known to exist in several phases, depending on the temperature. Of particular interest in nylon 6,6 is the phase change at a temperature of 435 K which occurs below the melting point of the nylon, 535 K. This was first observed by Brill,[86] who noticed that two of the strongest X-ray diffraction peaks merged at 435 K into a single peak. Such behaviour had not been observed for nylon 6. The crystalline packing of the nylon polymer chains at temperatures below the Brill transition, the so-called low temperature form, is a triclinic unit cell with sheets of hydrogen bonded chains. Above the Brill transition, the high temperature form has a higher symmetry, apparently a hexagonal unit cell, sometimes called pseudo-hexagonal. It has been proposed[87] that at the higher temperature the hydrogen bonds are constantly breaking and reforming, thus breaking up the sheet-like low temperature structure, by forming temporary hydrogen bonds between any of the nearest neighbour chains. However, other studies have suggested that a three-dimensional network of hydrogen bonds could not exist,[88] but that the Brill transition could be ascribed to increased torsional motion of the methylene groups.

Molecular modelling studies of this system were performed by Wendoloski et al.,[89] using molecular dynamics to study both the low, room temperature form and two high temperature structures at 503 K and 513 K. The high temperature forms corresponded to a crystal which had undergone the Brill transition. They used periodic boundary conditions with a repeat unit consisting of a cluster of nylon 6,6 with four triclinic unit cells along each axis, making 64 unit cells in all. They employed the AMBER[5] force field with electrostatic potential defined by atomic charges deduced from *ab initio* calculations on the monomer units. The hydrogen atoms were assumed to be bonded to the backbone atoms by bonds of fixed length, a constraint imposed by the SHAKE algorithm.[90] The detailed atomistic, high temperature structures were developed from the low temperature form by rescaling it in five uniform steps, to the required high temperature unit cell parameters, and re-minimizing the atom positions at each increment. Starting from the energy minimized structures the molecular dynamics calculations were equilibrated for 10 ps before accumulating data for 50 ps.

The predicted atomic positions of the non-hydrogen atoms in the energy minimized low temperature structure agree very well with the positions determined experimentally from fibre diffraction studies. The MD calculations at room temperature showed that the hydrogen bonded sheets were always maintained, but with large thermal motion of the amide linkage perpendicular to the hydrogen bonded sheets. At temperatures close to the melting point of nylon, the hydrogen bond sheets were found to be maintained, and there was an increase in the thermal motions of all groups, particularly the methylene units. This agrees with the suggestions put forward by Starkweather and Jones[91] and by Colclough and Baker,[88] who found no evidence for a dynamic three-dimensional network of hydrogen bonds. It would appear that the Brill transition does not rely on the formation of such a three-dimensional network. The molecular dynamics trajectories were also used by Wendoloski *et al.* to calculate ^{2}H NMR line shapes which were compared with experimental work. Very good agreement was found for the low temperature simulation, with qualitative agreement for the high temperature ones.

4.8.2 Modelling of poly(p-phenyleneterephthalamide)

The PPTA chain (alternatively known as poly(*p*-phenyleneterephthalamide) consists of a series of phenylene rings joined by amide linkages. The π electron system along the polymer chain is extremely delocalized and leads to a strong preference for planar conformations, which manifests itself in the observation of extremely high persistence lengths. This propensity for high extension with a low degree of conformational flexibility for the chain also leads to an unusual, but characteristically high, melting point. Indeed, so high is the melting point for PPTA that the polymer decomposes as it melts. X-Ray diffraction studies have suggested that PPTA may have two different packing modes[92,93] in the crystalline form, depending upon the conditions of the crystallization.

In work reported by Rutledge and Suter,[51–53] the PPTA chain repeat unit was taken to be as illustrated in Fig. 4.5. Bond lengths and bond angles, apart from the backbone bond angles, were held fixed. The phenylene ring units were kept as rigid entities. The force field for the simulation was mainly taken from previous work, with the atomic charges being chosen to reproduce dipole and quadrupole data and to ensure local charge neutrality of the chemical groups. The phenyl–amide group torsional potential was investigated using AM1 calculations and

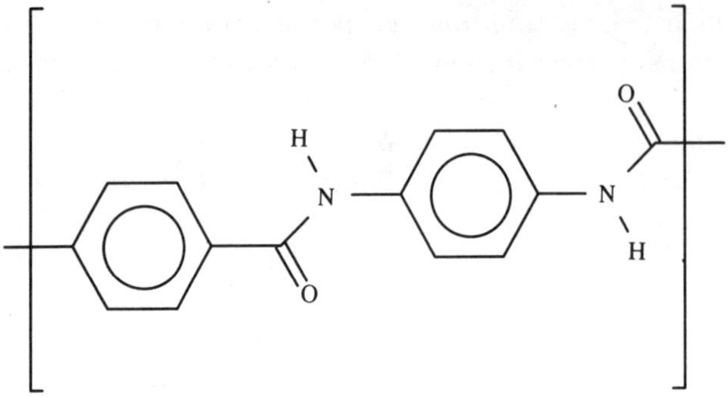

Figure 4.5 Poly(p-phenyleneterephthalamide) chemical repeat unit.

compared with the previous published calculations. The calculations on the packing of the polymer chains used the pseudo-crystalline method developed by Rutledge and Suter, described briefly above.

They assumed that the problems of determining optimum chain configurations and chain packing can be considered independently. Single chain simulations were used to determine the lowest energy conformations. As the fibre spinning process is likely to favour elongated conformers, low helical pitch chains were excluded. Then a grid search of the chain packing parameters was performed using fixed geometry chains to identify the potential optimum packing configurations. Finally full optimization of the potential low energy minima was carried out, allowing simultaneous optimization of both packing and chain geometry parameters. Such a procedure was developed in an attempt to address the problem of predicting the optimum crystal structure from first principles. Typical simulations of this type often use experimental information to make assumptions about the nature of the optimum crystal structure. This obviously biases the conformational and packing space that the simulation explores and in most cases, perhaps unsurprisingly, the predicted structure will agree well with the experimental one. In their simulation Rutledge and Suter made no such assumptions and suggested up to eight distinct polymer morphologies with cohesive energies within a range of 2.4 kcal/mole.

In a combined experimental and theoretical paper Rutledge et al.[53] went on to measure the wide-angle X-ray scattering, WAXS, and infrared absorption of several drawn PPTA samples, and to compare the calculated WAXS reflection intensities with the experimental values. The simulations suggest a low energy structure consistent with the X-ray data produced by modification I, and suggest that modification II is consistent with several polymorphs having similar packing.

PPTA and polybenzamide were also studied by Yang and Hsu,[68] since they are structures with strong interchain interactions, and so formed a useful contrast with their studies on polyethylene. Their work involved two distinct components: the prediction of crystalline structures, and the characterization of deformation mechanisms. The simulation was undertaken with the Dreiding force field[9] with Ewald summations for treating the charges. The unit cell parameters were most accurate for the c-repeat, and poorer for the stacking distances, particularly of the H-bonded sheets in PPTA. Errors of about 6% were obtained. Nonetheless, the calculated diffraction pattern for PPTA was in good agreement with experiment.

The calculated modulus did not agree so well. This was attributed to several factors including degree of crystallinity. Surprisingly, though, the values calculated for bulk PPTA were consistently higher than those from a single-chain approximation. It is likely that considerable refinement of the force field would be required, with particular attention to the atomic charges and to the electrostatic, van der Waals and hydrogen-bonding parameters, so that moduli be calculated reliably.

4.9 Conclusions

Atomistic modelling has always played an important role alongside experiment in the field of crystalline polymers. Single-chain geometries can be calculated reliably, and their packing into crystals can be in large part understood. Recent work has gone a long way to addressing issues of how interchain packing can affect the chain geometry, with the development of new techniques which give more realistic models of the system.

In this chapter, we have given an overview of the molecular simulation of crystalline polymers. Clearly, a number of hypotheses and methodologies have been proposed over the years, and with the advent of ever faster computers it will prove possible to test many of the hypotheses through simulation. Much work remains to be done in developing accurate force fields for a wide range of polymers, although the procedure seems now to be substantially established so that the task should be straightforward. Work also remains to be done in clarifying the methodologies underlying the simulations – for example, the use of periodic boundary conditions versus cluster methods, and the choice of cutoffs which are employed. Already, however, simulation is an essential partner for experiment in understanding X-ray diffraction and other spectroscopic techniques.

Molecular dynamics looks set to play a significant role in understanding the dynamic processes within crystals, including the onset of disorder and the role of defects. The use of MD will accelerate rapidly as faster, inexpensive computers become available.

As we said at the outset, crystalline polymers are of special interest because of their unique mechanical and barrier properties. It is a distinct possibility that, using simulation, the design of novel polymers with specific crystallinities will soon become reality.

Acknowledgement

We would like to thank Dr D J Blundell for useful discussions and a critical reading of the manuscript.

References

1. Goldbeck-Wood G in *Computer Simulation of Polymers*, Colbourn E A (ed), Longman, 1993, Ch. 5.
2. Bowen J P, Allinger N L in *Reviews in Computational Chemistry*, Vol. 2, Lipkowitz K B, Boyd D B (eds), VCH Publishers, New York, 1991, Ch. 3.
3. Karasawa N, Dasgupta S, Goddard W A III *J. Phys. Chem.* **95**: 2260 (1991).
4. Morley S D, Abraham R J, Haworth I S, Jackson D E, Saunders M R, Vinter J G *J. Computer-Aided Molecular Design* **5**: 475 (1991).
5. Weiner S J, Kollman P A, Case D A, Singh U C, Ghio C, Algona C, Profeta S Jr, Weiner P *J. Am. Chem. Soc.* **106**: 765 (1984); Weiner S J, Kollman P A, Nguyen D T, Case D A *J. Comput. Chem.* **7**: 230 (1986).
6. Brooks B R, Bruccoleri R E, Olafson B D, States D J, Swaminathan S, Karplus M *J. Comput. Chem.* **4**: 187 (1983).
7. Lifson S, Hagler A T, Dauber P *J. Am. Chem. Soc.* **101**: 5111 (1979).
8. *Discover Version 2.8 User Manual*, CFF91 Force Field, Biosym Technologies, San Diego, CA 92121, USA.
9. Mayo S L, Olafson B D, Goddard W A III *J. Phys. Chem.* **94**: 8897 (1990).
10. Sprague J T, Tai J C, Yuh Y, Allinger N L *J. Comput. Chem.* **8**: 581 (1987); Liljefors T, Tai J C, Li S, Allinger N L *J. Comput. Chem.* **8**: 1051 (1987).
11. Allinger N L, Yuh Y H, Lii J-H *J. Am. Chem. Soc.* **111**: 8551 (1989); Lii J-H, Allinger N L *J. Am. Chem. Soc.* **111**: 8566 (1989); Lii J-H, Allinger N L *J. Am. Chem. Soc.* **111**: 8576 (1989).
12. Dinur U, Hagler A T, in *Reviews in Computational Chemistry*, Vol. 2, Lipkowitz K B, Boyd D B (eds), VCH Publishers, New York, 1991, Ch. 4.
13. Kendrick J, Fox M, *J. Mol. Graphics* **9**: 182 (1991).
14. Maple J R, Dinur U, Hagler A T *Proc. Natl. Acad. Sci. U.S.A.* **85**: 5350 (1988).
15. Dinur U, Hagler A T *J. Am. Chem. Soc.* **111**: 5149 (1989).
16. Williams D E, in *Reviews in Computational Chemistry*, Vol. 2, Lipkowitz K B, Boyd D B (eds), VCH Publishers, New York, 1991, Ch. 6.
17. Cox S R, Williams D E *J. Comput. Chem.* **2**: 304 (1981).
18. Stone A J, Alderton M *Mol. Phys.* **56**: 1047 (1985).
19. Price S L *Chem. Phys. Lett.* **114**: 359 (1985).
20. Stouch T R, Williams D E *J. Comput. Chem.* **13**: 622 (1992).
21. Williams D E *Biopolymers* **29**: 1367 (1990).
22. Ferenczy G G, Reynolds C A, Richards W G *J. Comput. Chem.* **11**: 159 (1990).
23. Gasteiger J, Marsili M, *Tetrahedron* **36**: 3219 (1980).

24. Gasteiger J, Saller H *Angew. Chem.* **97**: 699 (1985).
25. Abraham R J, Griffiths L, Loftus P *J. Comput. Chem.* **3**: 407 (1982); Abraham R J, Hudson B *J. Comput. Chem.* **5**: 562 (1984).
26. Rappe A K, Goddard W A III *J. Phys. Chem.* **95**: 3358 (1991).
27. Coulter P, Windle A H *Macromolecules* **22**: 1129 (1989).
28. Tabor B J, Magre E P *Die Makromoleculäre Chemie* **126**: 130 (1969); Tabor B J, Magre E P, Boon J *Eur. Polym. J.* **7**: 1127 (1971).
29. Blundell D J, Newton A B *Polymer* **32**: 308 (1991).
30. Blundell D J, D'Mello J *Polymer* **32**: 304 (1991).
31. Abraham R J, Haworth I S *J. Computer Aided Molecular Design* **4**: 283 (1990).
32. Abraham R J, Haworth I S *Polymer* **32**: 121 (1991).
33. Powell B M, Dolling G, Bonadeo H *J. Chem. Phys.* **69**: 2428 (1978).
34. Abraham R J, Haworth I S *J. Chem. Soc. Perkin Trans.* **2**: 1429 (1988).
35. Flory P J *Statistical Methods of Chain Molecules*, Interscience, New York, 1969.
36. Tadokoro H *Structure of Crystalline Polymers*, R. E. Krieger, Malabar, 1990.
37. Yokouchi M, Chatani Y, Tadokoro H, Tani H *Polym. J.* **6**: 248 (1974).
38. Yokouchi M, Chatani Y, Tadokoro H *J. Polym. Sci. Polym. Phys. Ed.* **41**: 81 (1976).
39. Sakakihara H, Takahashi Y, Tadokoro H, Oguni N, Tani H *Macromolecules* **6**: 205 (1973).
40. Kusunagi H, Tadokoro H, Chatani Y, Suehiro K *Macromolecules* **10**: 405 (1977).
41. Corradini P, Avitabile G *Eur. Polym. J.* **4**: 385 (1968).
42. Corradini P, Napolitano R, Petraccone V, Pirozzi B, Tuzi A *Eur. Polym. J.* **17**: 1217 (1981).
43. De Rosa C, Napolitano R, Pirozzi B *Polymer* **26**: 2039 (1985).
44. Corradini P, De Rosa C, Gong Zhi, Napolitano R, Pirozzi B *Eur. Polym. J.* **21**: 635 (1985).
45. Napolitano R, Pirozzi B *Makromol. Chem.* **187**: 1993 (1986).
46. King M A, Blundell D J, Howard J, Colbourn E A, Kendrick J *Molecular Simulation* **4**: 3 (1989).
47. Colbourn E A, Kendrick J, unpublished results.
48. Sorensen R A, Liau W B, Boyd R H *Macromolecules* **21**: 194 (1988).
49. Sorensen R A, Liau W B, Kesner L, Boyd R H *Macromolecules* **21**: 200 (1988).
50. Liau W B, Boyd R H *Macromolecules* **23**: 1531 (1990).
51. Rutledge G C, Suter U W *Macromolecules* **24**: 1921 (1991).
52. Rutledge G C, Suter U W *Polymer* **32**: 2179 (1991).
53. Rutledge G C, Suter U W, Papaspyrides C D *Macromolecules* **24**: 1934 (1991).
54. Ewald P P *Ann. Physik* **64**: 253 (1921).
55. Karasawa N, Goddard W A III *J. Phys. Chem.* **93**: 7320 (1989).
56. Born M, Huang K *Dynamical Theory of Crystal Lattices*, Oxford University Press, London, 1954.
57. Keller A, O'Connor A *Discuss. Faraday Soc.* **25**: 114 (1958).
58. Müller A *Proc. Roy. Soc. A* **120**: 437 (1928).
59. Bunn C W *Trans. Faraday Soc.* **35**: 482 (1939).
60. Avitabile G, Napolitano R, Pirozzi B, Rouse K D, Thomas W W, Willis B T M, *J. Polym. Sci. B* **13**: 351 (1975).
61. Seto T, Hara T, Tanaka K *Japan J. Appl. Phys.* **7**: 31 (1968).

62. Yemni T, McCullough R L *J. Polym. Sci. Polym. Phys. Ed.* **11**: 1385 (1973).
63. Tai K, Kobayashi M, Tadokoro H *J. Polym. Sci. Polym. Phys. Ed.* **14**: 783 (1976).
64. Kobayashi M, Tadokoro H *Macromolecules* **8**: 897 (1975).
65. Brown D, Clarke J H R *J. Chem. Phys.* **84**: 2858 (1986).
66. Kobayashi M, Tadokoro H *Macromolecules* **8**: 897 (1975).
67. Williams D E, Cox S R *Acta Crystallogr.* **90**: 404 (1984).
68. Yang X, Hsu S L *Macromolecules* **24**: 6680 (1991).
69. Schachtschneider J H, Snyder R G *Spectrochem. Acta* **19**: 117 (1963).
70. Boyd R H *J. Polym. Sci. Polym. Phys. Ed.* **13**: 2345 (1975).
71. Reneker D H *J. Polym. Sci.* **59**: 539 (1962).
72. Reneker D H, Fanconi B M, Mazur J *J. Appl. Phys.* **48**: 4032 (1977).
73. Mansfield M, Boyd R H *J. Polym. Sci., Polym. Phys. Ed.* **16**: 1227 (1978).
74. Mansfield M L *Chem. Phys. Lett.* **69**: 383 (1980).
75. Skinner J L, Wolynes P G *J. Chem. Phys.* **73**: 4022 (1980).
76. Reneker D H, Mazur J *Polymer* **29**: 3 (1988).
77. Noid D W, Pfeffer G A *J. Polym. Sci. B Polym. Phys.* **27**: 2321 (1989).
78. Noid D W, Sumpter B G, Wunderlich B *Polym. Commun.* **31**: 304 (1990).
79. Noid D W, Pfeffer G A, Cheng S Z D, Wunderlich B *Macromolecules* **21**: 3482 (1988).
80. Sumpter B G, Noid D W, Cheng S Z D, Wunderlich B *Macromolecules* **23**: 4671 (1990).
81. Sumpter B G, Noid D W, Wunderlich B, in *Computer Simulation of Polymers*, Roe R D (ed), Prentice-Hall, Englewood Cliffs, NJ, 1990.
82. Sumpter B G, Noid D W, Wunderlich B *Macromolecules* **24**: 4148 (1991).
83. Noid D W, Sumpter B G, Wunderlich B *Macromolecules* **23**: 664 (1990).
84. Sumpter B G, Noid D W, Wunderlich B *J. Chem. Phys.* **93**: 6875 (1990).
85. Wunderlich B, Moller M, Grebowicz J, Baur H *Conformational Disorder in Low and High Molecular Mass Crystals*, Springer, Berlin, 1988.
86. Brill R *J. Prakt. Chem.* **161**: 49 (1942).
87. Schmidt G F, Stuart H A *Z. Naturforschung* **13a**: 222 (1958).
88. Colclough M L, Baker R, *J. Mater. Sci.* **13**: 2531 (1978).
89. Wendoloski J J, Gardner K H, Hirschinger J, Miura H, English A D *Science* **247**: 431 (1990).
90. Weiner P, Kollman P A *J. Comput. Chem.* **2**: 287 (1981).
91. Starkweather H W Jr, Jones G A *J. Polym. Sci. Polym. Phys. Ed.* **19**: 467 (1981).
92. Northolt M G *Eur. Polym. J.* **10**: 799 (1974).
93. Haraguchi K, Kajiyama T, Takayanagi J *J. Appl. Polym. Sci.* **23**: 915 (1979).

CHAPTER 5

Computer simulation of polymer crystallization

G GOLDBECK-WOOD

5.1 Introduction

Polymers with a regular chemical repeat are in principle able to crystallize by parallel alignment of at least sections of the long chains.[1,2,3,4,5] In the liquid phase the chains form random coils which can be envisaged as somewhat like a heap of spaghetti. On crystallization the chains tend to organize into thin platelets. The molecules traverse these lamellae perpendicular to the basal plane and have a tendency to fold back on themselves. Under certain conditions, in particular in crystallization from dilute solution or at an early stage of the solidification, these lamellae can be observed as separate entities. They then tend to form stacks of alternating crystalline and amorphous layers. In crystallization from the melt the lamellae eventually grow radially outwards and form macroscopic spherical semi-crystalline objects, the so-called spherulites.

Thus we find characteristic structures on two different length scales. The macroscopic level is defined by spherulites which have a typical size in the region of μm to cm. The microscopic level is given by individual lamellae with a typical thickness of 10 nm.

In theoretical investigations of polymer crystallization the macroscopic and the microscopic levels have generally been treated separately. The emphasis in studies of lamellar growth[6] has been on understanding the origin of the limited thickness. In this context not much notice is generally taken of the nature of chain-folding and the amorphous regions. These are again subject of separate studies. The impetus on the macroscopic level has come mainly from the processing side with an emphasis on the overall crystallization of a sample. However, the presence of lamellae within crystalline polymers has major consequences for macroscopic properties.[3] Hence, an integrated view is called for.

In principle, computer simulation could, with the processing power now available to us, be a significant aid in deciding about appropriate models of crystallization, thereby furthering our understanding of the often complex experimental evidence, as well as making predictions in new areas, hitherto inaccessible by experiment. However, the application of computer simulation methods in this area is still in its early stages but a number of good, promising ideas and models have been put forward in recent years. Unfortunately, not all of them have been published in international journals.

I have endeavoured to bring together the various strands in the hope that this will, apart from being a source of references, help the reader to gain a more integrated picture of the subject and develop links between its various parts in order to reach better and more realistic models.

Following the broad division into a macroscopic and a microscopic level this chapter consists of two parts. The first part (section 5.2) deals with the different aspects and models of lamellar crystallization, and the second part (section 5.3) with the overall crystallization behaviour.

5.2 The microscopic level

5.2.1 Physical background

From a supercooled melt or a supersaturated solution, regular, flexible long-chain molecules crystallize into lamellae of about 10 nm thickness by folding up and down the thin dimension with a large degree of adjacent or near-adjacent re-entry of chains.

Depending on conditions the crystals may occur either in isolation or as aggregates. In the latter the crystalline regions are separated by amorphous layers. A comprehensive study of polymer crystallization needs to address the formation of both phases, acknowledging their interdependence. In spite of this, models of the crystallization process (see sections 5.2.3 and 5.2.4) are usually limited to the crystalline lamella. Separate studies deal with chain folding and the amorphous phase (see section 5.2.5). With few exceptions, the latter are based on equilibrium thermodynamics, despite the recognition that polymer crystallization must be understood as a kinetic phenomenon.

The reason for the success of the simple lamellar models (shown schematically in Fig. 5.1) is that they describe fairly well the general kinetic and morphological laws which are found experimentally in a wide range of materials and conditions. The growth rate, G, and the lamellar thickness, l, generally obey the following dependencies on the supercooling $\Delta T = T_m^0 - T_c$:

$$G = \beta \exp[-K_G/(T_c \Delta T)] \qquad [1]$$

$$l = a/\Delta T + b \qquad [2]$$

where K_G, a and b can in a first order approximation be treated as

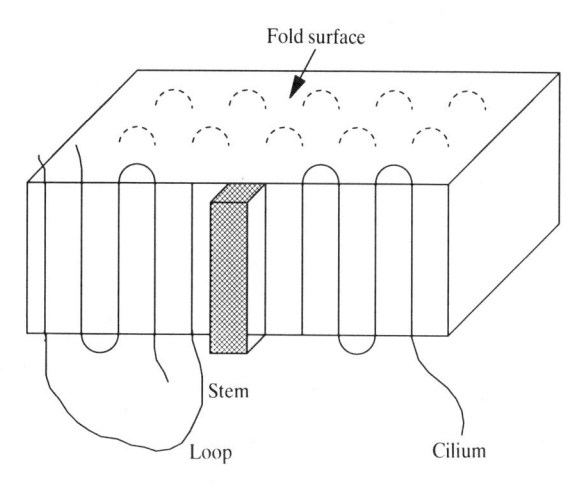

Figure 5.1 Schematic model of a polymer single crystal. The flexible chain traverses the thin dimension of the lamella, forming 'stems'. It folds back on itself by adjacent folds and loose loops. The chain ends form cilia in the amorphous phase.

constants. The prefactor β in the growth rate expression incorporates the laws governing the incident flux of material onto the growth face. The physics behind this transport term is a wide area which is not very well understood. Although it plays a significant role in the crystallization and even dominates the growth rates at temperatures close to the glass transition temperature it will not be our concern in this chapter. The interested reader is referred to the literature, e.g. Ref. 5.

In general terms, and without proposing any concrete model, the growth kinetics can be understood as a competition between a driving force and a barrier.[6] The origin of the driving force is the supercooling which takes the liquid into a metastable state with respect to an infinite crystal. A finite object like a lamella has a large surface free energy σ_e arising from the fold surfaces. In that case the crystal with heat of fusion Δh is in equilibrium with the surrounding liquid at a thickness

$$l_{min} = \frac{2\sigma_e T_m^0}{\Delta h \, \Delta T} \qquad [3]$$

Any thinner crystal will dissolve, but for thicker crystals the free energy driving force will increase with l.

The driving force is not the only force acting on the crystallization process. There is also a barrier force which exponentially disfavours thick crystals. The exact origin of this term has been a matter of great dispute in the past two decades. We can broadly distinguish two different approaches: the first is based on a surface nucleation concept.[7] An energy

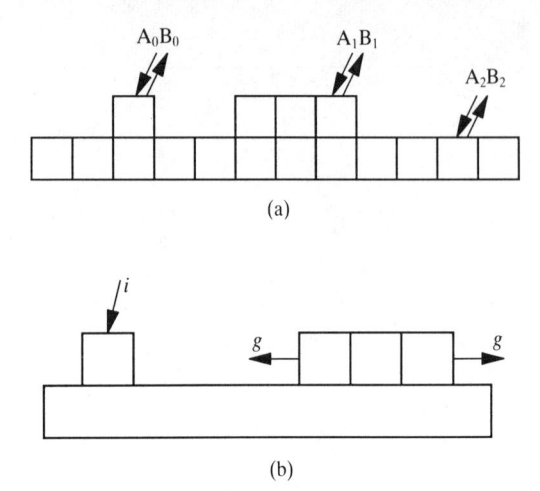

Figure 5.2 Projection views of crystallization. (a) On and off rate constants for stems with 0, 1 and 2 neighbours are specified in the stochastic model. (b) The mean-field models are based on a nucleation rate i and a spreading rate g.

barrier due to the creation of extra surface must be overcome via a random fluctuation as a chain segment is trying to attach to the growth face. The crucial simplification which is made at this stage is that this chain segment is treated as a rigid 'stem'. Furthermore a single stem is regarded as the nucleus which has to pass a critical length l^*, such that lateral deposition of further stems leads to a reduction in free energy. The view of the growth face is therefore that of a cliff edge of height l with no further structure in this thickness dimension. This amounts to a quasi two-dimensional 'projection' view (see Fig. 5.2). Computer simulations based on such models will be discussed in section 5.2.3.

The second, alternative explanation of the growth barrier rests on the consideration that the flexible chain will undergo a complex attachment process during which it explores a large number of configurations most of which will be unviable for further growth.[8,9] This barrier is then basically given by the loss in entropy involved in straightening out a section of the chain. The resulting picture of the growth face is one of a rounded or tapered edge. The thickness dimension is essential to the model, so that polymer crystallization is regarded as an inherently three-dimensional phenomenon. Computer simulations of such models will be discussed in section 5.2.4.

Apart from the main area of crystallization with a fixed thickness l there is a special mode which involves liquid-crystal-like phases.[3,10] The longitudinal mobility of the chains is strongly enhanced which inevitably leads to an at least partial suppression of the barrier force, and therefore to thickening as well as lateral growth, In *in-situ* studies of crystallization of polyethylene under high pressure it has been observed that the material can only crystallize while in this mobile phase.[10] These are quite unexpected discoveries for which a satisfactory explanation has yet to be found. Inevitably this will involve computer simulations. In section 5.2.6 simulations of the chain dynamics in a polymer crystal will be reviewed since these are of great relevance to further investigations of thickening growth.

5.2.2 Simulation methods

Simulations of polymer crystallization on the microscopic level are largely based on the great wealth of knowledge which has been gained in the area of crystal growth of low molecular weight substances.[11,12] The basic model is that of a two-dimensional or a three-dimensional Ising lattice of cells which represent a crystalline unit when they are occupied or a part of the liquid phase when unoccupied. Attractive interactions extend only to nearest neighbours, and the potential energy is the sum of all such pair interaction energies. The kinetics of crystal growth are then simulated by two basic events: impingement and evaporation. In order to determine the relevant rates, the structure of the surface and the associated probability distribution must be known. If the surface is in a configuration ψ then its equilibrium probability $P(\psi)$ can be determined from Boltzmann statistics.[13] The evolution of the system is then assumed to be based on the principle of microscopic reversibility even under the influence of a driving force. The justification for this rests on the regression axiom of Onsager[14] which states that the way in which the system goes to its equilibrium state is independent of how the initial state was reached. In the present case this means that the system will behave the same under a non-equilibrium chemical potential caused by external influences as under spontaneous fluctuations. Therefore the ratio of impingement and dissolution rate constants k^+ and k^- is given by

$$\frac{k^+}{k^-} \equiv \frac{p(\psi \to \psi')}{p(\psi' \to \psi)} = \frac{P(\psi')}{P(\psi)} \qquad [4]$$

This has been worked out in specific cases like a simple cubic lattice by van der Eerden *et al.*[11]

These considerations and equations form the basis of the microscopic models and simulations of polymer crystallization. Two methods have been used to compute the evolution of the system: the rate equation method and the Monte Carlo algorithm.

The rate or Master equation[15] follows directly from the consideration that the time evolution of the occupation of state ψ_j is given by the sum of all processes leading into and out of ψ_j:

$$\frac{dP(\psi_j, t)}{dt} = \sum_k [p(\psi_k \to \psi_j)P(\psi_k)] - \sum_k [p(\psi_j \to \psi_k)P(\psi_j)] \quad [5]$$

If an explicit expression for this equation can be found then the evolution of the system, and in particular its steady state, can be determined by numerical integration. The advantage of this method is its precision which is only limited by the numerical algorithm but not by any further stochastic errors. The disadvantage is that with increasing complexity of the system it becomes increasingly impractical to write down the associated rate equations.

Another method of 'solving' the master equation is the dynamic Monte Carlo technique.[16] It is of much greater, nearly unlimited, versatility. The algorithm which is most commonly used has the following general form. The surface configuration ψ is stored in the computer. Then a random number is used to determine a process which would lead to a configuration ψ'. Comparison of the transition probability $p(\psi \to \psi')$ with a further random number determines whether the process is actually carried out. If so, the surface configuration is updated and finally the cycle is again initiated. After a certain number of cycles average values of growth rates and other observables are obtained. A flow diagram of a typical Monte Carlo simulation of crystal growth can be in found in van der Eerden *et al.*[11] The sequence of configurations which is generated by this method is meant to simulate the time evolution of the system.[16] This is an important point to bear in mind in the choice of free energy barriers between the states.[17] Further, the following factors can affect the result of the simulation:

1. Convergence to a steady state distribution is only guaranteed asymptotically.
2. The initial configuration leads to a transient which must be discarded in any averaging.
3. Even with an efficient algorithm only systems which are small compared with macroscopic ones can be simulated. Therefore suitable boundary conditions must be used. Both system size and boundary conditions can affect the results.
4. Extreme precision is hard to achieve. Several runs must be carried out, and each extra significant figure in the accuracy of the result requires about a further 100 times of computing time.

5.2.3 Two-dimensional projection models

In this section we are going to examine models which treat the formation of a crystalline 'stem' as a one-step process.[7] This amounts to studying a two-dimensional 'projection' model of a polymer crystal as shown in Fig. 5.2 in a discrete, square lattice representation. Other types of lattice (e.g. hexagonal), or a continuum representation (where steps on the surface can occur at any irrational coordinate x) have also been studied. In the most general case which we shall call the 'Stochastic Model'[18-20] the growth units should be able to attach to or detach from any surface site with appropriate rate constants (see Fig. 5.2a) as long as they cover another occupied cell in the layer underneath (solid-on-solid (SOS) restriction). Surface migrations are neglected.

Three different sites are distinguished depending on the number i of neighbours. Their respective rate constants for addition and removal, A_i and B_i (Fig. 2.2a), are derived from the principle of detailed balance:

$$A_i/B_i = \exp[\Delta F_i/kT]; \quad i = 0, 1, 2 \qquad [6]$$

where ΔF_i is the total free energy difference between the two states. In the two-dimensional model this is

$$\Delta F_i = \delta f - 2\sigma_e + (i - 1)2\sigma_n; \quad i = 0, 1, 2 \qquad [7]$$

where δf is the bulk free energy loss per stem, σ_e the surface free energy gain per stem end and σ_n is the lateral surface free energy gain, which can be interpreted as the 'bond-energy' in the two-dimensional model.

The absolute values of the rate constants can in principle be chosen in any way which satisfies the above ratios. The choice, or so-called 'apportioning', which is made in nucleation theories[7] is

$$A_0 = \beta \exp[-(2\sigma_n - \psi\delta f)/kT] \qquad [8]$$
$$B_0 = \beta \exp[-(1 - \psi)\delta f/kT] \qquad [9]$$
$$A_1 = \beta \exp[-(2\sigma_e - \psi\delta f)/kT] \qquad [10]$$
$$B_1 = \beta \exp[-(1 - \psi)\delta f/kT] \qquad [11]$$
$$A_2 = A_1 \qquad [12]$$
$$B_2 = 0 \qquad [13]$$

where ψ is the apportioning factor which can in principle be chosen to lie between 0 and 1 (for further discussion see Armitstead and Goldbeck-Wood[6]). Notice that the creation of cavities is neglected as a possible process, hence the principle of microscopic reversibility is violated in that case.

This choice is in contrast to that generally made in growth models of atomic or low molecular weight materials, where the on-rate constants are set to unity.[11,12] Hence, there seems to be a certain degree of arbitrariness. However, the apportioning has an influence on the predicted behaviour. If, for example,[18] all the on rates are set equal to a constant flux, then units will be added to the surface sites completely at random, with no penalty for the creation of steps of any height. The result will be a series of columns whose heights increase according to a random walk. A quite different result is obtained if $B_0 = B_1$. Then the additions would tend to reduce the number of steps. The further from equilibrium, the more important the apportioning effect will be. We must therefore attempt to choose the physically most realistic apportioning. As a general approach consider a reaction between two states with a free energy difference ΔF and an activation barrier F_A. In the downwards reaction thermal fluctuations of the system must be equal to F_A before the transition can take place. In the backward transition, on the other hand, the fluctuations must surmount $\Delta F + F_A$. Hence, the change in free energy should be associated with the upward rather than the downward step. This can be shown to hold no matter how complex the reaction.[21]

It is therefore useful[20] to consider the following general expressions for the rate constants with the three apportioning factors ϕ_0, ϕ_1, ϕ_2, which act on the total free energy difference between the two connected states, rather than just on the bulk portion as above:

$$A_i = \beta \exp[\phi_i(\Delta F_i)/kT] \qquad [14]$$

$$B_i = \beta \exp[-(1 - \phi_i)(\Delta F_i)/kT] \qquad [15]$$

$$i = 0, 1, 2$$

Associating the positive change in free energy with the upwards step requires for $0 \leqslant \delta f - 2\sigma_e < 2\sigma_n$ the choice $\phi = (1, 0, 0)$ for the apportioning vector.

Most kinetic theories do not treat the full stochastic growth process. The 'nucleation-and-spreading' models (Fig. 5.2b)[22-24] are mean-field treatments. Following Sadler,[18] it is assumed that it is possible to separate the growth into the initiation of patches (which is generally called 'nucleation') at a rate of i patches per site and unit time and the spreading of these, once they are established, at a speed g sites per unit time. Removal from type 2 sites is neglected (see Eqn [13]). The net initiation rate i is then derived from a set of flux equations for the sequence of states which occur during the initiation of patches and averaging over all sequences of events. Two more assumptions enter at this stage. Firstly, during the initiation process, units can only be removed in the order in which they were added, and secondly, there are long stretches of smooth face over which the partially attached patches can fluctuate in size. The result is the nucleation rate

$$i = A_0(1 - B_1/A_1)/(1 - B_1/A_1 + B_0/A_1) \qquad [16]$$

Once a patch is initiated its extremities are presumed to move at a constant speed. The rate of step spreading is given by

$$g = (A_1 - B_1) \qquad [17]$$

The substrate is assumed to have a constant width L which has been related to some lattice coherence length.[7] The full width of the growth face, however, generally increases with time. It was only discovered later that this is significant for the explanation of curved faces (see below). When a niche reaches the boundary, it is generally assumed that it will be absorbed. Cyclic[24] and reflecting boundary conditions[25] have also been studied.

A continuous version of the nucleation-and-spreading model with absorbing boundary conditions was solved analytically by Frank[22] who used the mean-field approximation that steps moving in opposite directions are uncorrelated. He found two asymptotic regimes (generally known as Regime I and Regime II): (a) 'mononucleation' growth (i.e. not more than one nucleus on the substrate at any time) with the growth rate

$$G_M = biL \qquad [18]$$

and (b) 'polynucleation' growth (i.e. several nuclei at any time) with the growth rate

$$G_P = b(2ig)^{1/2}. \qquad [19]$$

At the crossover the dimensionless parameter $z = L^2 i/2g$, the so-called Lauritzen number,[26] is equal to one.

Hoffman[27] realized that the discreteness of the actual physical situation will limit the polynucleation regime as soon as the rate of nucleation i dominates over the rate of spreading g. In this so-called 'Regime II' he conjectured that the growth rate is again proportional to the nucleation rate i. This should then more properly be called 'aggregation rate'.

These results have been checked by computer simulations without the restrictions necessary for analytical treatments. Firstly, simulations of the spreading model[23,24] allowed step correlations. A tailored Monte Carlo algorithm, the so-called 'step-gas-model', speeded up the simulation in cases of relatively small nucleation rates. In each unit-time interval all left- and right-facing steps are moved by one site in the respective directions, collision and boundary checks are performed and a number n of new nuclei are placed at random positions on the surface. This number n is determined by comparison of the probablity $P(n) = \sum_{m=0}^{n} C_m i^m (1-i)^{L-m}$ with a random number. Secondly, the effect of the mean-field treatment on the attachment process was studied by simulations of the stochastic model[18,19] In the following we are going to discuss the main results.

5.2.3.1 The concentration of niches

The concentration of niches essentially governs the type of growth in a two-dimensional model and describes the character ('flat' or 'corrugated') of the interface. An analytical expression on the basis of Frank's mean-field model can easily be derived as

$$C_{\text{mean-field}} = (2i/g)^{1/2} = 2(z)^{1/2}L^{-1} \qquad [20]$$

Simulations of the spreading model[23,24] gave good agreement with this formula in Regime I, but deep in Regime II the concentration rises more slowly and eventually saturates. This is not unexpected, however, since the Frank model breaks down for high step densities and according to Hoffman[27] Regime III sets in about $C \approx 1/3$.

The equilibrium concentration of niches in a two-dimensional SOS model was derived by Sadler[18] on the basis of the fundamental treatment of the equilibrium structure of a one-dimensional surface by Leamy et al.[28] For large $\sigma_n/2kT$ the result is

$$C_{\text{equil}} = 2\exp\left(-\frac{\sigma_n}{2kT}\right) = 2(i/g)^{1/2} = (2)^{1/2}C_{\text{mean-field}} \qquad [21]$$

The discrepancy of $\sqrt{2}$ has been explained[18] by the fact that the stochastic model permits the creation of cavities, and therefore has an additional mode of niche creation.

The simulation of the stochastic mode (with $\phi = (0, 0, 0)$) agrees with the equilibrium formula for low supercoolings. A rising niche concentration with ΔT is interpreted[18] as a result of kinetic roughening. However, Salt[20] found that the dependency of the niche concentration on the supercooling varies with the apportioning. The increase found for $\phi = (0, 0, 0)$ turns into a decrease for $\phi = (1, 1, 1)$. The interpretation that there is a kinetic roughening effect can therefore not be held up. This is in agreement with the observation that the factor which brings about higher step densities at lower temperatures is σ_n in Eqn [21] which in nucleation theories depends on $l \propto \Delta T^{-1}$, i.e. C increases primarily because of decreasing thickness and not because of increasing growth rate. The mean-field concentration was reached asymptotically in the case $B_2 = 0$ for large free energy differences δf. This corroborates the interpretation that the combination of the prohibition of cavity creation and the neglect of fluctuations (small for sufficiently large δf) accounts for the difference between the stochastic and the mean-field models.

5.2.3.2 Fluctuation of niches

The simulation of the stochastic model allows direct visualization of the limits of the nucleation-and-spreading description (Fig. 5.3). As the step free energy is decreased, i.e. the ratio i/g increased (Eqn [21]), the niche separation becomes smaller than the size of the fluctuations and a coherent spreading process can no longer be observed.

5.2.3.3 Growth rates and regimes

According to surface nucleation theory the logarithm of the growth rate is a piece-wise linear function of the logarithm of the nucleation rate. There are two slope changes marking the Regime I–II and II–III transitions. Note that this interpretation rests on the assumption that the spreading rate g remains approximately constant over a large range of supercoolings.

Guttman and DiMarzio[23] claim to have verified all three regimes. However, the mean-field and constant spreading-rate assumptions discussed above are already built into the model. Regimes I and II agree well with the analytical results of Frank[22] even for very small substrate lengths ($L = 10$ sites). The location of the Regime II–III transition was expected to lie around the point where the nucleation rate is equal to the spreading rate. In the simulation, however, the upswing does not occur until considerably larger nucleation rates which are needed to generate the necessary niche concentration of about one-third. (Guttman and DiMarzio falsely state that the nucleation rate at the transition is much smaller than the spreading rate, in disagreement with their figures.) Such high nucleation rates create an unphysical situation, at least in the present context. They relate to the case where the proliferation of lateral surface area is favoured ($i > g$ means negative σ_n according to Eqn [21]). As, additionally, hole creations are not allowed ($B_2 = 0$) the resulting morphology on a square lattice is not space filling, as in diffusion limited aggregation. Guttman and DiMarzio resort to a hexagonal lattice symmetry in which such 'skyscrapers' are not allowed, only to find that there is no longer an upswing in the growth rate. Simulations of the stochastic model on a square lattice also yield a levelling off at the same step free energy (see Fig. 6 in Sadler[19]). This combined evidence strongly suggests that the growth rate is basically determined by the niche concentrations: as C saturates so does G. The question therefore remains as to what is the reason for the experimentally observed upswing. The answer given by Guttman and DiMarzio is that growth occurs in the hexagonal lattice on more than one plane such that the number of growth sites can keep on rising in Regime III to about 1.6 growth sites per substrate site (Fig. 5.4).

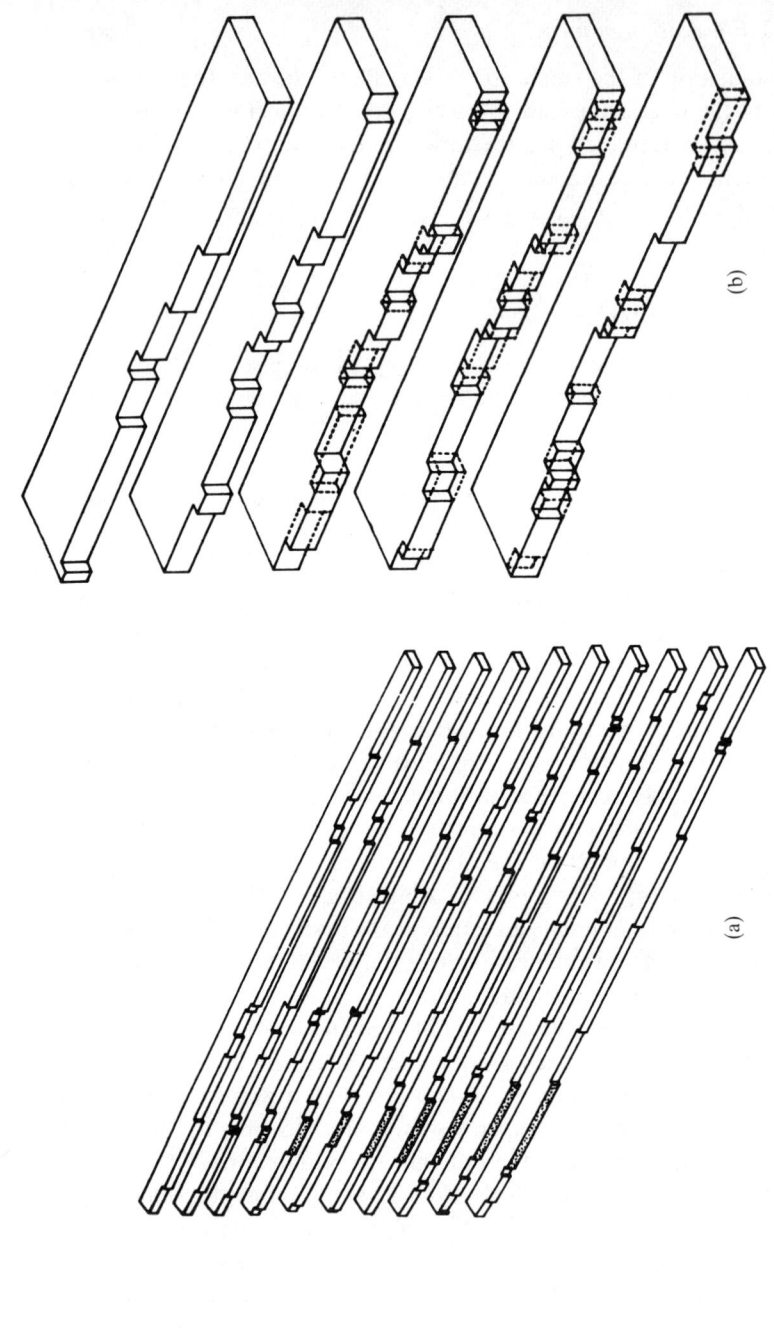

Figure 5.3 Stages of growth in the stochastic model showing (a) 'nucleation and spreading', and (b) strongly fluctuating niches. From Sadler.[18]

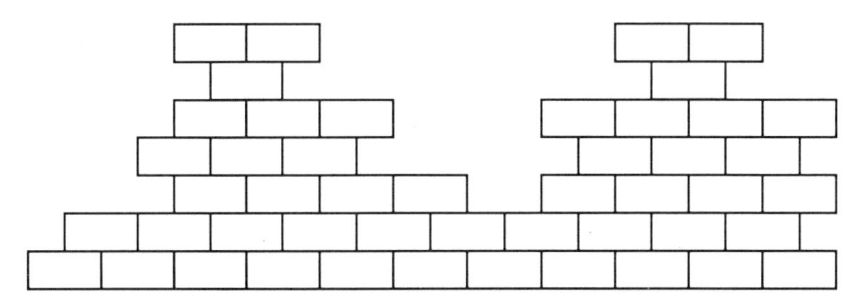

Figure 5.4 Expected morphology in Regime III resulting from growth on several planes on a hexagonal lattice.

Although agreement is claimed with interpretation of SANS data it remains debatable whether this is a realistic model. Further work is called for, and other effects, like for example the dependency of the surface free energy on the supercooling and lamellar thickness, should also be taken into account.

5.2.3.4 Lateral shapes

The observation of curved crystals in experiments, in particular at high crystallization temperatures,[29,30] posed a serious challenge to nucleation theories. The Frank model[22] predicts the faces to be flat or nearly flat in all cases, and in particular in Regime I. The obvious question to ask is whether any of the simplifications and assumptions which enter the model are the reason for this deficiency. These are in particular the restriction to two dimensions (i.e. stems are treated as rigid rods) and the mean-field assumptions discussed above.

As a possible starting point Sadler[31] noted the similarity between polyethylene single crystal habits and the increase in curvature observed in low molecular weight crystals which is caused by surface roughening.[32] In three dimensions this is a sharp transition. Above a temperature T_R the free energy per length of an infinite isolated step is zero, and below T_R it is finite. A change from faceted to curved crystals takes place over a small temperature interval around T_R. In two dimensions, however, the roughening temperature is 0 K. Therefore the density of steps is always finite and increases continuously with temperature, and so does the curvature of the equilibrium shapes.[18] The observed rounding in polyethylene, however, is much more abrupt. In order to assess whether this is a non-equilibrium effect Sadler constructed crystal outlines from growth rates on substrates oblique to principal crystallographic directions. The resulting shapes (Fig. 5.5) can still not explain the rather dramatic

[Shape with $\sigma_n/kT_m = 1$]

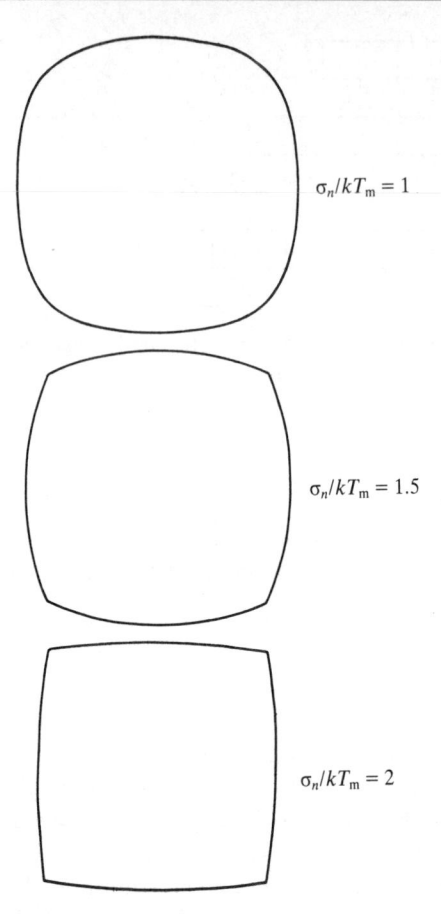

[Shape with $\sigma_n/kT_m = 1.5$]

[Shape with $\sigma_n/kT_m = 2$]

Figure 5.5 Rounding of habits in the stochastic model with fixed substrate length. From Sadler.[18]

change in curvature observed experimentally. Sadler concluded that a proper three-dimensional model with a fine-grained view of the thickness dimension of the lamella (see Fig. 5.11) was required.

However, the Sadler analysis did not consider the effect of the type of lattice, nor of the type of boundary condition. The substrate length was kept constant, whereas in reality the crystals' faces are growing, and are bound by adjacent sectors which might have the effect of absorbing or reflecting incoming steps.

Moving boundary effects were first introduced into the mean-field model by Mansfield.[33] The substrate length increases linearly with time: $L = 2ht$, and no steps enter from outside these limits. Computer simulation led Mansfield to an asymptotic analytical solution for long times. The shape of the interface turned out to be a section of an ellipse. In

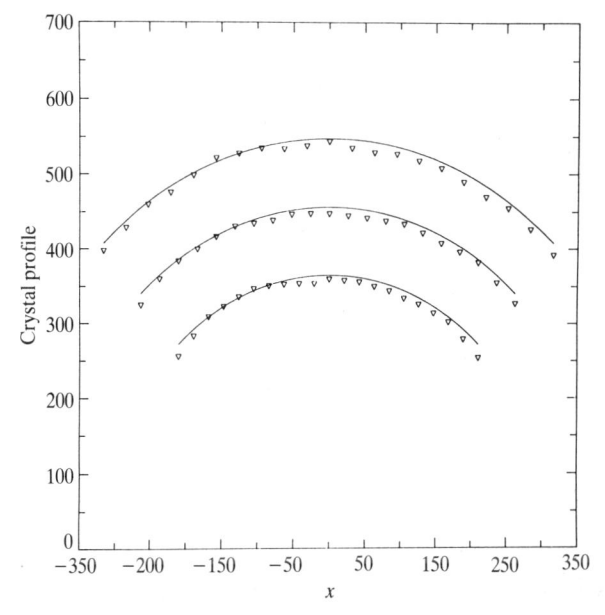

Figure 5.6 Crystal profiles in the nucleation-and-spreading model with moving boundaries, $g/h = 1.5$. From Mansfield,[34] by permission of the publishers, Butterworth Heinemann Ltd. ©.

support of his calculations Mansfield also published the computer simulation study.[34] The method used is basically that of a step-gas model. The resulting crystal profiles agree well with the analytical solution (Fig. 5.6). The deviations are explained by the asymptotic nature of the analytical result.

Toda[35] used the SOS Monte Carlo method on a square lattice to simulate the moving boundary model not only in the case $h < g$ but also for $h > g$. In the latter case the so-called lenticular crystals, which have been observed in melt-crystallization[29,36] are generated. He found very good agreement between the analytical and the simulation results in the limit of very large time. These could be reached because the SOS method is more efficient than the step-gas method used by Mansfield in the present case of high step densities. The deviation was generally less than 1% and highest close to the points of highest step density ($x = \pm ht$). However, in the case of large nucleation rates the deviation was systematic and reached about 3% close to the points $x = \pm ht$. This is not unexpected as the continuous model reaches its limits (see above). In particular it allows for any curvature, whereas in reality this is limited by the lattice spacing.

A corollary of this is that the type of lattice could also affect the results.

The definition of a moving boundary is not so obvious in the case of the more realistic hexagonal lattice.[35] Finally, the interaction between the two adjacent growth sections (110) and (200) in polyethylene could be crucial. This has not been studied so far. Further simulations seem feasible and should also check the effect of a full stochastic treatment.

5.2.4 Three-dimensional models

5.2.4.1 Introductory remarks

It was recognized early on in the development of polymer crystallization theory that the deposition of a chain segment onto the lateral growth face cannot without further comment be treated as a simple laying down of rigid stems of equal length. This issue was addressed by Frank and Tosi.[21] They allowed fluctuating fold lengths and further showed that the formation of stems out of smaller 'flexibility units' can indeed be substituted by a deposition of complete stems if there is a high free energy barrier between any two stems. In the polymer case this is generally assured because of the fold energy. The crucial premise of their analysis, however, is that a new stem cannot start to form unless the previous stem is completed, i.e. has reached the steady state length l. Figure 5.7 shows this process, but also further deposition paths which are not allowed. Hence, there is no configurational path degeneracy in the Frank and Tosi model, or in other words, it is a zero-entropy model.

The issue was left unattended until firstly Binsbergen[37] and then in more detail Point[38] proposed models which allowed the wider class of deposition processes, and therefore a configurational entropy which adds to the free energy barrier for stem attachment. The basic concept of a nucleation step was still maintained by Point. In his analytical model Point had to make the restriction that, once the length has been chosen in the multipath initiation process, it remains fixed during the spreading process. It was possible to lift this restriction in a computer simulation, which was carried out by Dupire.[39] This work will be described in more detail in section 5.2.4.2 along with a recent computer study of the same model using rate equations.[40]

A second group of models which allows a wider class of deposition processes was put forward by Sadler.[31] He invoked the existence of a roughening transition on the lateral growth faces. According to the Jackson criterion[41] this necessitates small growth units (of the order of five CH_2 in polyethylene) which should be able to attach anywhere on the lateral growth face.[31] Sadler and Gilmer formulated a Monte Carlo algorithm for this three-dimensional entropy-barrier model[8] and a rate equation analysis for a scaled-down version[42] (see section 5.2.4.3).

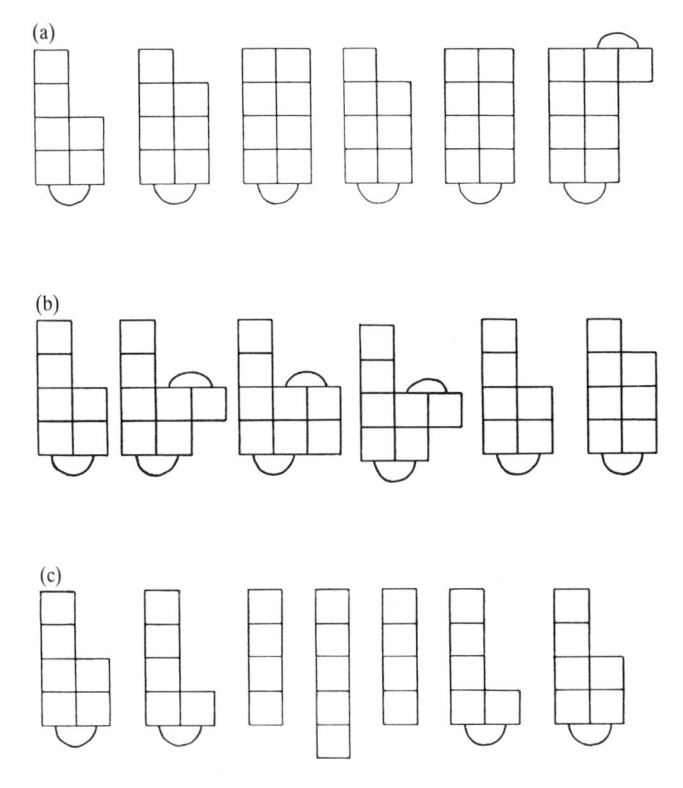

Figure 5.7 Stem attachment as a multistep process. (a) Folding is allowed only if the specified length is reached. (b) and (c) Folding allowed at any stage. From Goldbeck-Wood,[51] by permission of the publishers, Butterworth Heinemann Ltd. ©.

5.2.4.2 The multipath model

The basic picture of this model is one of chains which deposit as a series of small units, each representing only a few monomers. Because of the flexibility of the chain and its random coil state in the melt it is thought to be unlikely that these segments will immediately form straight stems of matching lengths. More likely there will be partial attachments and subsequent rearrangements before a stem is finally incorporated into the crystal.

Dupire[39] simulated the formation of a secondary nucleus by adjacent re-entry of the chain. At each stage a segment of length Δl is either deposited on top of the previously attached segment or adjacent laterally, thereby initiating a new stem, or removed from the last position.

The growth is governed by rate constants (see section 5.2.2). In the present model we can distinguish three different configurations (Fig. 5.8):

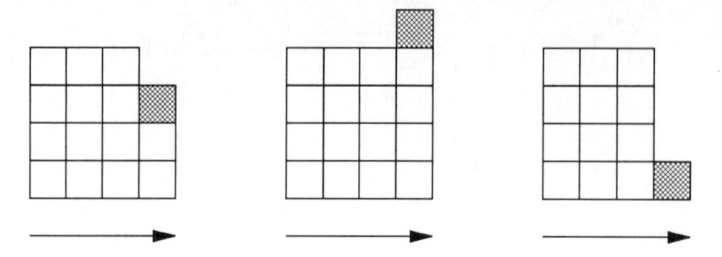

Figure 5.8 The different configurations for the attachment of a growth unit in the multipath model: 'niche', 'overhang' and 'new stem'.

niche, overhang and new stem. The respective ratios of rate constants are

$$\frac{k^+}{k^-} \text{(niche)} = \exp(H_f \, \Delta T/T) \qquad [22]$$

$$\frac{k^+}{k^-} \text{(overhang)} = \exp(H_f \, \Delta T/T - E_n T_m^0/T) \qquad [23]$$

$$\frac{k^+}{k^-} \text{(new stem)} = \exp(H_f \, \Delta T/T - E_s T_m^0/T) \qquad [24]$$

where

$$H_f = \frac{ab \, \Delta l \, \Delta h}{kT_m^0} \qquad E_l = \frac{2b \, \Delta l \sigma_n}{kT_m^0} \qquad E_s = \frac{2ab\sigma_e}{kT_m^0}$$

and a, b are the lattice constants. Some caution may be expressed at this point about the use of thermodynamic potentials in a statistical model. The free energies are the result of microscopic interaction energies and a stochastic process which contributes an entropy. The surface and bulk interaction energies in a microscopic model are not independent quantities, e.g. in a square lattice model an interaction energy ε yields a 'broken bond' or surface energy $\varepsilon/2$, and an enthalpy or bulk energy per unit of 2ε.

As above, there is still a degree of freedom in the apportioning of the free energy barriers and, as in the projection models the particular choice affects the behaviour far away from equilibrium.[40]

The kinetics can be solved either by Monte Carlo or by numerical integration of rate equations,[40] equivalent to those of the 'row-of-stems' model.[42] The advantage of the latter method is its precision. The disadvantage is that the transient states have no relation to the time evolution of the actual system as in the Monte Carlo. Therefore it is very important to reach convergence. This, however, is very slow in the case

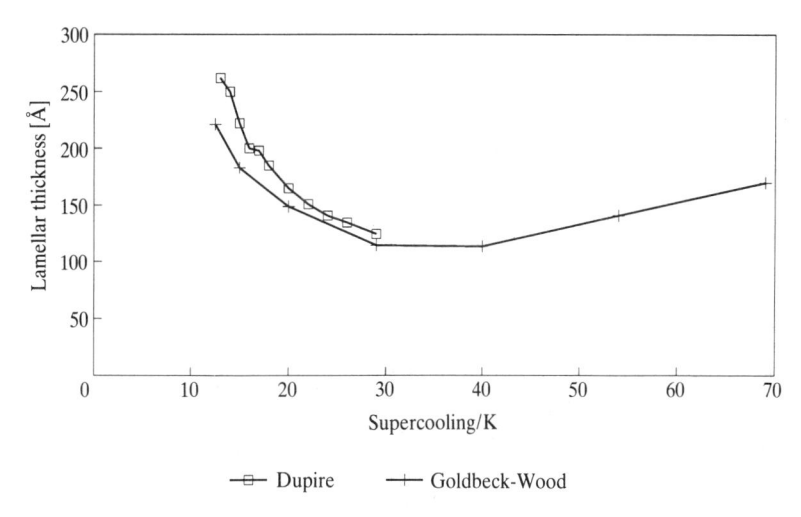

Figure 5.9 Lamellar thickness of polyethylene as simulated by the multipath model.

of large thicknesses and small growth rates. The Monte Carlo method can give a rough estimate of the steady state more quickly.

In the following the main results of a simulation using data typical for polyethylene ($\Delta h = 280$ J cm^{-3}, $\sigma_n = 12 \times 10^{-7}$ J cm^{-2}, $\sigma_e = 90 \times 10^{-7}$ J cm^{-2}) will be discussed.

(a) *Average stem length.* The average stem length agrees reasonably with experimental values and follows the typical ΔT^{-1} behaviour for small supercooling (Fig. 5.9). The influence of a variation in Δh, σ_n, and σ_e was also studied showing that an increase in σ_e or a decrease in σ both lead to an increase in the average lamellar thickness.[39,40]

It was further found[40] that the length starts rising again at large supercoolings combined with increasing fluctuations. This phenomenon is, however, quite different from the δl catastrophe, since l still remains finite in all cases. The increase is only obtained if the fold surface energy is larger than the overhang (i.e. lateral) surface energy, so that stem extensions are more favourable than the initiation of new stems. The effect of this is most noticeable at large attachment rates, i.e. at large ΔT.

(b) *The average spreading rate.* The spreading rate in the multipath model depends strongly on the supercooling (Fig. 5.10). This is in sharp contrast to the constant g assumption which enters the regime analysis in nucleation theory. If the average spreading rate is really supercooling dependent, then it is not valid to explain the changes in slope in the growth rate plots simply by the changing power of the nucleation rate alone (see also Ref. 43).

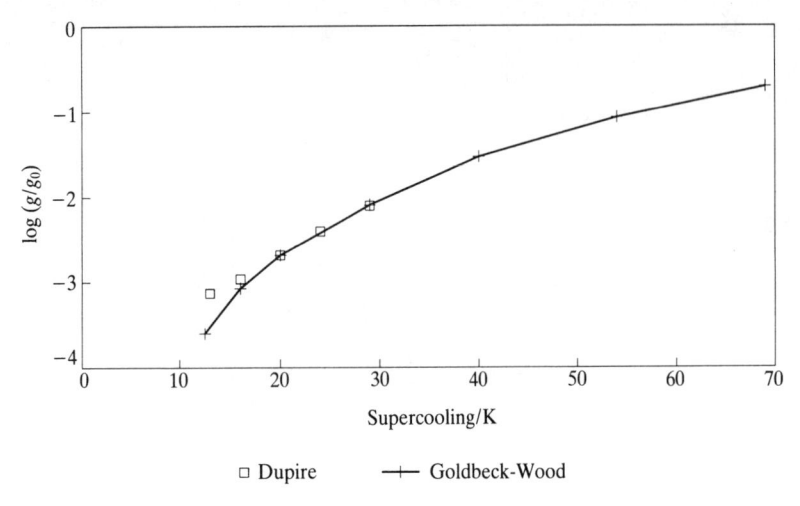

Figure 5.10 Spreading rate of a substrate layer in polyethylene simulation.

5.2.4.3 The rough surface model

The observation of rounded lateral shapes led Sadler[31] to the conjecture that there is a roughening transition on the (110) face. If this was the case then there would be no nucleation barrier to growth and at first sight a linear increase of the growth rate with supercooling would be expected. This is not observed, however. The growth rates seem unaffected by the changing habits, and always show the classical $\exp(K_G/T\Delta T)$ dependency. This means that there must still be a barrier. At this point the entropy of chain attachment comes into play.[4,8] Segments of the chain attach 'blindly' at several surface sites. Because of the randomly coiled conformation of the chain in the melt and the connectivity this leads to conformations which are not viable for further growth. Rearrangements then become necessary to form a straight stem. These are equivalent to stretching a part of the chain and removing all intervening entanglements. This requires an activation entropy proportional to the sequence length which in turn leads to the exponential growth law. Based on this concept simulations have been carried out by Sadler and Gilmer[8,42,45] and the author.[40]

In the model of lamellar growth[8] shown in Fig. 5.11 the crystallizing entities are simple blocks which represent small segments of the chain. They interact with their nearest neighbours only in a simple cubic lattice. These interactions are governed by bond energies ε_x, ε_y and ε_z between units in the x-, y- and z-directions of the lattice, i.e. between stems in different layers, in the same layer and between neighbouring units along a crystalline stem. These are the only parameters required to specify this

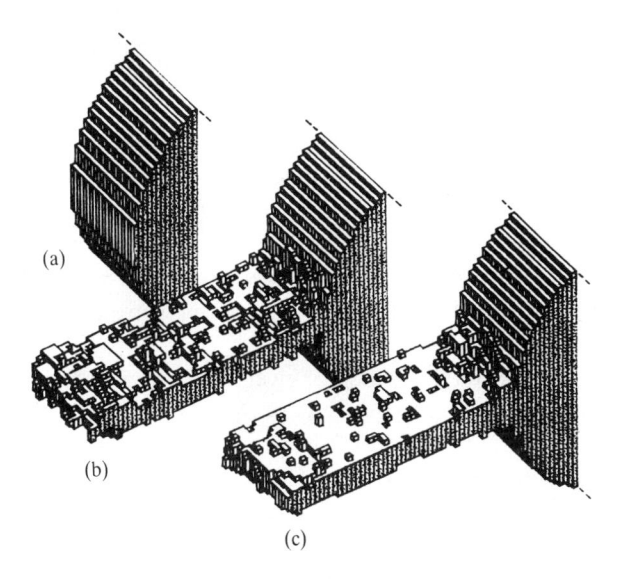

Figure 5.11 Computer generated lamellae (a) seed crystal (b) with $\varepsilon/kT_m^0 = 1.4$, (c) with $\varepsilon/kT_m^0 = 1.8$. From Sadler and Gilmer,[8] by permission of the publishers, Butterworth Heinemann Ltd. ©.

microscopic model. No macroscopic potentials, like lateral and fold surface free energies, enter at this level. The crystallization is envisaged as the attachment and rearrangement of a coiled chain. Loops are already present. The end result of the solidification process may be a chain-folded crystal in which the folds represent high free energy states. Their formation, however, is the result of a highly stochastic process. It is therefore considered not to be appropriate to model chain-folding directly by means of an extra fold energy at this level.

The chain-connectivity is accounted for by a set of so-called 'pinning' rules which basically limit the stem extension once a stem has been covered by the next layer in the x-direction. A further requirement due to the connectivity of the chain is that there must not be any interruptions along crystalline stems. No distinction is made between adjacent stems that belong to the same or to different molecules. The model effectively considers only one lateral growth face, namely that perpendicular to the x-axis. In the y-direction periodic boundary conditions are applied to mimic an infinitely wide crystal.

The simulation proceeds as follows. At the start a seed crystal is set up on the cubic lattice (Fig. 5.11a). At each lattice point the number of neighbours and bonds in each direction are stored. Then the Boltzmann factors for the various adding and removing processes are calculated from the interaction energies. The categories are: adding at the top or bottom of an existing stem, subject to the pinning rules; creating a new stem by adding a unit to a position with at least one lateral neighbour, and which is part of an unoccupied column; removing from the various lateral surface sites. At any stage of the crystallization each site is classified as either active or inactive with respect to the various processes stated above. An action category is chosen by comparison of the probability distribution with a random number. Next, of all sites in this category one is chosen by another Monte Carlo decision. The action is then carried out and the categorization of sites is updated. This ends the cycle and the process is started again.

The main results of this model concern (a) the morphology, (b) the growth rate and (c) the lamellar thickness.

(a) *Morphology.* The effect of the bond energies on the morphology was investigated in analogy with the case of low molecular weight materials where a rounding due to the roughening transition occurs at $kT_m^0/\varepsilon = 0.6$. A noticeable difference in surface roughness was indeed found (Fig. 5.11) between lamellae grown with isotropic interaction energy $\varepsilon/kT_m^0 = 1.4$ and 1.8. The lower value of ε results in a 'rounding off' of corners.

(b) *Growth rates.* The linear rate of advance of the lamella along x depends exponentially on the supercooling in the same way as in nucleation theory (Fig. 5.12). Additional restrictions designed to model the effect of an attachment of one molecule in two locations on the growth face were also studied. Some new stems were given a maximum and a minimum length value beyond which they were not allowed to extend.

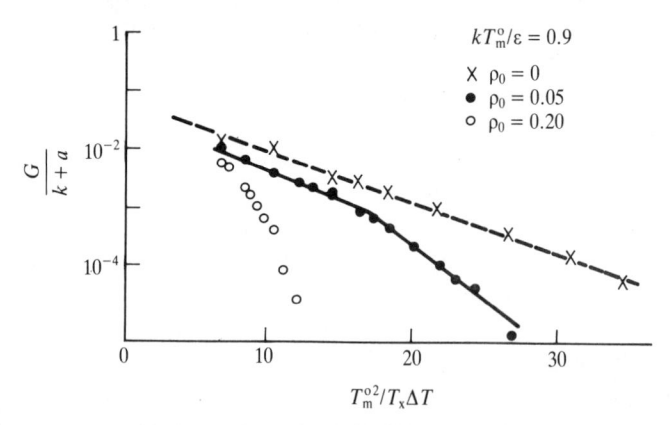

Figure 5.12 Dimensionless growth rates of simulated lamellae for different values of the density of pinning sites ρ_0. From Sadler and Gilmer,[8] by permission of the publishers, Butterworth Heinemann Ltd. ©.

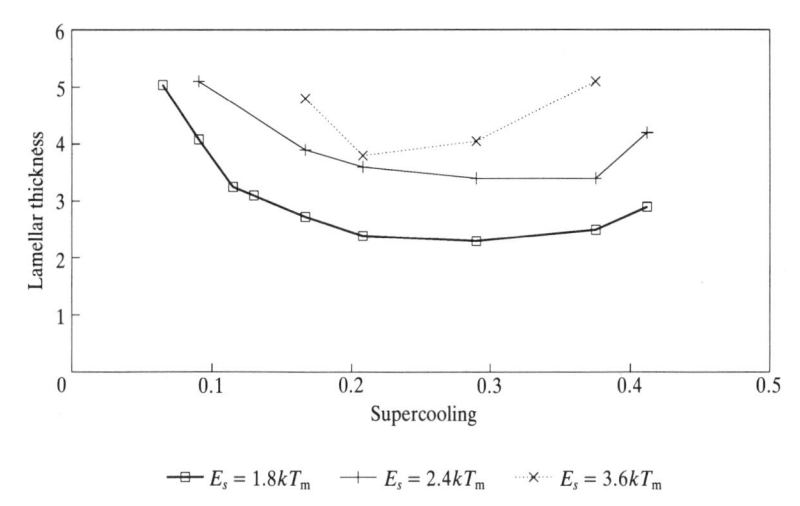

Figure 5.13 Thickness of simulated lamellae with $\varepsilon/kT_m^0 = 1.8$ and extra fold surface energy $\varepsilon - E_s$.

Legend: $E_s = 1.8kT_m$; $E_s = 2.4kT_m$; $E_s = 3.6kT_m$

With increasing density of such pinning sites the growth rates decrease more strongly at small supercoolings because it is more difficult to form long stems. The downturn of the slope is similar to that observed experimentally and which is generally interpreted as the Regime I–II transition.

(c) *Lamellar thickness.* The average lamellar thickness (Fig. 5.13) follows the ΔT^{-1} law of the minimum thickness for small supercoolings. This is the result of the pinning rules which tend to minimize l. At large supercoolings a plateau is reached. This agrees well with experimental evidence.[44] The absolute thickness values, however, are small in comparison with actually observed fold lengths. For polyethylene, for example, the average lamellar thickness predicted by the simulation is about 30 CH_2, in comparison to about 200 in reality. This level can only be slightly influenced by a different choice of interaction energies, a decrease in ε leading to an increase in l.[40] The main reason for the small l is a small minimum thickness which in turn is a result of the fact that there is no high fold energy in the model. If, despite our arguments above, such an extra fold energy is introduced into the model, the lamellar thickness can be somewhat increased (see Fig. 5.13). The side-effect of this is, however, that there is an upswing at large supercoolings, when stem extension becomes more favourable than the initiation of new stems. This leaves the statistical modelling of polymer crystallization in somewhat of a dilemma.

In the limit of growth above the roughening transition there is no correlation between attachments at different sites. In the present case of

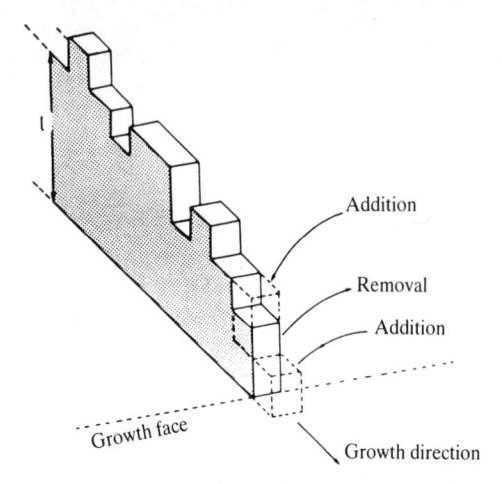

Figure 5.14 'Row-of-stems' model, representing a slice cut out of the crystal perpendicularly to the growth face.

polymers one could envisage that there is no regular spreading of patches by chain-folding in this regime (see Fig. 5.3). Under such conditions it seems reasonable further to simplify the model[42,45] down to a single row of stems perpendicular to the growth face (Fig. 5.14). This 'slice' then represents the full lamellar crystal (rather than a layer on the growth face as in the multipath model). Additions and removals are permitted only at the outermost end of the row. The growth is then a strictly sequential process and is governed by the same set of rate equations as in the layer model. From the microscopic interaction energy ε follows the heat of fusion $\Delta h = 2\varepsilon$ and, as in the three-dimensional case, there is no extra fold energy.

Despite its simplicity the row-of-stems model shows the same features as the full model, and furthermore makes more detailed analysis[45] and special applications feasible. Firstly, it was applied[46] to the case of short chain poly(ethylene oxide) (PEO) which shows distinct growth branches in which the lamellar thickness remains constant at the extended or n-times folded chain length. The extended chain mode was modelled by setting a limit on the stem length. Preferred fold lengths were simulated by associating a favourable interaction energy $\delta\varepsilon$ with adjacent stems of equal lengths for certain selected lengths. The model does indeed reproduce the linear increase of the growth rate with supercooling which was observed experimentally.

The row-of-stems model with preferred fold lengths[47] exhibits growth rate minima at those crystallization temperatures where the thermodynamic stability requires a transition to a thickness above the previous preferred level l_p. These minima arise because of a 'poisoning' mechanism whereby a series of stems of the now substable l_p tends to attach and has to come off again before growth can finally proceed. Such growth rate minima have indeed been observed in crystallization of paraffins.[48] The strength of the model lies in the fact that it provides a picture of the growth process in the transition region. Its weakness is that it has not yet been fitted to experimental data.

The only case so far in which the rough surface model has been applied to experimental data is poly(aryl ether ether ketone) (PEEK).[49] The experimental lamellar thickness and growth rate could be fitted reasonably well by a choice of the interaction energies which was guided by the experimentally determined heat of fusion. The low values of the lamellar thickness inherent to the model corresponded well to the small fold length of PEEK.

The concept of an entropy barrier which is due to the probability of finding a matching segment of the chain molecule is of particular relevance when it comes to copolymers.[50,51] If the comonomer units cannot be incorporated into the lattice the entropy barrier will increase. In particular, the crystallizable sequences are of finite length and therefore the on-rate constants become decreasing functions of the stem length:

$$k^+(i) = k_0^+ p^{(i-1)} \qquad [25]$$

where p is the probability of crystallizable units. Incorporating this into the row model leads to a strong depression of growth rates, a melting point significantly below the equilibrium value calculated by Flory[52] and a finite lamellar thickness at the melting point. The case of incorporation of comonomers has also been simulated[51] (Fig. 5.14). The inclusion concentration was found to be kinetically determined.

Finally, the row of stems model was extended to allow the simulation of non-isothermal crystallization.[49,53] It was found that in crystallization during heating the lamellar thickness keeps on increasing through recrystallization processes (frequent on/off events). This is limited, however, by the increasing entropy barrier and eventually leads to melting well below the equilibrium melting point. An approach to equilibrium (i.e. to an infinite crystal) by lowering the heating rates was found to be more than exponentially suppressed.

5.2.5 Chain-folding and the amorphous phase

Models of the amorphous layer have played an important role in forming a mental picture of polymer crystallization. The debate about the two diametrically opposed views of the so-called 'switch-board model'[54] and the model of regular adjacent chain-folding[55] centred largely around the chain-conformation in the amorphous layer. Computer simulation models have been an indispensable aid in the interpretation of experimental data, in particular from small angle neutron scattering (SANS). The simulation method was generally a Metropolis Monte Carlo sampling of the conformations generated by either a 'statistical weight matrix'[54] of *trans*, *gauche*$^+$ and *gauche*$^-$ bonds in a continuous space, or the same on a lattice,[55] or a method of breaking and re-forming chains except within the same chain (so-called 'bond-flips').[56] In any of these methods thermodynamic equilibrium conditions have to be maintained, i.e. the temperature is kept just above T_m^0. Therefore predictions can only be applied with great caution to materials which have been solidified under highly non-equilibrium conditions, as is the case in quench-crystallization of blends of protonated and deuterated polyethylene used for neutron scattering studies. The off-lattice study of Yoon and Flory[54] deduces from comparison with SANS data that the type of re-entry is more random than regular. However, it is often overlooked that 70–80% of the chains are actually predicted to form an interfacial layer and re-enter the same crystalline lamella. The deficiency of the Yoon and Flory model is an unphysically high density. In fact, if overcrowding is to be avoided, a high degree of the predicted re-entry must be in the near vicinity of the exit points.

This consideration actually brings the result much closer to that of Guttman *et al.*[55] and Mansfield.[56] They carried out simulations on a lattice to enforce the correct density and arrived at a probability for adjacent re-entry of 60–80% and a degree of combined adjacent and next-nearest-neighbour re-entry of more than 50% In the lattice models the entropic forces actually act to favour nearby re-entry because this leaves the system as a whole with a chance to access a larger configuration space. However, a lattice representation may not be adequate for the amorphous phase, e.g. because of the artificial bond and dihedral angles, in which case the respective results would be irrelevant. We conclude in agreement with Keller[1] that, considering the respective shortcomings and error margins, the actual numerical results of the opposing studies (Yoon–Flory and Guttman–Mansfield) do not justify any polarized interpretation of a random switchboard or a regular chain-folding picture of polymer crystallization.

The lattice model of Mansfield has also been applied to study the distribution of short branches in the amorphous region.[57,58] It was found that branches which occupy two sites (equivalent to 1-alkene) segregate near the crystal-amorphous boundary. They tend to lie parallel to the boundary thereby decreasing the amount of re-entrant loose loops which in turn is compensated by an increase in tight folds.

The interrelation between the formation of amorphous and crystalline layers during solidification has so far not received much attention. Toda et al.[59] recognized that cilia can effect the nucleation. They amended the Frank equations[22] by a cilia nucleation term. Under this premise a new growth mode can exist between Regimes I and II. This was confirmed by simulation.[24] In the crossover regions the agreement was poor, however, according to the authors possibly because the correlation between steps and cilia was neglected.

Some new kinetic studies[60,61] have emerged recently which incorporate the entropy contribution to the free energy due to the chain segments in the amorphous regions. They are assumed to perform non-reversal random walks on a three-dimensional cubic lattice. A Grand Canonical Monte Carlo simulation is carried out on the basis of the following free energy of the polymer state:

$$F_s = E_s - kT \left[\sum_k \ln N_l(l_k, r_{k,k+1}) + \sum_{0,N} \ln N_t(l) \right] \quad [26]$$

where E_s is the energy contribution of the (rigid) stems due to their chemical potential and interaction energy, N_l is the number of states for a flexible piece of length l and end-to-end distance r and N_t is the number of states for a flexible piece of chain at the tails. Preliminary results have been reported which include the degree of adjacent re-entry. This is found to be about 80% at low densities (equivalent to solution growth) but decreasing with supercooling to less than 20% at melt densities. This is interpreted as a tendency towards a random switchboard behaviour. However, more detailed analysis of the loop sizes is required in this very promising model.

A similar model is applied to copolymer crystallization[61] with the additional refinement of treating attachment as a multipath process. This probably makes it the most comprehensive model of polymer crystallization yet.

5.2.6 Thickening growth and internal dynamics

So far, we have considered models which allow growth only on the lateral faces. The top and bottom faces are thought to be fixed by the folds and loops. This is justified by the experimental evidence that the lamellar thickness is at least initially fixed by the supercooling at which the crystallization is carried out. Annealing processes lead to a thermally activated thickening which is not described by the crystallization models, but separate models have been proposed.

Additionally, there is evidence for a special type of growth in which the lamella grows laterally as well as in its thickness dimension simultaneously.[3,10] A schematic representation is shown in Fig. 5.15 which is based

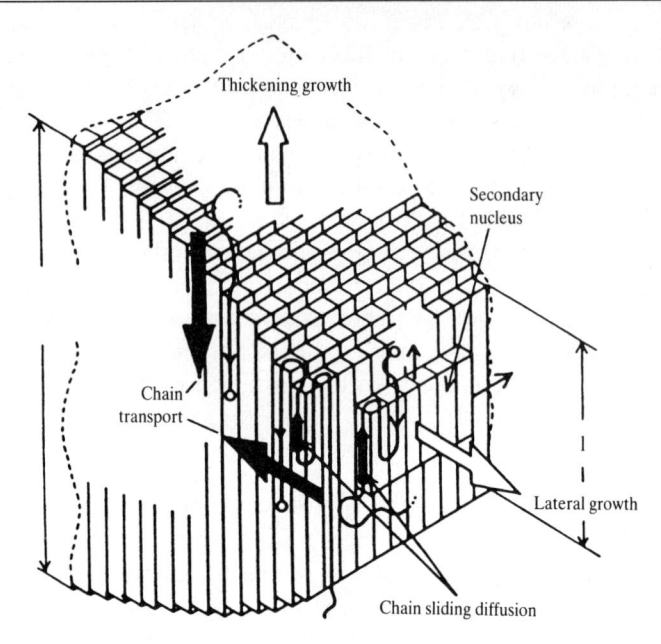

Figure 5.15 Development of a crystal via chain-folding and subsequent chain-sliding. From Rastogi et al.[63]

on a model by Hikosaka.[62,63] Preliminary computer simulations of this model have recently been carried out by the author.[64] The agreement with the calculations by Hikosaka is good. Furthermore the simulation allows an investigation of a wider range of parameters than has been feasible analytically.

It is believed that the reason for the thickening growth lies in the presence of a 'mobile' phase in the crystalline region. There is considerable argument about the exact nature and classification of this phase. But as far as the crystallization process is concerned the most important aspect is the chain dynamics and in particular the extent of longitudinal motion. Theoretical models of the microscopic motion have recently become possible through accurate molecular dynamics simulations.

Ryckaert and Klein[65] modelled solid n-alkanes with flexible backbones and infinite chain-length. Starting off from an orthorhombic packing, the unit cell was found to expand with increasing temperature and tended towards the hexagonal packing at 400 K. Translational diffusion occurred already above 250 K first without any accompanying rigid body rotation but becoming more liquid-like at higher temperatures with rotational diffusion between four well defined sites.

Sumptner et al.[66] carried out a detailed study of the internal dynamics of polyethylene. They argue that the dynamic disorder in a crystal involves large-amplitude, anharmonic motion which is affected by dynamic processes. These can only be effectively studied by molecular dynamics. They discovered that rotational isomeric defects are created gradually at temperatures above 300 K. These could account for many of the observed properties like, for example, lamellar thickening. The simulation showed a single chain can slide through the crystal more easily at higher temperatures corresponding to greater defect densities.

These results are very encouraging for a further development of molecular dynamics simulations in studies of polymer crystallization. Even if the crystallization process cannot be treated we can nevertheless gain much more insight into the chain-dynamics and deduce relationships which can then be fed into other models and Monte Carlo simulations.

5.3 The macroscopic level

5.3.1 Introduction

On the macroscopic level the growing lamellae organize into superstructures which often have a spherical shape. Under isothermal conditions the radius of these spherulites increases at a constant rate, the so-called 'linear growth rate', until growth is terminated by impingement of neighbouring spherulites. In a coarse grained picture crystallization can therefore be regarded as a two-stage process of primary nucleation and subsequent linear growth. This, of course, is not specific to polymers, and indeed the macroscopic development of the phase transition was first described quantitatively by several studies with a metallurgical context,[67-70] (see Wunderlich).[2] They assume an infinite crystallizing volume, a uniform distribution of potential nuclei and a constant nucleation and growth rate. The transformation kinetics can then be solved for objects which are symmetrical in all directions, i.e. spheres in three dimensions, disks in two dimensions and rods in one dimension. The transformed volume fraction $\alpha(t)$ develops according to the well-known Avrami equation

$$-\ln(1 - \alpha(t)) = Kt^n \qquad [27]$$

where the rate parameter K and exponent n depend on the geometry and mode of nucleation.[71] In the polymer case $\alpha(t)$ differs from the crystalline volume fraction because the spherulites are not compact crystalline objects but are formed by the microscopic lamellae interspersed with amorphous material.

Several modifications and extensions of the original treatment have been proposed in order to study *inter alia* non-isothermal crystallization and the effects of a non-homogeneous distribution of potential nuclei (see Billon[72]). Further generalization was achieved by geometrical models of nucleation and growth generated on a computer. One of the significant advantages of simulation is that it can give more detailed information on the morphology: the sizes, the shapes of the spherulites and their respective distributions can be determined and also visualized.[72-74]

5.3.2 Models and simulation methods

It is well known that crystallization can be initiated by either homogeneous or heterogeneous nuclei. The latter tend to nucleate almost simultaneously, giving rise to the so-called 'athermal' mode of nucleation, whereas the former lead to a roughly constant nucleation rate per unit volume – the so-called thermal mode. Both modes as well as mixed mode nucleation have been studied. The thermal mode was first modelled by Galeski[73,74] by alternating a growth time step with a nucleation step in which a fixed number of nuclei was introduced thus giving a discrete representation of the crystallization process. The nuclei start growing if their location has not been overlapped by a previously nucleated spherulite which ensures a number of new growth centres which is on average proportional to the unoccupied volume. Billon[72] solved the activation problem on a continuous time scale. The activation time of each nucleus was determined and nuclei introduced one by one at random positions.

The development of the transformed area or volume is generally monitored by scanning a fixed number of points which are either placed on a lattice[71,72] or randomly across the sample.[73,74] The precision of this method depends on the ratio of the number of scanning points to the number of entities. This has been increased from 150/200[71] (about 7% deviation from Avrami exponents when a ratio of 50/200 was used), to 2000/500,[73] and recently to 10000/15.[72]

For a characterization of the morphology Galeski determined the spherulite size distribution by scanning the sample with 40 000 randomly positioned points. Billon follows the evolution of the morphology by calculating point by point the development of the boundaries of each set of two disks which may impinge on one another.

5.3.3 Discussion of results

The average time dependences of the degrees of conversion in the standard cases (see Hay and Przekop[71]) were generally found to be in very good agreement with theoretical predictions. This indicates that even the less precise methods used are adequate but the higher precision used by Billon might be needed for special cases.

The first effect studied by simulation was that of non-symmetrical entities. As the analytical treatments require the growing entities to be symmetrical in all directions the greatest error can be expected if the growing regions are rodlike in a three-dimensional sample. Price and Thornton[75] simulated this case and reported that for a given volume the deviation from Avrami behaviour increased with the number, density and diameter of the rods and was commonly about 10–15%. In terms of Eqn [27] the effect on n was insignificant but typically one order of magnitude on K. Hay studied varying proportions of rods, disks and spheres in a sample. Even in the most extreme case of respective numbers of 125:50:25 Avrami behaviour was obtained, probably because the overall volume occupied by non-spherical entities was still small.

The degree of crystallinity per unit volume can depend on time because of impurity rejection and molecular weight segregation during melt crystallization. A not unrealistic variation of the density by 10–20% during the transformation process was found[71] to have a marked effect on the kinetics. Any Avrami fit had a poor correlation coefficient, indicating that the Avrami exponent varied during the transformation and therefore such a description was no longer appropriate. This strongly indicates that a simple Avrami description is probably not appropriate in many cases of polymer crystallization.

The general features of the sample morphology in two and three dimensions were investigated by Galeski. The size distributions have bell-like shapes with a distinct maximum in the athermal case but being more spread out in the thermal case. The largest spherulites were formed in the thermal mode. Similar morphologies were obtained by Billon but the size distributions have not been quantified.

During the course of the transformation the volume decreases due to at least partial crystallization of the transformed regions. Hay and Przekop conjecture that this leads to a proportional shift of the nucleation centres towards the centre of mass. In the simulation this had, however, no significant effect on the Avrami exponent. A more significant effect of volume changes are the so-called 'weak spots' modelled by Galeski and Piorkowska.[76,77] If part of the melt gets encircled by spherulites, the flow of material from the rest of the sample is cut off which on crystallization leads to a local thinning of films and the formation of holes. Patterns were generated in two dimensions and weak spots found to occupy more than 10% of the sample area. The data on number, total area, average size and largest size were found to be in very good agreement with experimental results obtained by scanning electron microscopy of poly-(ethylene oxide) films. In three dimensions the size distribution and total volume of weak spots depends strongly on the type of primary nucleation. A more sporadic nucleation reduces the total volume several-fold and leads to a decrease in the maximum size. These results are of great significance for mechanical and electrical properties.

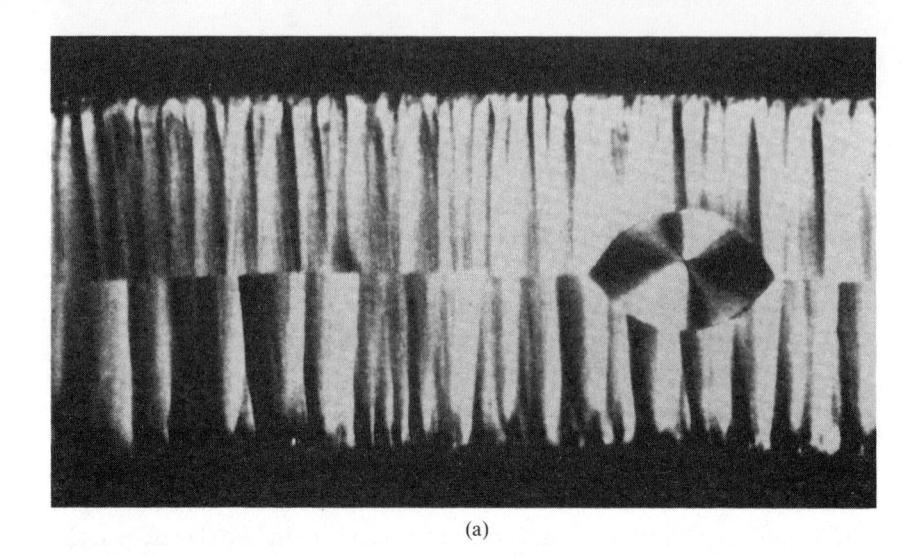

Figure 5.16 (a) Optical micrograph of a nylon 6-6 sample crystallized in a DSC. (b) Two-dimensional simulation of (a). From Billon and Haudin.[72]

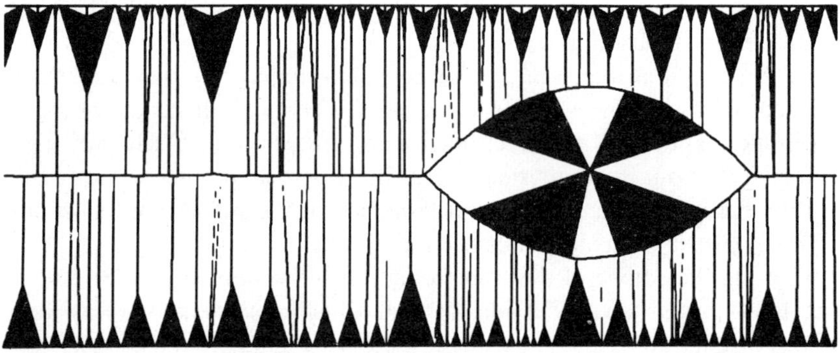

Surfaces due to tools or fibres introduced into the samples often act as nucleating agents. A slight non-uniformity in nucleation distribution of about 20% was found to have no significant effect.[71] If surface nucleation completely dominates, growth is entirely perpendicular to the surface and the so-called transcrystalline regions are produced[72] which resemble closely those observed in reality (Fig. 5.16). The simulation showed that transcrystallization can have a significant effect on the transformed volume fraction in rectangular samples.

Most recently Billon et al.[78] developed their model further to reconstruct the conversion process from a photograph of the final morphology of spherulitic growth under isothermal conditions. This yields the rate of appearance of spherulites dN_a/dt and the transformed surface fraction $\alpha(t)$ as a function of time. From the equation

$$dN_a/dt = SqN_0(1 - \alpha(t))\exp(-qt) \qquad [28]$$

N_0, the initial number of nuclei per surface area, and q, the activation frequency, can be determined.

We conclude that computer simulations of the overall crystallization process have been very helpful in assessing the limits of the analytical Avrami treatments and can provide further insight into morphological features which are of great importance for material properties. Still, a number of factors are neglected even in simulation, in particular the effect of a molecular weight distribution which is expected to yield different growth rates, fractionation effects and post-crystallization and annealing effects.

Acknowledgements

I would like to thank Dr A. Toda for discussions and communication of results prior to publishing. This work was supported by the SERC.

References

1. Keller A *Disc. Faraday Soc.* **68**: 145 (1979).
2. Wunderlich B *Macromolecular Physics*, Academic Press, London, 1980, Vol. 2.
3. Bassett D C *Principles of Polymer Morphology*, Cambridge University Press, London, 1981.
4. Strobl G in *Springer Proceedings in Physics*, Vol. 19, Springer, Berlin, 1987, p. 191.
5. Phillips P J *Rep. on Prog. Phys.* **53**: 549 (1990).
6. Armitstead K, Goldbeck-Wood G *Adv. Polym. Sci* **100**: (1992).
7. Hoffman J D, Davies G T, Lauritzen J I Jr in: *Treatise on Solid-State Chemistry, Vol 3*, Hannay N B (ed.) Plenum, New York, 1976, p. 497.
8. Sadler D M, Gilmer, G H *Polymer* **25**: 1446 (1984).
9. Sadler D M *Nature* **326**: 174 (1987).
10. Keller A in: *Sir Charles Frank, OBE, FRS: An eightieth birthday tribute*, Chambers G et al. (eds), Adam Hilger, Bristol, 1991.
11. van der Eerden J P, Bennema P, Cherepanova T A *Prog. Cryst. Growth, Charact.* **1**: 219 (1978).
12. Gilmer G H *Science* **208**: 355 (1980).
13. Bennema P, van der Eerden J P *J. Cryst. Growth* **42**: 201 (1977).
14. Onsager L *Phys. Rev.* **37**: 405 (1931); *Phys. Rev.* **38**: 2265 (1931).
15. Haken H *Synergetics. An Introduction*, Springer, Berlin (1978).
16. Binder K *Monte Carlo Methods in Statistical Physics*, Springer, Berlin, 1986.
17. Kang H C, Weinberg W H *J. Chem. Phys.* **90**: 2824 (1989).

18. Sadler D M *J. Chem. Phys.* **87**: 1771 (1987).
19. Sadler D M *Polymer* **28**: 1440 (1987).
20. Salt C D *Project No. T17*, University of Bristol (UK), Dept. of Physics, 1988.
21. Frank F C, Tosi M *Proc. Royal Soc. A* **263**: 323 (1961).
22. Frank F C *J. Cryst. Growth* **22**: 233 (1974).
23. Guttman C M, DiMarzio E A *J. Appl. Phys.* **54**: 5541 (1983).
24. Toda A, Tanzawa Y *J. Cryst. Growth* **76**: 462 (1986).
25. Point J J, Villers D *Polymer* **33**: 2263 (1992).
26. Lauritzen J I Jr, *J. Appl. Phys.* **44**: 4353 (1973).
27. Hoffman J D *Polymer* **24**: 3 (1983).
28. Leamy H J, Gilmer D H Jackson K A in *Surface Physics of Materials*, Blakely J M (ed.) Academic Press, London, 1975, p. 121.
29. Labaig J J Ph.D. Thesis, Université Louis Pasteur, Strasbourg, 1978.
30. Organ S J, Keller A *J. Polym. Sci. B* **24**: 2319 (1986).
31. Sadler D M *Polymer* **24**: 1401 (1983).
32. van Beijeren H, Nolden I in *Topics in Current Physics, Vol 43: Structure and Dynamics of Surfaces, II*, Schommers W. and von Blanckenhagen P (eds), Springer, Berlin, 1987, p. 259.
33. Mansfield M L *Polymer* **29**: 1755 (1988).
34. Mansfield M L *Polym. Commun.* **31**: 283 (1990).
35. Toda A private communication (1991); to be published.
36. Toda A *Colloid Polym. Sci.* **270**: 667 (1992).
37. Binsbergen F L *J. Cryst. Growth* **13/14**: 44 (1972).
38. Point J J *Macromolecules* **12**: 770 (1979), and *Disc. Faraday. Soc.* **68**: 167 (1979).
39. Dupire M Ph.D Thesis, Université de l'Etat à Mons, Belgium (1984).
40. Goldbeck-Wood G Ph.D Thesis, University of Bristol, UK (1992).
41. Bennema P, Hartman P (ed.), Gilmer G H in *Crystal Growth, An Introduction*, North-Holland, Amsterdam, 1973, p. 282.
42. Sadler D M, Gilmer G H *Phys. Rev. Lett.* **56**: 2708 (1986).
43. Point J J, Colet M-Ch *Ann. Chim. Fr.* **15**: 221 (1990).
44. Raoult J *J. Macromol. Sci. Phys. B* **15**: 567 (1978).
45. Sadler D M, Gilmer G H *Phys. Rev. B* **38**: 5684 (1988).
46. Sadler D M *J. Polym. Sci. Polym. Phys. Ed.* **23**: 1533 (1985).
47. Sadler D M, Gilmer G H *Polym. Commun.* **28**: 243 (1987).
48. Ungar G, Keller A *Polymer* **28**: 1899 (1987).
49. Goldbeck-Wood G, Sadler D M *Molec. Sim.* **4**: 15 (1989).
50. Goldbeck-Wood G, Sadler D M *Polym. Comm.* **31**: 143 (1990).
51. Goldbeck-Wood G *Polymer* **31**: 586 (1990).
52. Flory P J *Trans. Faraday Soc.* **51**: 848 (1955).
53. Goldbeck-Wood G *J. Polym. Sci. Polym. Phys. Ed.* **31**: 61 (1993).
54. Yoon D Y, Flory P J *Polymer* **18**: 509 (1977).
55. Guttman C M, Hoffman J D, DiMarzio E A *Disc. Faraday Soc.* **68**: 297 (1979).
56. Mansfield M L *Macromolecules* **16**: 914 (1983).
57. Mathur S C, Mattice W L *Macromolecules* **21**: 1354 (1988).
58. Mathur S C, Rodrigues K, Mattice W L *Macromolecules* **22**: 2781 (1989).
59. Toda A, Kiho H, Miyaji H, Asai K *J. Phys. Soc. Jpn.* **54**: 1411 (1985).
60. van Dieren F, Baumgärtner A *Proc. Intern. Disc. Rolduc* (1987).

61. van Ruiten J, van Dieren F, Mathot V B F in *Crystallization of Polymers*, Dosière M (ed.), NATO ASI-C Series, to appear 1993.
62. Hikosaka M *Polymer* **31**: 458 (1990).
63. Rastogi S, Hikosaka M, Kawabata H, Keller A *Macromol. Chem. Macromol. Symp.* **48/49**: 103 (1991).
64. Goldbeck-Wood G in *Crystallization of Polymers*, ed. M. Dosière, NATO ASI-C Series, to appear 1993.
65. Ryckaert J-P, Klein M L *J. Chem. Phys.* **85**: 1613 (1986).
66. Sumptner B G, Noid D W, Wunderlich B *J. Chem. Phys.* **93**: 6875 (1990).
67. Kolmogoroff A N *Izvest. Akad. Nauk. SSSR Ser. Math.* **1**: 335 (1937).
68. Johnson W A, Mehl R T *Trans. AIME* **135**: 416 (1939); Johnson W A, Mehl R T *Met. Technol. AIME*, Tech. Publ. No. 1089 (1939).
69. Avrami M *J. Chem. Phys.* **7**: 1103 (1939), and loc. cit. **8**: 212 (1940), and loc. cit. **9**: 177 (1941).
70. Evans U R *Trans. Faraday Soc.* **41**: 365 (1945).
71. Hay J N, Przekop Z J *J. Polym. Sci. Polym. Phys. Ed.* **17**: 951 (1979).
72. Billon N, Haudin J M *Ann. Chim. Fr.* **15**: 1 (1990).
73. Galeski A *J. Polym. Sci. Polym. Phys. Ed.* **19**: 721 (1981).
74. Galeski A, Piorkowska E *J. Polym. Sci. Polym. Phys. Ed.* **19**: 731 (1981).
75. Price F P, Thornton J M *Bull. Amer. Phys. Soc.* **18**: 295 (1973).
76. Galeski A, Piorkowska E *J. Polym. Sci. Polym. Phys. Ed.* **21**: 1299 and 1313 (1981).
77. Piorkowska E, Galeski A *J. Polym. Sci. Polym. Phys. Ed.* **23**: 1723 (1985).
78. Billon N, Haudin J M *Coll. Pol. Sci.*, to be published.

CHAPTER 6

Molecular models for polymer deformation and failure

Y TERMONIA

6.1 Introduction

Polymer deformation and failure is an extremely important problem of great experimental as well as theoretical interest. A theoretical approach to the problem is, however, greatly complicated by the need to consider many structural parameters such as free volume, trapped entanglements, amorphous and crystalline regions, molecular weight and its distribution, etc. For that reason, all previous models were mainly phenomenological or semi-empirical and no attempt was made to relate the mechanical properties to molecular structure. These models range from descriptions based on principles of fracture mechanics to static calculations of stress–strain curves of aggregate models, and from merely pictorial views to accounts of morphological changes on a 10 nm level. (For an exhaustive overview the reader is referred to the excellent texts of Refs 1–2.) All these models, to our knowledge, are incapable of simultaneously providing quantitative information on (i) stress–strain behaviour, molecular extension, chain fracture and slippage; (ii) morphological changes that occur during deformation; (iii) effects of testing variables, such as temperature and strain rate, and (iv) importance of the molecular weight and its distribution.

A model study that attempts to tackle all those aspects of polymer deformation and fracture evidently requires that dynamic events on a molecular level be included. It is the purpose of the present chapter to review a series of models that explicitly take into account the roles of chain extension, chain slippage through entanglements, primary and secondary bond breaking during failure, etc. In these models, the polymer network is represented by an array of bonds on a lattice. Depending on the application of interest, these bonds represent atomistic linkages, e.g. C–C covalent bonds, or full molecular chain strands between entanglements and/or crosslinks. The molecular weight and its distribution is easily accounted for by breaking a certain number of bonds or chain strands between lattice sites prior to deformation. These lattice networks are then strained and a set of thermally activated processes allows for chains to slip through entanglements and/or bonds to break at maximum extension. These processes are simulated by a Monte Carlo approach using Eyring's chemical activation rate theory. At regular time intervals the lattice sites are relaxed towards mechanical equilibrium by a series of fast computer algorithms which steadily reduce the net residual forces acting on those sites.

This chapter will be organized as follows. Section 6.2 is entirely devoted to linear flexible polymers with particular emphasis on semi-crystalline polyethylene (PE). The first part deals with the ultimate mechanical properties expected in perfectly ordered and extended arrays of macromolecules. The second part dwells on the factors controlling the drawability and orientation of those flexible macromolecules when still in the isotropic state. In section 6.3, we turn to crosslinked networks with particular emphasis on the role of the entanglements that are inevitably present in those systems prior to crosslinking. The model is then extended to the case of multiaxial deformation in an attempt to isolate the effects of entanglements on the form of the strain energy density function. We conclude in section 6.4 and present an outline for future research.

6.2 Linear polymers

6.2.1 Tensile strength of oriented chains

The present section deals with the mechanical properties of polymer fibres in which the macromolecules are perfectly ordered and oriented along the testing axis with no defect other than chain-ends resulting from the finite molecular weight of the sample. As a result, our study deals with the maximum axial tensile strength which should help guide experimental research towards the development of stronger polymeric fibres.

Figure 6.1 shows the model representation of those fibres.[3] The nodes in the array represent the elementary repetition units of the polymer chains, i.e. methyl units for polyethylene. The nodes are joined in the x- and z-directions by secondary bonds having an elastic constant K_2. These bonds account for the van der Waals forces in polyethylene (PE). Only nearest neighbour interactions are considered. In the y-direction, stronger forces with elastic constant K_1 account for the primary C–C bonds in PE.

In the kinetic theory of fracture,[4,5] a bond, in the absence of external mechanical forces, breaks whenever it becomes excited beyond a certain level, U, the activation energy of the bond. For chains in the process of being strained, the activation energy barrier is linearly decreased by the

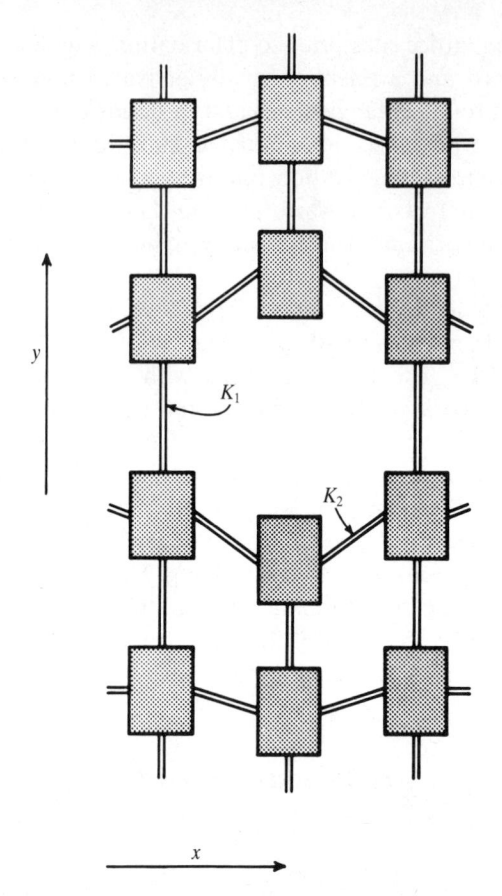

Figure 6.1 Two-dimensional representation of the array of nodes and of bonds for a perfectly ordered and oriented polymer fibre. K_1 and K_2 are the elastic constants for the primary and for the secondary bonds, respectively. From Termonia et al.[3]

local stress.[5,6] Thus, in the presence of stress, the rate of breakage of a bond i is given by

$$v_i = \tau \exp[(-U_i + \beta_i \sigma_i)/kT] \quad [1]$$

where τ is the thermal vibration, T is the absolute temperature, β_i is the activation volume and σ_i is the local stress

$$\sigma_i = K_i \varepsilon_i \quad [2]$$

Here K_i and ε_i are the elastic constant and the local strain, respectively.

The simulation is started from an unstrained array in which all the bonds are unbroken, except for a few broken primary bonds along the y-axis which account for the finite molecular weight of the polymer chains.

The unbroken bonds are visited at random by a Monte Carlo lottery which breaks a bond according to a probability

$$p_i = v_i/v_{max} \qquad [3]$$

where v_{max} is the rate of breakage of the mostly strained bond in the array. After each visit of a bond, the time t is incremented by $1/[v_{max} \cdot n(t)]$, where $n(t)$ is the total number of unbroken bonds at time t. After a small time interval δt has elapsed, the bond breaking process is stopped and the sample is strained along the y-axis by a small constant amount depending on the rate of strain. The network is then relaxed towards its minimum energy configuration by using a set of fast computer algorithms which steadily reduce the net residual force on each node.[3]

Results of a simulation for a two-dimensional array of nodes are illustrated in Fig. 6.2 for a small (case a) and for a large (case b) elongation, respectively. Note the development of a large fracture focus in Fig. 6.2(b), which appears as a result of an acceleration of its growth due to the high concentration of stresses at its boundary.

The model described above has been applied with success to studies of the effects of molecular weight and testing conditions (temperature and rate of deformation) on the ultimate tensile strengths of polyethylene[3,7] and poly(p-phenylene-terephthalamide)[8] fibres. The importance of the molecular weight distribution has been investigated by Termonia et al.[9] The approach has been later refined to include possible imperfections in the fibre, such as weak-link chain defects[10] and chain-end segregation.[11] For the purpose of illustration of the potential of our approach, we now turn to present results obtained for the effect of molecular weight on the ultimate strength of oriented PE fibres. Values of the parameters for Eqns [1] and [2] can be found in Termonia et al.[3] Note that these parameters are difficult to measure experimentally and their values are not established with great accuracy. Our model results therefore give only expected trends and they should not be interpreted too quantitatively.

6.2.1.1 Effect of molecular weight

Figure 6.3 shows a series of stress–strain curves for various (monodisperse) molecular weights ranging from 1.4×10^3 to 3×10^5. At low molecular weights ($M < 8 \times 10^4$), the results show that intermolecular slippage involving rupture of secondary bonds occurs in preference to molecular fracture (i.e. primary bond breakage). Under such circumstances, plastic deformation is observed and the curves are bell-shaped with a very slow decrease of the stress towards the breaking point. At higher molecular weights ($M > 8 \times 10^4$), primary as well as secondary bond rupture is observed and the fracture of the sample appears much more brittle. Note

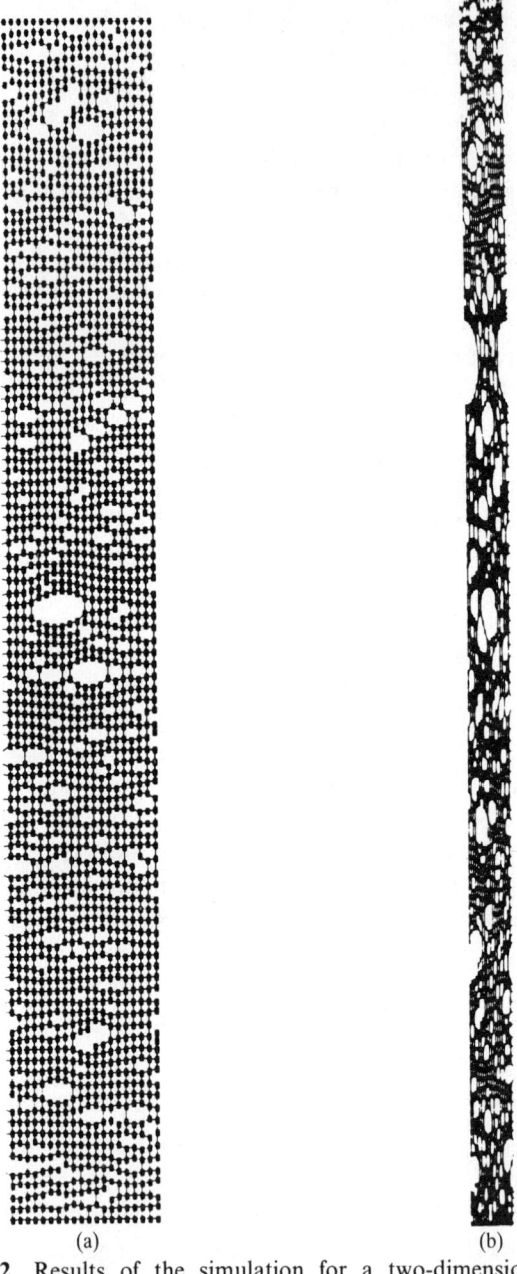

Figure 6.2 Results of the simulation for a two-dimensional array of 20 × 100 nodes: case (a), 150% strain; case (b), 300% strain. The probabilities for breaking primary and secondary bonds in these highly stretched samples were chosen equal to $\exp(\varepsilon^2)$ for simplicity. Cylindrical boundary conditions along the x-axis were imposed. From Termonia et al.[3]

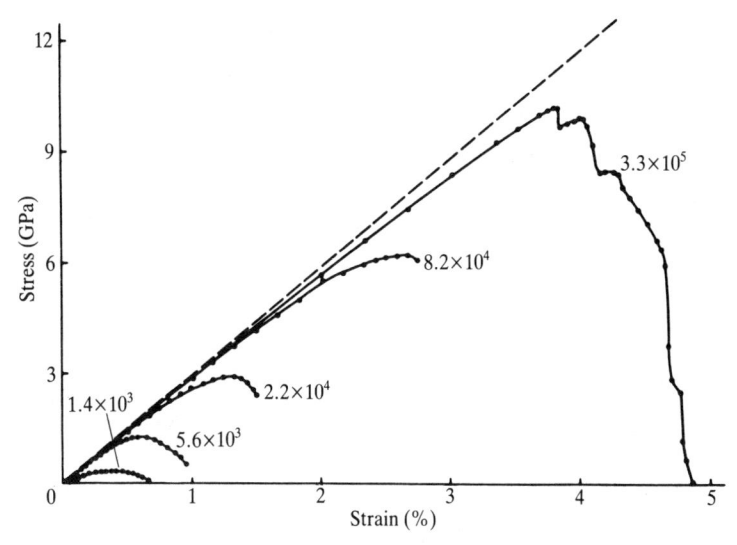

Figure 6.3 Stress–strain curves calculated for several oriented polyethylene fibres of various (monodisperse) molecular weights. The strain rate is 100% min. Three-dimensional simple cubic arrays of up to $6 \times 6 \times 1000$ nodes were used in the simulations. Cylindrical boundary conditions along the x-axis were imposed. Curves for $M = 2.2 \times 10^4$ and $M = 8.2 \times 10^4$ have not been calculated in their entirety, due to the lack of computer time. The dashed line indicates a slope equal to the theoretical modulus ($=300$ GPa). From Termonia et al.[3]

that at those higher molecular weight values, a renormalization scheme has been adopted. Nodes were made to correspond to $m > 1$ methyl groups each and the secondary bonds linking two neighbouring nodes were renormalized accordingly in groups of m units with elastic constant $K_2 \times m$. The validity of that procedure has been discussed in Termonia et al.[3]

6.2.2 Drawability of unoriented chains

Section 6.2.1 has shown the potential of our approach in predicting the ultimate mechanical properties of polymer fibres in which the macromolecules are perfectly ordered and oriented along the fibre axis. That state of perfect orientation and extension is usually achieved by drawing the unoriented material to very high draw ratios at temperatures close to its melting point. From an experimental point of view it is therefore of primary importance to have a detailed knowledge of the factors controlling the tensile deformation of those materials. The goal of the present section is to show the potential of molecular models in providing such information.

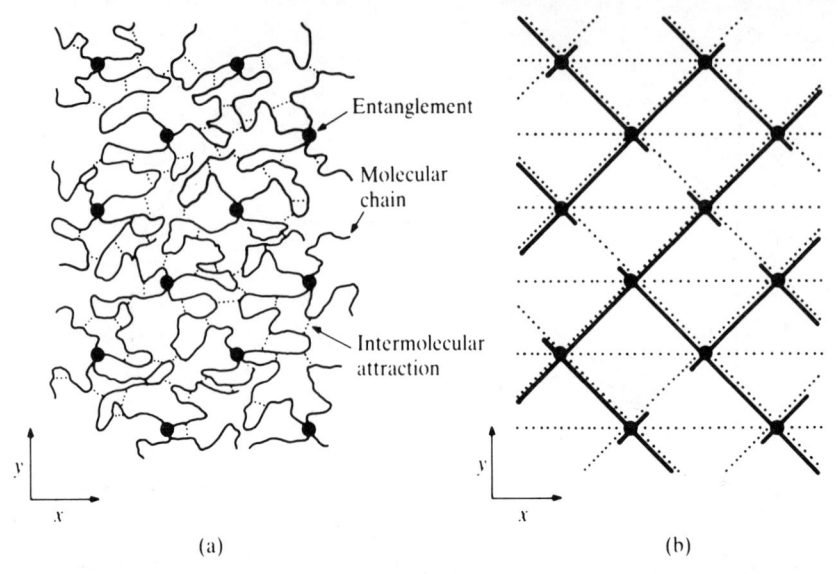

Figure 6.4 (a) Dense system of unoriented polymer chains. The heavy black circles represent entanglement loci and the dotted lines denote the Van der Waals bonds. (b) More schematic representation of the network in (a). The heavy solid lines indicate chain vectors between entanglements; chain ends are represented by shorter solid lines. Individual VdW bonds have been replaced by 'overall' bonds (dotted lines) joining each entanglement to its nearest neighbours. From Termonia and Smith.[13]

6.2.2.1 Model

In our model,[12–14] the undeformed (semi)-crystalline polymer solid is represented by a loose network of entangled macromolecules. Prior to deformation, the chain strands between the entanglement loci are coiled and are tied together through weak inter- and intramolecular bonds, e.g. Van der Waals forces. These forces represent the initial stiffness of the polymer. This representation of the polymer solid is illustrated in the schematic of Fig. 6.4. The heavy black circles in this figure denote the entanglement loci. The dotted lines represent the weak attractive (Van der Waals) bonds connecting sections of either the same chain, or of different chains. Since the coordination number of an entanglement is assumed to be 4, the actual three-dimensional network for convenience has been given a planar (x–y) configuration. The y-axis is chosen as the direction of draw. Periodic boundary conditions are imposed along the transverse x-axis.

In our approach, the above polymer solid is further simplified as shown in Fig. 6.4(b). Here, the details of the chain configurations are omitted altogether. The coiled chain strands between entanglements are replaced by chain vectors denoted by heavy solid lines. Chain strands that do not connect two entanglement loci (i.e. chain end sections) are indicated by shorter solid lines. In actual undeformed solids the chain vectors are randomly oriented in three-dimensional space, i.e. $\langle \cos^2 \theta \rangle = \frac{1}{3}$, where θ is the vector's angle with the draw axis. Accordingly, in our two-dimensional lattice representation of Fig. 6.4(b) we take the value $\theta = 54.7°$ for all vector orientations along the draw (y-)axis. The various Van der Waals bonds are replaced by 'overall' bonds (dotted lines in Fig. 6.4(b)) joining each entanglement point to its neighbours. Only overall bonds between the two nearest neighbours are taken into account.

At the start of the simulations, the entanglement lattice is filled in with macromolecules of monodisperse molecular weight M. This is a rather complex orientation that is realized as follows. For given values of M and M_e (the molecular weight between entanglements) random chains of M/M_e strands are constructed using a step-by-step procedure based on examining, at each step, for possible continuations of the chain in the subsequent steps.[15] This procedure allows the lattice to be filled in with macromolecules of a length distribution as close as possible to monodisperse.

This macromolecular network is then deformed at constant temperature T and rate of elongation ε. This leads to straining of the Van der Waals bonds which are broken, according to the Eyring kinetic theory of fracture, i.e. at a rate given by Eqn [1] (see section 6.2.1). These VdW bond breakages lead to a release of the chain strands, which are now to support the external load. Once broken, VdW bonds are assumed not to re-form. The present model thus ignores the possible influence of the crystalline phase in the deformation process past its initial stage. Experimental evidence has accumulated[16] that plastic deformation of weakly bonded polymer systems is dominated not by crystallites but by chain entanglements; this justifies the present simplification.

As the stress on the 'freed' chain strands increases, slippage through entanglements is assumed to set in at a rate that has the same functional form as that for VdW breakings (Eqn [1]), but with different values for the activation energy U and volume β. In addition, σ now denotes the difference in stress in two consecutive chain strains that are separated by an entanglement. This stress difference is calculated using the classical treatment of rubber elasticity.[17] According to this theory, the stress on a stretched chain strand having a vector length r is given by

$$\sigma = \alpha k T \mathscr{L}^{-1}(r/nl) \qquad [4]$$

Here n is the number of statistical chain segments of length l in a strand. $\mathscr{L}^{-1}$ is the inverse Langevin function and α is a proportionality constant which depends on the number N of chain strands per unit volume.

Following Treloar,[17]

$$\alpha = (N/3)\sqrt{n} \qquad [5]$$

In accordance with reported experimental data for viscous flow of paraffins,[18] we assume the average length of a hydrocarbon chain capable of a coordinated movement, at a rate given by Eqn [2], is approximately 25 carbon atoms. This slippage process leads to a change in the number of statistical chain units between entanglements and, occasionally, to chain disentanglement. If the rate of chain slippage is too low, maximum elongation of the chain strands between entanglements can be reached and chain fracture occurs at a local draw ratio $\lambda = \sqrt{n}$. In view of the high stress required for chain rupture and the ensuing important retraction of the broken molecular ends, fractured chains are assumed not to re-form.

The simulation of the VdW bond breakings is performed using the same Monte Carlo procedure as that described in section 6.2.1. Simulation of chain slippage is executed using a similar technique in which $n(t)$ now denotes the total number of entanglements left at time t. After a small time interval δt has elapsed, the VdW bond breaking, chain slippage and fracture processes are halted. The network is then relaxed towards mechanical equilibrium and elongated along the y-axis using the same procedure as that described in section 6.2.1. Values of the model parameters for polyethylene can be found in Refs. 12–14.

Termonia and Smith[12] studied in detail the effect of molecular weight on the stress–strain curves and morphology of deformation of semi-crystalline linear polyethylene. The dependence of drawability on testing conditions (rate and temperature of deformation) has been studied by Termonia et al.[14] Here, we limit ourselves to a detailed investigation of the effect of entanglement spacing (i.e. the molecular weight between entanglements) on the deformation of high molecular polyethylene. More details can be found in Termonia and Smith.[13] Apart from being academically interesting, that study has become a topic of industrial significance with the commercialization by DSM/Toyobo and, under licence, Allied Signal Corporation of high performance polyethylene fibres[19] (sold under the respective trade names Dyneema SK60 and Spectra 900/1000). These fibres are manufactured according to a solution spinning/drawing technique known as 'gel'-spinning, which is founded on control of the entanglement spacing through solution processing.

6.2.2.2 Effect of density of entanglements

To systematically examine the effect on the deformation behaviour of the entanglement spacing, i.e. of the number of statistical chain segments

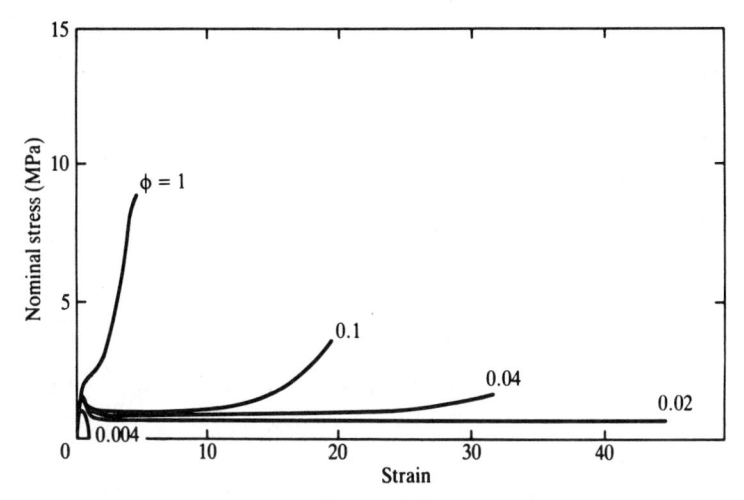

Figure 6.5 Calculated nominal stress–strain curves for monodisperse linear polyethylene with $M = 475\,000$ at four different values – indicated in the graph – of the entanglement spacing factor ϕ (see text). The deformation temperature was 109 °C and the rate of elongation was 500%/min. From Termonia and Smith.[13]

between entanglements, we introduce the spacing factor ϕ defined as[13, 20]

$$\phi = (M_e/1900)^{-1} \qquad [6]$$

in which M_e is the chosen value for the molecular weight between entanglements, whereas the numerical scaling factor 1900 is the reference value for the melt itself.

Figure 6.5 shows a series of nominal stress–strain curves calculated for monodisperse polyethylene of $M = 475\,000$ at five different values of the entanglement spacing factor ϕ (1, 0.1, 0.04, 0.02 and 0.004). The deformation temperature was taken to be 109 °C. This figure reveals the dramatic effect of the entanglement spacing on the deformation characteristics, notably on the post-yield strain hardening and on the strain at break. At decreasing values of ϕ, the rate of strain hardening, i.e. the post-yield modulus, rapidly drops to reach a negative value at $\phi = 0.004$. The strain at break, on the other hand, drastically increases from 4.5 to 45 when ϕ decreased from 1 to 0.02. At much lower values of ϕ, e.g. 0.004, the plastic deformation leading to high values of the strain at break no longer occurs as a result of continued strain softening and ductile failure is observed. The latter result is, of course, due to the fact that at $\phi = 0.004$ the molecular weight between entanglements is $1900/0.004 = 475\,000$, which equals the molecular weight of the polyethylene in the simulation. Accordingly, transfer of applied load, in this special case, occurs only from one

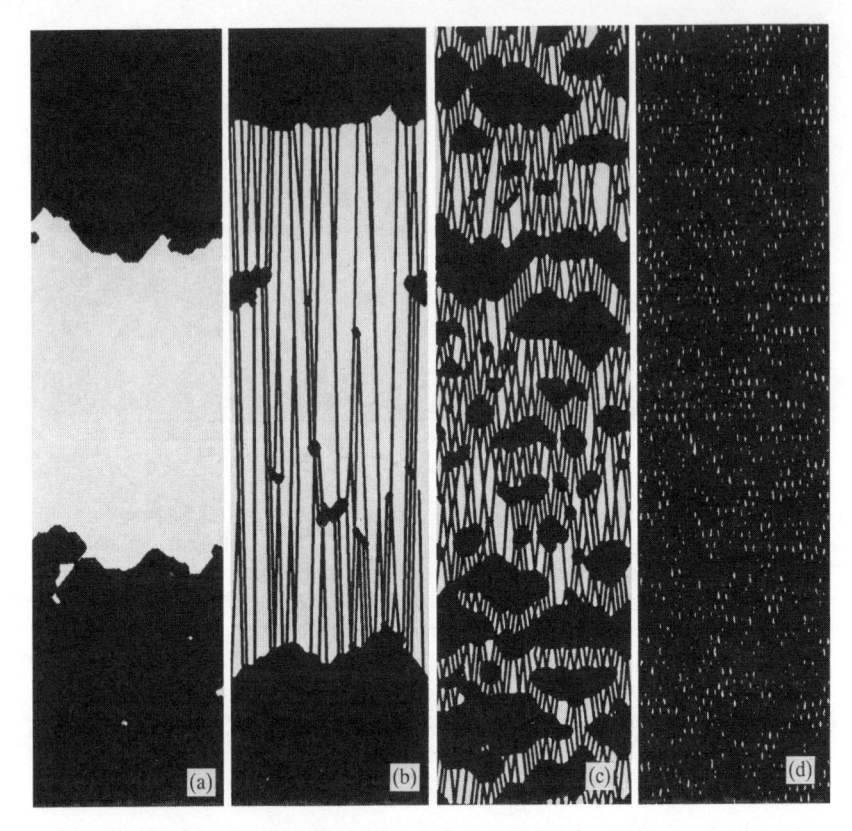

Figure 6.6 Typical 'morphologies' obtained with the model for the materials of Fig. 6.5 with ϕ: (a), 0.004; (b), 0.02; (c), 0.1; (d), 1. In cases (b)–(d) the draw ratio $\lambda = 2.7$. The widths of the 'samples' are, respectively, 3.2, 1.6, 0.7 and 0.2 µm. From Termonia and Smith.[13]

chain to its nearest neighbours through weak VdW bonds and stress concentrations arising from VdW-bond breaks are not distributed uniformly throughout the entanglement network. As a result, very little deformation occurs and failure is highly localized (see also the morphology in Fig. 6.6).

The calculated stress–strain curves of Fig. 6.5 show a remarkably good accord with those recorded of ultra-high molecular weight (UHMW) polyethylene crystallized from solutions of various initial polymer concentrations.[16] Thus, it is clear that the current model predicts very well the dramatic effect of the initial polymer volume fraction on the deformation behaviour of UHMV PE, particularly the spectacular increase of the strain at break at decreasing polymer concentration. Originally, this effect was explained in terms of reduced entanglement densities in solution crystallized polymer solids.[16] The present calculations, which are based on this

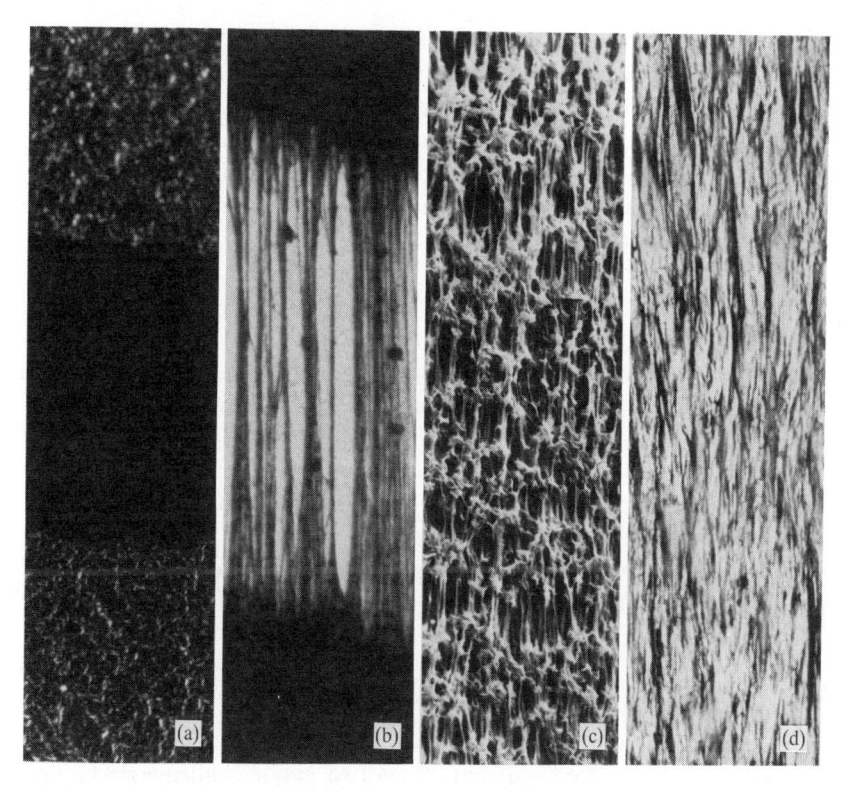

Figure 6.7 Micrographs of drawn samples of polyethylene films of $M_w = 1.5 \times 10^6$ and $M_n \sim 2 \times 10^5$ crystallized from solutions in decalin, and from the melt (see Smith et al.[21] for experimental details); these samples were drawn to a macroscopic draw ratio of approximately 3 at 100 °C. The initial polymer volume fractions were, respectively, (a), $\phi = 0.005$; (b), $\phi = 0.02$; (c), $\phi = 0.1$; (d), $\phi = 1$. Prints (a), (b) and (d) are optical micrographs taken under crossed polarizers (except b) and (c) is a scanning electron micrograph. The width of all strips shown is 0.1 mm. From Termonia and Smith.[13]

concept, of course, strongly reinforce this hypothesis. In turn, the excellent agreement between calculated and measured stress–strain curves provides great confidence in the model.

It was pointed out in the introduction that the present model also provides schematic 'morphologies' of the deformed polymer solids by plotting the positions of the entanglement loci and the connecting chain vectors. Figure 6.6 shows these morphologies for four entanglement spacing factors $\phi = 1$, 0.1, 0.02 and 0.004, at an 'overall' draw ratio of 2.7 (except for $\phi = 0.004$). This figure displays the dramatic effect of the entanglement spacing on the fractography, the deformation behaviour and particularly on the necking phenomena. At the very low value of $\phi = 0.004$ brittle-like fracture is observed. At the higher value of 0.02 a well-defined neck appears, which upon subsequent deformation travels through the entire specimen. For $\phi = 1$ essentially homogeneous deformation is observed, whereas $\phi = 0.1$ represents an intermediate case between homogeneous deformation and necking. In the latter case multiple micronecks are formed.

For the purpose of comparison, micrographs of actual samples of solution-crystallized UHMW PE are presented in Fig. 6.7. This figure displays photographs of polyethylene films of $M_w = 1.5 \times 10^6$ and $M_n \sim 2 \times 10^5$ crystallized from respectively 0.5, 2 and 10% v/v solutions in decalin, and from the melt (see Smith et al.[21] for experimental details); the latter three samples were drawn to a macroscopic draw ratio of approximately 3 at 100 °C. The resemblance between these micrographs and the 'calculated morphologies' of Fig. 6.6 is truly remarkable. It illustrates that the present model indeed is capable of handling the very complex issue of connecting events on molecular level to macroscopic properties and features.

6.3 Crosslinked polymers

A theoretical description of the factors controlling the performance properties of crosslinked polymers faces two major challenges. The first is to devise a realistic model for network formation during crosslinking. Previous network models can be divided into two categories:[22] (a) Flory–Stockmayer type of approaches which neglect cyclic bonds and excluded volume effects[23,24] and (b) computer simulations in space which allow the formation of rings of any size.[25,26]

The second problem is to relate the elastic properties of those networks to their molecular structure. There is at present no firm molecular model for describing the role of entanglements on the stress–strain behaviour of elastomeric systems. Previous approaches were only phenomenological or semi-empirical and have led to two opposing views. Earlier studies by Flory[27] and Ronca[28] limit the role of chain–chain interactions to that of restricting fluctuations in network junctions. A more concise treatment of these constraining effects can be found in Flory and Erman.[29] At small strains, all junction fluctuations are suppressed and the shear modulus, defined through

$$G = vkT \qquad [7]$$

takes on its 'chemical' value for which v equals the number of starting chains per unit volume.[27–29] The above point of view has been challenged by many other researchers[30–35] who contend that chain–chain interactions are also present along the chain contour and add a significant contribution to the modulus value given by Eqn [7]. Entanglements along chain contours have been modelled as confining tubes,[36,37] hoops[38] or slipping links.[39]

The object of the present work is to extend the molecular models described in section 6.2 to the deformation behaviour of elastomeric networks, in order to specifically address the two problems raised above.[40–43] Thus, in analogy with section 6.2, the network formation is simulated on a lattice with the help of a Monte Carlo procedure, which allows the formation of rings of any size. However, in contrast to the works of Refs 25 and 26, our approach also takes into account the presence of entanglements latent in the polymer prior to crosslinking. For simplicity, the model has been applied to elastomeric networks formed through end-linking of difunctional polymer chains with plurifunctional junction sites.

6.3.1 Model

In our model, the elastomeric network prior to crosslinking is represented by an entangled network of (bifunctional) macromolecules which are in a random coil configuration. That network is built on a computer, as follows. We start with a regular empty lattice of nodes which represent potential elastically active junctions. For simplicity, we restrict ourselves to the case of macromolecules having a molecular weight M larger than the molecular weight between entanglements, M_e. This implies that the distance between nodes be of the order of that between entanglements. These nodes are then connected with macromolecules having a prescribed molecular weight distribution (for details, see Termonia and Smith[12,13]). In the present section, we will deal solely with monodisperse molecular weight distributions for the starting polymer. Figure 6.8 depicts a small section of our network for a polymer having a molecular weight (M) equal to four times that between entanglements (M_e). Each macromolecule in Fig. 6.8 is thus defined through a random walk connecting three nodes (entanglements) on the lattice. Details of the configuration of a molecular chain strand between two entanglements are omitted in the model and only end-to-end vectors (wiggling lines) are being considered. Since the coordination number of an entanglement is only 4, the actual three-dimensional network has been given a planar $(x - y)$ configuration. Periodic boundary conditions have also been imposed along the transverse x-axis.

The nodes in Fig. 6.8 are of three different types: nodes with 4 chain strands originating from them (type I); nodes with 2 strands (type II);

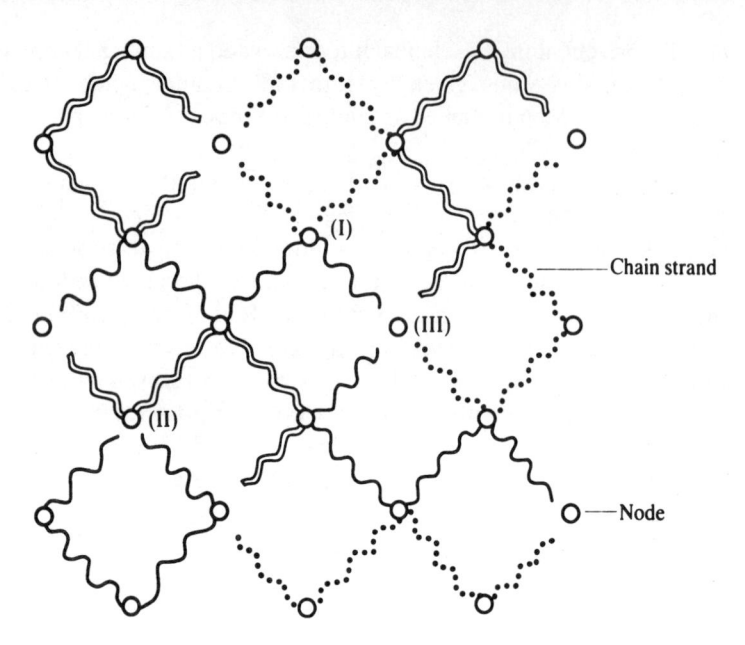

Figure 6.8 Schematic lattice representation of an entangled (x–y) network prior to crosslinking. The lattice nodes (symbol ○) are distributed on a regular lattice and they represent potential elastically active junctions. The polymer has a monodisperse molecular weight (M) equal to four times that between entanglements (M_e). Details of the configuration of a molecular chain strand between entanglements are omitted and only end-to-end vectors (wiggling lines) are represented. Different line types are for different molecules. The lattice nodes are of three types: nodes with 4 chain strands originating from them (type I); nodes with 2 strands (type II); nodes with 0 strand (type III). From Termonia.[40]

nodes with 0 strand (type III). Nodes of type (I) represent true entanglements (denoted by a filled-in circle in Fig. 6.9, later). Nodes of types (II) and (III), on the other hand, are vacant and the crosslinking procedure is initiated by filling them in with tetra- or tri-functional monomers (see Fig. 6.9). These monomers react – with probability p – with the neighbouring chain ends. Nodes of type (II) with two reacted chain ends lead to the formation of an additional entanglement (nodes type (IV), see Fig. 6.9); those with only one or zero reacted ends are not elastically active and are therefore removed from the lattice (node (V)). Note that the local removal of nodes requires a careful updating of the number of statistical segments along neighbouring chain strands. Similarly, nodes (III) with three or four reacted ends are considered as elastically active crosslinks (nodes VI and VII), whereas the others are removed.

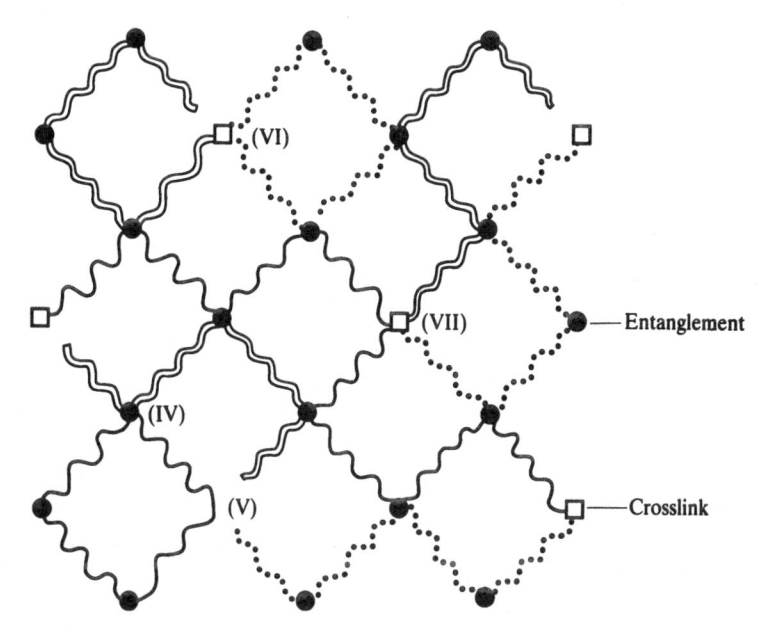

Figure 6.9 Same network as Fig. 6.8 but, after crosslinking with tetrafunctional monomers. Entanglements are designated by symbol ●, whereas crosslinks are denoted by symbol □. For further details, see text. From Termonia.[40]

The procedure outlined above leads, for a given value of the molecular weight M and advancement of the reaction p, to the formation of a network of junction points (entanglements and crosslinks), all of which are elastically active. Among the active entanglements, however, only those trapped between crosslinks are permanent. Since we are, in the present work, interested in measuring the effect of entanglements at elastic equilibrium, this leads to the need to further remove from the lattice all those entanglements – like node IV in Fig. 6.9 – which are not independently connected to the network through 4 intact chain strands.

The network, at this stage of the simulation procedure, is quite different from its initial representation in terms of a regular lattice of nodes. The removal of all elastically inactive nodes – described above – has led to a very irregular structure whose deformation behaviour is expected to be quite non-affine. A correct description of the global stress–strain behaviour thus requires knowledge of the local stress distribution around individual nodes. For a given value of the external stress, this is obtained through a detailed relaxation of every single junction point towards mechanical equilibrium with its connected neighbours. That relaxation is performed with the help of a series of computer algorithms already described in section 6.2.1 (see also Termonia et al.[3]). The force f on a particular chain strand between two junctions is calculated by using the classical treatment of rubber elasticity, i.e. employing an equation similar to Eqn [4]

$$f = (kT/l)\mathscr{L}^{-1}(r/nl) - f_0 \qquad [8]$$

in which we introduce the quantity f_0 to denote the local force in the absence of strain, i.e. that for which $= n^{1/2}l$. This prevents collapse of the network in the state of rest.

The non-affine displacements of the various junctions, obtained through network relaxation, create large force gradients on chains passing through trapped entanglements. These gradients lead to the possibility of chain slippage. In the present case, which assumes total mechanical equilibrium between successive strain increments, slippage is allowed until the difference in force in the two strands of a chain separated by an entanglement falls below an entanglement friction force f_e (compare with slippage rules in section 6.2.2). The latter, which is left as a free parameter, thus effectively prevents an entanglement from moving freely along the chain contour. Note also that the above treatment neglects the effects of chain–chain interactions along the length of the chain. The justification for that neglect is that friction at entanglement junctions essentially dominates the rheological behaviour. Values of the model parameters can be found in Termonia.[40]

The approach described above has been used to study in detail the effects of entanglements on the structure and mechanical properties of polydimethyl-siloxane (PDMS) networks.[40] It has been extended to bimodal PDMS networks[41] in an attempt to elucidate their unusual toughness properties found experimentally.[44] Termonia[42] investigated the changes in mechanical properties in networks crosslinked in a state of strain, whereas Termonia[43] dealt with interpenetrating networks in which a given polymer is crosslinked in the immediate presence of another. Here, we shall limit ourselves to a brief summary of the results of Termonia[40] on the effect of entanglements on the mechanical properties of monodisperse PDMS networks.

6.3.2 Network properties and role of entanglements

Figure 6.10 shows a typical network structure, as obtained from our computer model, prior to deformation. The starting molecular weight equals 4 times the molecular weight between entanglements, $M = 4M_e$, and the degree of advancement of the reaction $p = 0.96$. The functionality of the crosslinker is taken to be 4. The figure shows only the gel fraction with its elastically active junctions and their connecting strands. Dangling ends have not been represented. Elastically active crosslinks (symbol □) are connected to the network through at least 3 intact chain strands, whereas 4 strands are required for trapped entanglements (symbol ●).

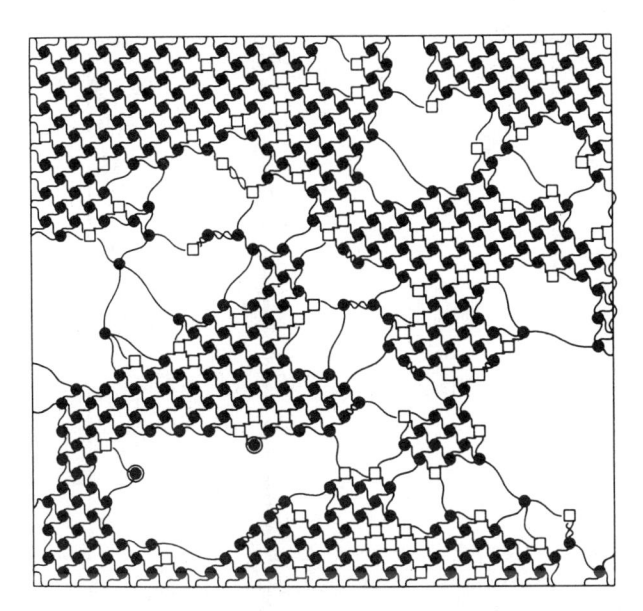

Figure 6.10 Actual network prior to deformation, as obtained from the computer model for a polymer with $M = 4M_e$ and crosslinked with tetrafunctional monomers. The fractional conversion of crosslinker groups is $p = 0.96$. The figure shows only the gel fraction with its elastically active junctions (crosslinks □ and entanglements ●) and their connecting strands. Dangling ends have not been represented. A circle around an entanglement point denotes the presence of an internal loop. From Termonia.[40]

Note that chain vectors, particularly near removed nodes, do not necessarily have the same number of statistical segments. A circle around an entanglement point denotes the presence of an internal loop. The network of Fig. 6.10 is in local mechanical equilibrium that is, for every strand $f = f_0$ (see Eqn [8]).

The effect of molecular entanglements on network properties is studied in Fig. 6.11. The figure is for a starting molecular weight $M = 4M_e$ crosslinked with tetrafunctional monomers. The first effect of the presence of entanglements is found in the gel fraction which, at any $p > 0.85$, is much higher than that for low molecular weight unentangled polymers.[40] Entanglements, however, also lead to much lower values of the density of elastically active strands v. This is due to the presence in the gel of a large number of non-permanent entanglements which do not contribute to the elastic properties at equilibrium. The figure indeed also shows that the sharp drop in v closely follows that of the density v_t of permanent entanglements trapped between crosslinks (symbol △). Thus, even though

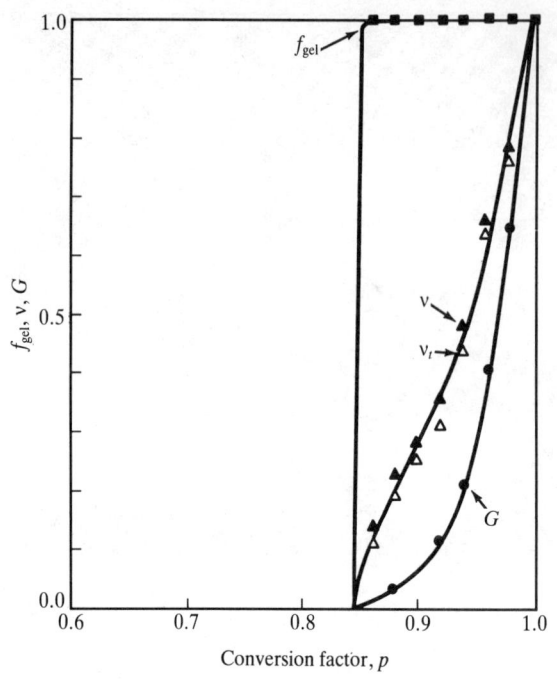

Figure 6.11 Network properties for the case $M = 4M_e$ (entanglements present) as a function of the conversion factor, p. The crosslinker is taken to be tetrafunctional. The figure shows theoretical results for the gel fraction f_{gel} (symbol ■), density of elastically active chain strands v (▲), trapped entanglements v_t (△) and modulus G (●) as a fraction of their respective values for the case $p = 1$. The results are for a network of 60×60 nodes. From Termonia.[40]

their gel fractions are close to 100%, the networks in Fig. 6.11 contain a number of quite long dangling ends which have slipped through untrapped entanglements and do not contribute to the elastic properties. This is clearly exemplified by the modulus results G which closely follow our v_t (or v) curves.

We now turn to study the effects of entanglements on the deformation behaviour. The results are presented in Fig. 6.12 for a value $p = 0.99$. Figure 6.12(a) shows a typical configuration of the array of active crosslinks (□) and trapped entanglements (●) for those networks prior to deformation ($\lambda = 1$). Again, dangling ends and elastically inactive junctions have not been represented for simplicity of the representation. This allows one to clearly identify in the figure the presence of two elastically inactive regions which act as network defects. Figs 6.12(b, c) show the deformation behaviour of that structure at $\lambda = 3$, for different

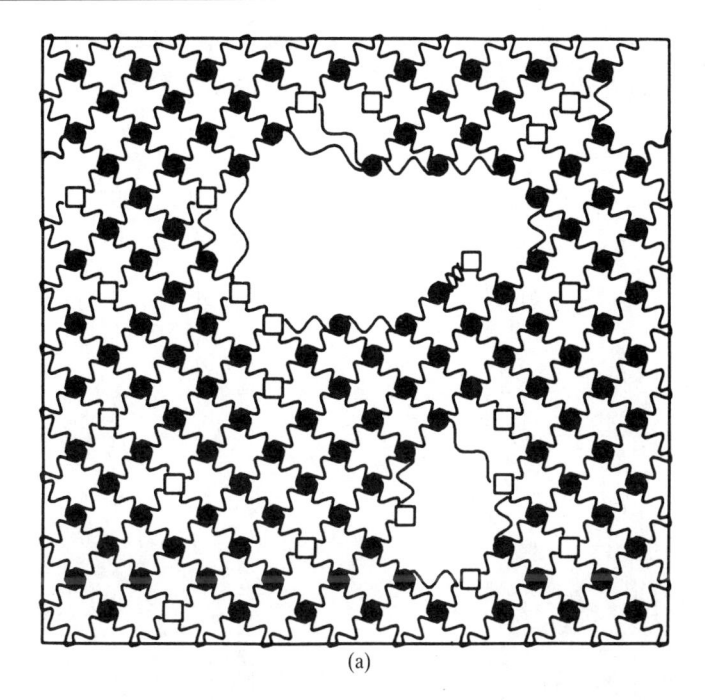

Figure 6.12 Networks of active crosslinks (□) and trapped entanglements (●) for $p = 0.99$ at different states of deformation. (a), Network prior to deformation ($\lambda = 1$). See over for part (b).

values of the friction factor f_e ($f_e = 0.02$, Fig. 6.12b and $f_e = 100$, Fig. 6.12c). In the presence of chain slippage (Fig. 6.12b), we observe an important increase in the distance between entanglements in the high stress regions (left and right boundaries of the defects). This is accompanied by a densification of entanglement points in the low stress regions (upper and lower boundaries of the defects). When slippage is absent (Fig. 6.12c), the deformation behaviour is seen to be more affine. As a result, the defect opening is much less pronounced than in Fig. 6.12(b).

6.3.3 Multiaxial deformation

A complete understanding of the behaviour of polymers under multiaxial deformation is a very complex problem of great experimental as well as theoretical interest. Rubberlike materials are ideal candidates for those studies because of their large extensibility and weak time dependence of their stress–strain curves. The deformation of those materials has been usually described in terms of the strain energy density function, W.[45] That function plays in continuum mechanics a role equally as important as

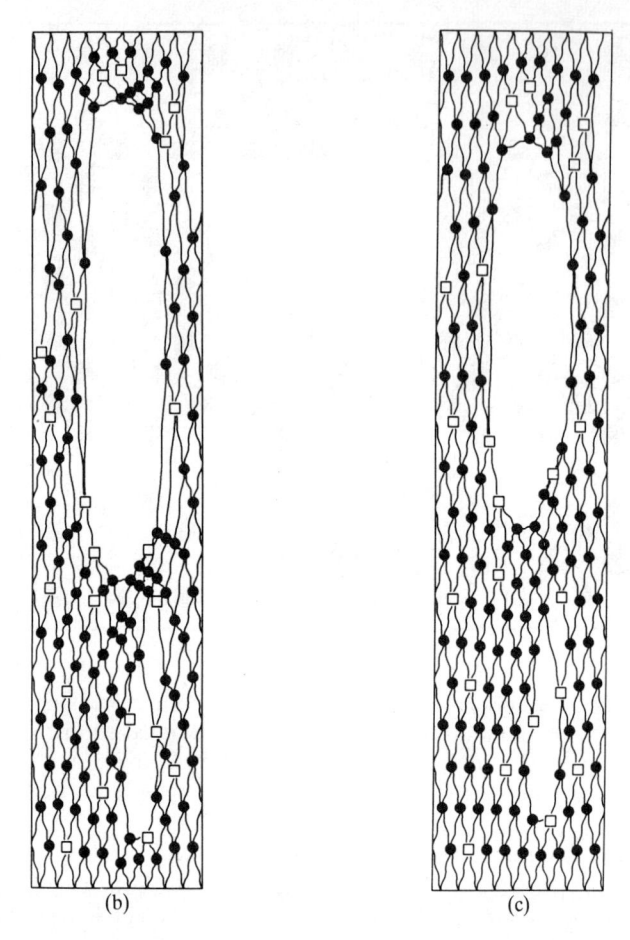

Figure 6.12 (*continued*) (b) Network at $\lambda = 3$ in the presence of entanglement slippage ($f_e = 0.02$); (c), same as in (b), but without slippage ($f_e = 100$). From Termonia.[40]

that of the free energy in thermodynamics. Thus, a knowledge of the dependence of W on strain allows a complete determination of the state of stress at any given point in a deformed body. Also, the precise form of the energy density function for polymeric materials must be obtained from a detailed molecular study of the effect of deformation on the macromolecular chains, just as the determination of the free energy functional for a ferromagnet cannot be derived from classical thermodynamics and requires a separate model for the local fields acting on a single atom.

Several models have been proposed for describing the dependence of the energy density on strain for rubberlike materials. The phenomenological approach of Riblin and Saunders[46] led to the first constitutive relation for W. Some alternative semi-empirical relationships were proposed more recently.[47–49] Other researchers have focused on more comprehensive models, looking in detail into the effects of entanglements on deformation behaviour. These models can be grouped into four categories: constrained fluctuation,[50] primitive path,[51,52] sliplink[53,54] and tube[55,56] models. The relative merits of these various approaches in fitting experimental data on biaxial deformation have been recently reviewed by Gottlieb and Gaylord.[57] None of the models, however, was found to be successful in predicting experimental observation over the entire range of investigation.

In this section we extend the models described in section 6.3.1 to the case of multiaxial deformation. Our approach thus explicitly takes into account the role of entanglements latent in the polymer prior to crosslinking. It also allows for chain slippage through a detailed analysis of the force gradients near entanglement points. The model has been quite successful at correctly describing the peculiar behaviour of the strain energy derivatives observed experimentally. The results are presented in detail in Termonia.[58] Here, we shall limit ourselves to describing the model and illustrating its potential with a few examples.

Figure 6.13 describes our model set-up for biaxial deformation of polymer networks. The network was obtained (see section 6.3.1) through end linking of difunctional polymer chains with tetrafunctional crosslinks (symbol □). The study is restricted to starting polymer chains having a monodisperse molecular weight distribution with M equal to 4 times the molecular weight between entanglements, M_e. Each macromolecule in Fig. 6.13 is thus defined through a random walk connecting three nodes (entanglements, symbol ●) on the network. Again, details of the configuration of a molecular chain strand between entanglements are omitted, and only end-to-end vectors (wiggling lines) are being considered. Upon completion of that process, the chains are crosslinked at their ends with tetrafunctional monomers, taking a degree of advancement of the reaction p equal to 0.98. Chain ends in the network have not been represented and this allows one to identify in Fig. 6.13(a) the presence of several defects due to incomplete reaction ($p < 1$). The network of Fig. 6.13(a) contains 31 junctions (entanglements + crosslinks) in both the x- and y-directions. The film is assumed to be infinitesimally thin with a thickness of the order of the distance between nearest-neighbour entanglements (Fig. 6.13b).

The film is entangled at its four sides with rigid bars which are assumed to be frictionless so that the chain 'hooks' at the edges of the network are allowed to run freely on the rail bars. The corners of the film, however, are clamped in order to minimize non-uniformity of the deformation in those regions. The film is stretched biaxially along the x- and y-directions by displacements of two rail bars (Fig. 6.13a) that can move back and

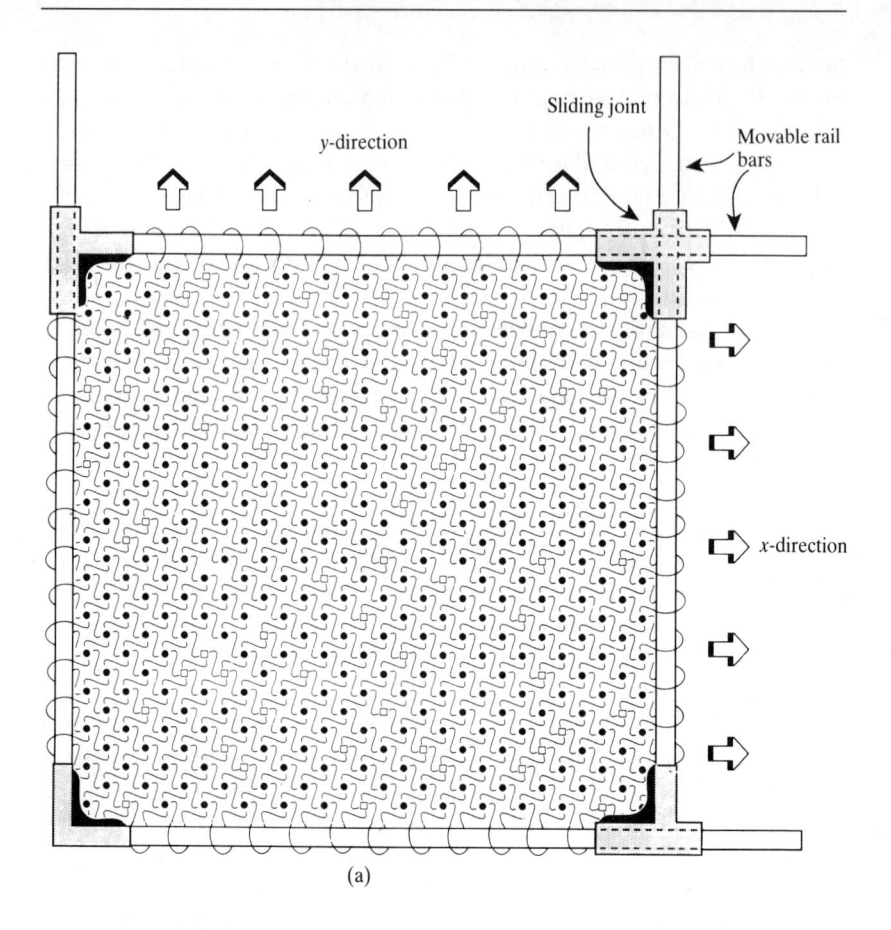

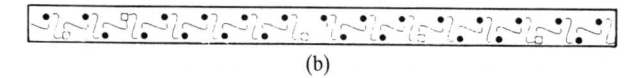

Figure 6.13 Model set-up for biaxial deformation of polymer networks. The network was obtained through end linking of difunctional polymer chains with tetrafunctional crosslinks (symbol □). The chains have a molecular weight M close to four times that between entanglements, M_e, and thus have roughly four entanglements (symbol ●) along their contour. The degree of advancement of the crosslinking reaction is set equal to $p = 0.98$. Chain ends have not been represented. The network comprises 31 junctions (entanglements + crosslinks) in both the x- and y-directions. The film is assumed to be infinitesimally thin with a thickness of the order of the distance between nearest-neighbour entanglements. From Termonia.[58]

forth through frictionless sliding joints. The forces exerted by the bars on the film lead to displacements of the network nodes in the x-, y- and z-directions. These displacements are calculated by minimizing the net residual stresses acting on each of the nodes in turn. The relaxation towards mechanical equilibrium was considered to be complete when the largest residual stress on a node fell below 0.2% of the average stress per strand. For simplicity, we do not explicitly calculate displacements of the nodes along the z-direction and assume that the network junctions move affinely in that direction so as to keep the total volume constant.

The state of stress of a particular chain strand between junction nodes (entanglements or crosslinks) is obtained as follows. Let r and $w(r)$ denote respectively the end-to-end vector length and free energy of deformation for that chain. The engineering stresses σ_x and σ_y along the x- and y-directions, are given by[59]

$$\left.\begin{aligned}\sigma_x &= \partial w(r)/\partial \lambda_x = [dw(r)/dr][\partial r/\partial \lambda_x] \\ \sigma_y &= \partial w(r)/\partial \lambda_y = [dw(r)/dr][\partial r/\partial \lambda_y]\end{aligned}\right\} \quad [9]$$

in which λ_x and λ_y are the draw ratios along the x- and y-directions, respectively. Equation [9] can be rewritten in simpler form

$$\left.\begin{aligned}\sigma_x &= f[\partial r/\partial \lambda_x] \\ \sigma_y &= f[\partial r/\partial \lambda_y]\end{aligned}\right\} \quad [10]$$

in which

$$f = dw(r)/dr \quad [11]$$

is the force on the chain. The latter is obtained from Eqn [8] in which we now omit the term $(-f_0)$ which was artificially introduced in sections 6.3.1 and 6.3.2 in order to prevent collapse of the network at zero stress. Here, in contrast, an exact description of the stress–strain behaviour is required. Thus, Eqns [9] and [10] prevent any network collapse and exactly describe the coupling of the stresses σ_x and σ_y in the two directions of strain.

To summarize: The local stresses acting on a chain strand are evaluated numerically from Eqns [8], [10] and [11]. These equations are used at two different stages in our simulations: (i) in our estimation of the nodes displacements in the x- and y-directions (see above) and (ii) in our measurements of the forces acting on the rail bars. The latter have been obtained from the means of the tensions measured in the central 12 chains hooked to the appropriate rail (see Fig. 6.13a). Our model also explicitly allows for chain slippage through entanglements. For a given value of the external strain, this process is performed, as in sections 6.3.1–6.3.2 until the difference in force f in the two strands of a chain separated by an entanglement falls below the entanglement friction force, f_e.

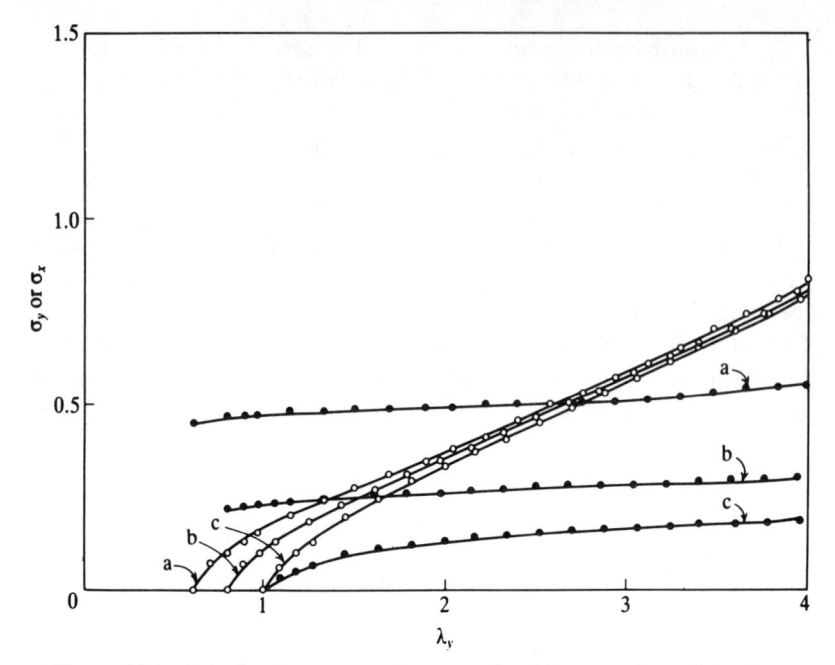

Figure 6.14 Calculated stress–strain curves for different values of the draw ratio λ_x in the x-direction: curves a, $\lambda_x = 2.68$; b, $\lambda_x = 1.56$; c, $\lambda_x = 1$. σ_y (symbol ○) and σ_x (symbol ●) denote the engineering stresses. From Termonia.[58]

Figure 6.14 shows a series of calculated stress–strain curves obtained for general biaxial deformation. In those experiments, the samples were first brought to constant and different values of the draw ratio λ_x in the x-direction. They were then stretched in small increments along the y-axis and the stresses σ_y and σ_x recorded. In order to facilitate the interpretation of our models' results, the figure is for the case of no slippage of chains through entanglements. The results show that the σ_y curves (symbol ○) are shifted towards higher values by an increase in the initial draw λ_x. Similarly, the stress σ_x (symbol ●) measured along the x-axis is seen to increase with the strain along the transverse y-axis. These data thus clearly illustrate the coupling between the stresses in the three directions. Representation of biaxial stress data in the form of Fig. 6.14 is, however, not very enlightening since its essential features are easily reproduced by all previous models.

An alternative representation, due to Jones and Treloar,[60] consists in plotting the differences $\sigma_y - \sigma_x$ vs. λ_y (or $-\sigma_x$ vs. λ_z). The curves at different λ_x are then shifted vertically so as to coincide with the pure shear curve ($\lambda_x = 1$) at the point $\lambda_y = 1$. The results are presented in Fig. 6.15. All our data for extensions λ_x ranging from 1 to values as high as 6, are

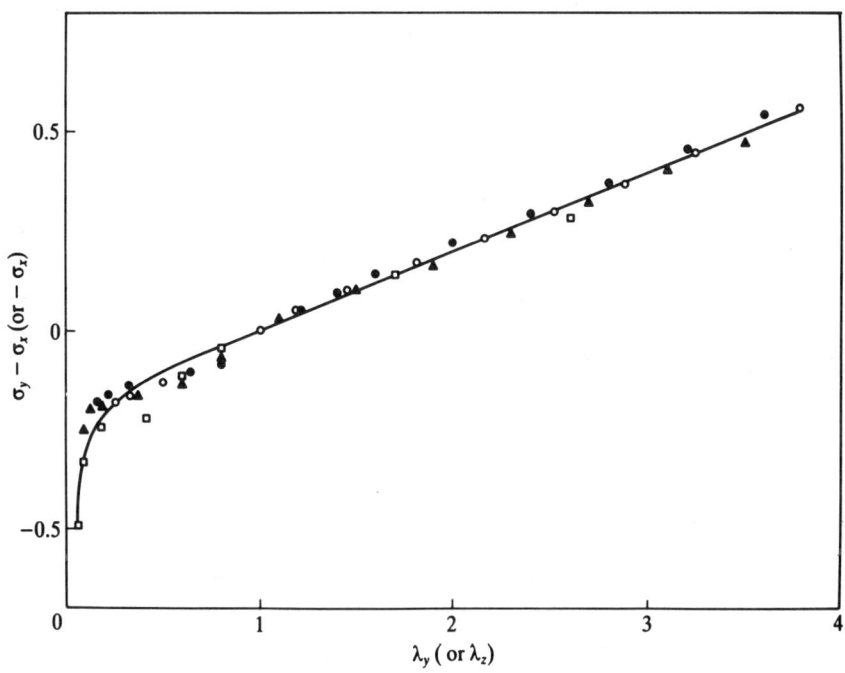

Figure 6.15 Jones and Treloar[60] representation of biaxial strain data. The figure shows the differences $\sigma_y - \sigma_x$ vs. λ_y (or $-\sigma_x$ vs. λ_z), all the curves at different λ_x being shifted vertically so as to coincide with the pure shear curve ($\lambda_x = 1$) at the point $\lambda_y = 1$. The symbols denote our calculated data for different values of λ_x: $\lambda_x = 1$ (○); $\lambda_x = 1.56$ (●); $\lambda_x = 2.68$ (▲); $\lambda_x = 6$ (□). From Termonia.[58]

seen to fall on a single master curve in agreement with experimental observation on rubber vulcanizates.[60] An essential feature of the Jones and Treloar representation in Fig. 6.15 is the sharp downturn of the results at low $\lambda_y < 0.1$. That trend can be reproduced by only three[57] of the previous models: the Mooney–Rivlin equation,[46] the Edwards[52] and the Treloar–Riding[59] models. As previously argued in Ref. 59, the molecular origin of that downturn is the finite extensibility of the chains which is explicitly taken into account in our approach through the use of an inverse Langevin expression for the stress (see Eqn [8] in that connection).

6.4 Conclusions

To conclude, we have presented a series of simple models for the description of the factors controlling the deformation behaviour of polymeric systems. The main advantage of these models is that they are molecular in nature but, at the same time, are not burdened with the need to provide a lot of details on structure and dynamics at the atomistic level. Rather, our approach is in terms of a few parameters such as elastic modulus, activation energies for rupture that can be rather easily extracted from experimental data. Admittedly, our models are only approximate since they, for example, neglect the effect of crystallinity and the temperature dependence of the activation energies. However, they bring very clearly to light the importance of many other factors such as the testing conditions, the molecular weight and its distribution, the density of entanglements, the mode and extent of crosslinking, etc. As such, our model results should not be interpreted too quantitatively but, rather, viewed as providing expected trends.

References

1. Ward I M *Mechanical Properties of Solid Polymers*, 2nd edn, Wiley, New York, 1983.
2. Kausch H H *Polymer Fracture*, 2nd edn, Springer-Verlag, Berlin, 1987.
3. Termonia Y, Meakin P, Smith P *Macromolecules* **18**: 2246 (1985).
4. Zhurkov S N, Sanhirova T P *Dokl. Akad. Nauk, SSSR* **101**: 237 (1955).
5. Reference 2, Chapter 3.
6. Zhurkov S N, Vettegren V I, Korsukov V E, Novak I I *Proc. 2nd Int. Conf. on Fracture*, Brighton, Chapman & Hall Ltd. (London) p. 545, 1969.
7. Termonia Y, Meakin P, Smith P *Macromolecules* **19**: 154 (1986).
8. Termonia Y, Smith P *Polymer* **27**: 1845 (1986).
9. Termonia Y, Greene W R, Smith P *Polymer Commun.* **27**: 295 (1986).
10. Termonia Y, Smith P *Polymer Commun.* **28**: 61 (1987).
11. Smith P, Termonia T *Polymer Commun.* **30**: 65 (1989).
12. Termonia Y, Smith P *Macromolecules* **20**: 835 (1987).
13. Termonia Y, Smith P *Macromolecules* **21**: 2184 (1988).
14. Termonia Y, Allen S R, Smith P *Macromolecules* 21: 3485 (1988).
15. Meirovitch H *J. Phys. A: Math. Gen. A* **15**: L735 (1982); *J. Chem. Phys.* **79**: 502 (1983).
16. Smith P, Lemstra P J, Booji H C *J. Polym. Sci.: Polym. Phys. Ed.* **19**: 877 (1981).
17. Treloar L R G *The Physics of Rubber Elasticity*, 2nd edn, Clarendon, Oxford, 1985.
18. Kauzmann H, Eyring H *J. Amer. Chem. Soc.* **62**: 3113 (1940).
19. Smith P, Lemstra P J to Stamicarbon B V (DSM), US Patent 4 344 908 1982; US Patent 4 430 383 1983.
20. Graessley W W *Adv. Polym. Sci.* **16**: 58 (1974).
21. Smith P, Lemstra P J, Pijpers J P L, Kiel A M *Colloid Polym. Sci.* **259**: 1070 (1981).
22. For an excellent review, see Dusek K and MacKnight W J in: *Cross-Linked Polymers: Chemistry, Properties and Applications*, Dickie R A, Labana S S and Bauer R S (eds), ACS Symposium Series, Washington, Ch. 1, 1988.
23. Flory P J *Principles of Polymer Chemistry*, Ithaca (NY), Cornell University Press, 1953.

24. Stockmayer W H *J. Chem. Phys.* **11**: 45 (1943); ibid. **12**: 125 (1944).
25. Leung Y K, Eichinger B E *J. Chem. Phys.* **80**: 3877 (1984); ibid. **80**: 3885 (1984).
26. Galiatsatos V, Eichinger B E *Rubber Chem. Techn.* **61**: 205 (1988).
27. Flory P J *J. Chem. Phys.* **66**: 5720 (1977).
28. Ronca G, Allegra G *J. Chem. Phys.* **63**: 4990 (1975).
29. Flory P J, Erman B *Macromolecules* **15**: 800 (1982).
30. Langley N R *Macromolecules* **1**: 348 (1968).
31. Ferry J D *Viscoelastic Properties of Polymers*, 2nd edn, Wiley, NY, 1979.
32. Dossin L M, Graessley W W *Macromolecules* **12**: 123 (1979).
33. Pearson D S, Graessley W W *Macromolecules* **13**: 1001 (1980).
34. Meyers K O, Bye M L, Merrill E W *Macromolecules* **13**: 1045 (1980).
35. Gottlieb M, Macosko C W, Benjamin G S, Meyers K O, Merrill E W *Macromolecules* **14**: 1039 (1981).
36. Marucci G *Macromolecules* **14**: 434 (1981).
37. Gaylord R J *Polym. Bull.* **9**: 181 (1983).
38. Ball R C, Doi M, Edwards S F, Warner M *Polymer* **22**: 1010 (1981).
39. Adolf D *Macromolecules* **21**: 249 (1988).
40. Termonia Y *Macromolecules* **22**: 3633 (1989).
41. Termonia Y *Macromolecules* **23**: 1481 (1990).
42. Termonia Y *Macromolecules* **23**: 1976 (1990).
43. Termonia Y *Macromolecules* **24**: 1392 (1991).
44. Mark J E *Adv. Polym. Sci.* **44**: 1 (1982).
45. Green A E, Zerna W *Theoretical Elasticity*, Oxford University Press, London, 1954.
46. Rivlin R S, Saunders D W *Phil. Roy. Soc. (London) A* **243**: 251 (1951).
47. Valanis K C, Landel R F *J. Appl. Phys.* **38**: 2997 (1967).
48. Ogden R W *Proc. Roy. Soc. A* **326**: 565 (1972).
49. Blatz P J, Sharda S C, Tschoegl N W *Proc. Natl. Acad. Sci. (USA)* **70**: 3041 (1973).
50. Flory P J, Erman B *Macromolecules* **15**: 800 (1982).
51. Graessley W W *Adv. Polym. Sci.* **46**: 67 (1982).
52. Edwards S F *Br. Polym. J.* **9**: 140 (1977).
53. Marrucci G *G. Rheol. Acta* **18**: 193 (1979).
54. Ball R C, Doi M, Edwards S F, Warner M *Polymer* **22**: 1010 (1981).
55. Marrucci G *Macromolecules* **14**: 434 (1981).
56. Gaylord R J *Polym. Eng. Sci.* **19**: 263 (1979).
57. Gottlieb M, Gaylord R J *Macromolecules* **20**: 130 (1987).
58. Termonia Y *Macromolecules* **24**: 1128 (1991).
59. Treloar L R G, Riding G *Proc. Roy. Soc. Lond. A* **369**: 261 (1979).
60. Jones D F, Treloar R L G *J. Phys. D.: Appl. Phys.* **8**: 1285 (1975).

CHAPTER 7

Monte Carlo simulations of the free energies and phase diagrams of macromolecular systems

S KUMAR

Abstract

A knowledge of the free energies of macromolecular condensed phase systems is important for studying the range of applicability of analytical theories, as well as for the prediction of their phase equilibrium behaviour. The calculation of free energies for truly macromolecular systems from simulation, however, is associated with serious sampling difficulties. Recently, two different techniques that facilitate the calculation of these quantities from simulation have been proposed and we shall begin by outlining these two approaches, the Rosenbluth method and the chain increment procedure. Subsequently, we show that, although the two methods are completely equivalent to each other from a mathematical viewpoint, the Rosenbluth method is useful for short chains while the chain increment is appropriate for truly macromolecular chains. We then illustrate the usefulness of these methods in understanding the free energies of isolated chains, polymer solutions and blends and pay special emphasis on delineating the effects of free volume on these properties. Finally, we shall examine simulations which attempt to calculate the phase equilibria for macromolecular chains and provide directions for future research.

7.1 Introduction

The knowledge of the phase behaviour of synthetic polymeric systems is important in several practical contexts.[1,2] For example, this information can aid in the design of miscible polymer blends,[3] understanding the swelling of rubber matrices with solvent,[4] the prediction of the θ-temperature of a dilute polymer–solvent mixture,[5-7] the depression of the melting point of a semicrystalline polymer due to the presence of diluents,[8] and the surface segregation of polymer melt chains due to molecular weight effects.[9-11]

The first theoretical effort at modelling the phase equilibria of polymeric systems was devised by Flory[12] and Huggins[13] and this mean-field lattice model, and its recent modifications,[14-16] have found extensive use in this context. While this theory is extremely robust, it cannot be used to obtain quantitative estimates of phase diagrams since the χ parameter, which is assumed to be a system specific constant in the original lattice theory, is normally a complicated function of the chain lengths of the constituent polymers and composition.[7,15–18] More sophisticated off-lattice RISM models[19,20] have recently been proposed to successfully overcome the many approximations associated with the original Flory lattice model. It should be realized, however, that such models have limitations since they invoke appropriate analytical approximations in the derivation of the thermodynamics of the systems of interest.[19]

In contrast to these analytical models computer simulation is a particularly valuable tool since it can be used, in principle, to calculate the dynamic and thermodynamic properties of a molecular system given a form for the interparticle interaction potentials without resort to mathematical approximations.[18,21] While this might imply an undue dependence of the final results on the chosen form of the potentials, it is emphasized that the general long wavelength phenomena observed (e.g. critical points, θ-temperatures, order–disorder transitions) will not depend on the exact details of the interaction potentials. This statement is consistent with the following results derived from simulations:

1. Recent experiments have shown that the χ parameter for a symmetric polymer blend must scale with $1/N$ at an upper critical solution temperature (UCST), where N is the chain length of the polymers in question.[22] This result, which is qualitatively consistent with the predictions of the Flory–Huggins theory, has also been reproduced by computer simulations[23–26] of lattice polymer blends. Since the scaling predictions of the simulations are in agreement with experiment, it is clear that the short range structure and interactions of the polymeric chains cannot play a significant role in determining the chain length dependence of an observable quantity such as the UCST.
2. Kremer and Grest[27,28] have performed extensive simulations to investigate the dynamics of entangled polymer melts. They have shown that the results obtained from the simulations were consistent with reptation. In addition, estimates for self diffusion coefficients, when scaled in an appropriate manner, were in quantitative agreement with those determined experimentally.[28] These results are compatible with this assertion since polymer chains were modelled in these calculations as 'soft-hard' beads connected by springs and all short range structure, such as bond angle bending, torsion angle rotations, as well as the atomistic nature of the monomers, were ignored.
3. Lattice simulations of microphase separations in amphiphilic systems[29–34] and in diblock copolymers[35,36] predict order–disorder

transitions and changes in morphology (i.e. from disordered to spheres, cylinders, lamellae or other bicontinuous structures) with variations in interactions and the chain length. It has been found that the results of these simulations, which ignore local structure by invoking a lattice model, are in qualitative agreement with theoretical and experimental findings on these systems thus emphasizing the relative unimportance of short range effects in this context.[33,35]

From these examples it is clear that simulations using simplistic models qualitatively capture the long wavelength properties of polymeric systems. They, however, cannot yield quantitative predictions of observable properties which are expected to be sensitive to the details in the short range potentials.[37] Recent work in this area[38–42] has shown that incorporation of these short range details do permit a quantitative atomistic modelling of polyethylene, polypropylene, polystyrene, poly(vinylchloride), polycarbonate and polysulphone systems. The point to be stressed here is that one can incorporate models of any desired accuracy in the simulations and the results obtained would be appropriate for the length scale of interest.

The simulations that have been discussed to this point have concentrated on calculating some of the thermodynamic properties of polymeric systems, such as internal energy, and have illustrated that the Monte Carlo method is an extremely efficient vehicle for obtaining these 'mechanical' properties. However, most of them do not attempt to model the phase equilibria of polymeric systems at liquid-like densities. The reason for this is the inherent difficult in obtaining 'non-mechanical' quantities[43] such as free energies, or component chemical potentials from simulation.[44] To illustrate this point we begin with the statistical mechanics definition for the Helmholtz energy for a molecular system,

$$-\beta A = \ln\left[\frac{Q_N}{\Lambda^{3N} N!}\right] \quad [1]$$

where β is the thermodynamic temperature, A is the Helmholtz energy, Λ is the deBroglie wavelength of the particles $[\equiv h/(2\pi m k_B T)^{1/2}]$[45] and Q_N is the contribution of energetic interactions to the partition function for a system of N particles,

$$Q_N = \int \exp[-\beta U(r^N)] \, dr^N \quad [2]$$

Here $U(r^N)$ is the potential energy of the system in configuration $r_1, r_2 \ldots r_N$. While one method for the calculation of free energies, which is very useful for isolated chains on a lattice, is to enumerate exactly the partition function in Eqn [1][46–48] it is clear that this technique is not useful for systems at liquid-like densities. A method to implement Eqn [1] in the context of a computer simulation experiment was illustrated by Mezei and Beveridge,[44] and this technique clearly defines the problems associated with this calculation. We start with the identity,

$$\langle \exp(\beta U_N) \rangle = \frac{\int \exp(\beta U_N) \exp(-\beta U_N) \, d\mathbf{r}_N}{\int \exp(-\beta U_N) \, d\mathbf{r}_N} = \frac{V^N}{Q_N} \quad [3]$$

From here it follows that the excess or configurational free energy is obtained from the relationship,

$$A_E = -k_B T \ln\left[\frac{Q_N}{V^N}\right] = k_B T \ln\langle \exp(\beta U_N) \rangle \quad [4]$$

where A_E is the Helmholtz energy difference between the real system and the ideal gas state at the same temperature and density. While this equation suggests a potential route for the calculation of A_E it also suggests that this procedure is weighted the largest by configurations with large repulsive energies which normally are not sampled in the framework of a typical Monte Carlo simulation utilizing important sampling. Thus this method will not allow for a reliable estimate of system free energy especially at liquid-like densities. In spite of this complication, however, there is a lot of interest in calculating these quantities since the knowledge of free energies is extremely important in determining equilibrium states of molecular systems, especially in the context of protein folding.[44,49,50] Several different methods have therefore been proposed to aid in these calculations, and all of them are related[49] to enumerating the free energy difference between two states 0 and 1, respectively, in a manner that is computationally more tractable than the calculation of absolute free energies.

$$\Delta A = A_1 - A_0 = -k_B T \ln\left[\frac{Q_1}{Q_0}\right] \quad [5]$$

Here Q_1 and Q_0 represent the partition functions corresponding to the two states of interest, and the methods, in general, allow for a detailed evaluation of free energy changes between the states 0 and 1, as long as they are reasonably close to each other. These free energy difference calculations have been performed in the context of a Monte Carlo simulation through thermodynamic integration,[44,51,52] free energy perturbation,[44,50] and the coupling parameter approach[44,53,54] and several recent papers have utilized these ideas to evaluate the free energy differences for biological systems.

While we have summarized an extremely large body of work in the general area of biopolymer simulations and have made no effort to be comprehensive, we would like to stress that all of these calculations have only enumerated the free energy difference between two thermodynamic states and that no attempt has been made to obtain estimates for the absolute values of these quantities. However, there is interest in enumerating the absolute free energies of complex systems since they aid in the critical evaluation of existing thermodynamic models for macromolecular condensed phases and also allow for a direct evaluation of related non-mechanical quantities, such as entropy, which are not easily obtained through the simulation procedures discussed above.

A method to efficiently calculate the absolute free energies for small molecule systems at liquid-like densities has been suggested by Widom who showed that Eqn [1] could be simplified to the form[55,56]

$$-\beta \mu_r = \ln \langle \exp(-\beta U_t) \rangle_N \qquad [6]$$

where μ_r is the residual chemical potential of the molecules ($\equiv \mu - \mu^{ID}$, where μ^{ID} is the chemical potential in an ideal gas at the same temperature and density), and $\langle .. \rangle_N$ represents a canonical ensemble average. The test bead insertion method involves the insertion of a ghost particle into a 'frozen equilibrium snapshot' of a system which contains N particles, and U_t represents the total interaction energy experienced by this test bead. The chemical potential is then obtained by averaging the appropriate Boltzmann factor over many different frozen snapshots of the system, and it has been shown that this method provides an accurate route to determining the chemical potentials of atomic and molecular fluids.[57-63]

This method can then be extended to the case of mixtures, and the chemical potentials of each component in the mixture obtained following a procedure similar to Eqn [6]. With a knowledge of the chemical potentials of the components the free energy of a multicomponent system could then be obtained from the simple relationship,

$$G = \sum_{i=1}^{N} \mu_i x_i \qquad [7]$$

where G is the Gibbs energy of the system, and the x_i's represent the mole fractions of the different species in the multicomponent system.

The calculation of the Gibbs energy of a polymeric system would, in principle, involve the direct extension of this bead insertion scheme to the macromolecular system of interest. This would entail the insertion of a polymer chain into the fluid of interest (which is in a frozen state) and calculating the test particle energy, U_t, associated with this randomly inserted chain. The chemical potential would then be computed following Eqn [6]. Such ideas have been implemented for solutions and melts of relatively short chain length polymers ($v \approx 20$) on a lattice,[64,65] as well as for melts of tangent hard sphere chains in free space ($v \leqslant 15$) by Hall and coworkers.[66–69] It is clear, however, that this method will fail even for relatively short chain length oligomers ($v \approx 15$) at liquid-like densities due to the increasing probability of overlap of the randomly inserted test chain with itself or with the other molecules in the system.[70]

Recently, two different approaches have been proposed to overcome the difficulties associated with the application of the Widom method to the polymer problem. Frenkel and Smit[71,72] and dePablo et al.[73] have proposed a configurational bias Monte Carlo scheme (CBMC), based on earlier work by Rosenbluth and Rosenbluth,[74] Meirovitch,[75] Harris and Rice[76] and Siepmann and Frenkel,[77] that allows for the insertion of chains into a fluid in a manner that samples favourable conformations preferentially. An extension of this CBMC procedure, termed the Rosenbluth method, then allows for the calculation of the chemical potentials of whole chains in a fluid. A complementary technique proposed by Kumar et al.,[79–81] termed the chain increment method, calculates the incremental chemical potential between a homopolymer chain of length v and one of length $v + 1$, $\mu_r(v + 1)$, at any temperature or density,

$$\mu_r(v + 1) = \mu^r_{chain}(v + 1) - \mu^r_{chain}(v) \qquad [8]$$

where $\mu^r_{chain}(v)$ is the residual chemical potential of a chain of length v ($\equiv \mu_{chain} - \mu^{ID}_{chain}$, where μ^{ID}_{chain} is the chemical potential of the chain in an ideal gas state). The residual chemical potential of a chain of length v is then obtained as a sequential sum of incremental chemical potentials of all chains shorter than v, i.e.

$$\mu^r_{chain}(v) = \sum_{i=1}^{v} \mu_r(i) \equiv \sum_{i=1}^{v} [\mu^r_{chain}(i) - \mu^r_{chain}(i - 1)] \qquad [9]$$

We begin here by presenting these two methods, and derive the appropriate statistical mechanics which illustrates that the system free energies computed by these two techniques should be identical in the absence of sampling problems. We will then present numerical examples to show the advantages and weaknesses associated with both methods. The usefulness of these methods in understanding the molecular thermodynamics of polymer melts, blends and solutions will then be examined. Subsequently, the utility of these concepts in delineating the phase equilibria of macromolecular systems will be discussed, and we shall conclude by presenting directions for future research.

7.2 Chain chemical potentials from simulation

7.2.1 The Rosenbluth method[71–73]

Rosenbluth and Rosenbluth[74] originally proposed a configurational bias procedure (CBMC) for lattice polymers which allowed for the calculation of the chemical potential of polymeric systems. This method was extended to free space systems by Meirovitch,[75] and later by Harris and Rice.[76] More recently Frenkel and Smit[71,72] and dePablo et al.[73] have extended these earlier works and presented a more generalized method to obtain chain chemical potentials. It is to be stressed that although previous workers have attempted to implement the Rosenbluth method[74] for off-lattice systems,[82] the work of these two different groups represents, to our knowledge, the first successful realization of this scheme for systems with arbitrary molecular structures and interactions. In this section we outline their procedure, and begin by considering the formal definition of the chemical potential of a chain of length v in a pure homopolymer phase,

$$-\beta\mu^r_{chain}(v) = \ln Z(N+1, v; \beta, V) - \ln Z(N, v; \beta, V) \quad [10]$$

Here μ^r_{chain} is the residual chemical potential of a chain of length v, β is the thermodynamic temperature $[\equiv 1/k_B T]$, V is the system volume, Z is the configurational partition function and N is the number of chains in the system. We now simplify Eqn [10] to a form similar to Eqn [6] following the Widom procedure,[55]

$$\exp(-\beta\mu^r_{chain}) = \left\langle \exp\left(-\beta \sum_{j=1}^{v} U_{j,N+1}\right) \right\rangle = \left\langle \prod_{j=1}^{v} \exp(-\beta U_{j,N+1}) \right\rangle \quad [11]$$

where U_{N+1} $[=\sum U_{j,N+1}]$ is the energy experienced by a test chain of length v that is inserted into a frozen snapshot of the system. Equation [11] suggests that one technique to calculate the chemical potential of a polymer chain is to insert it bead by bead as a test molecule. If one then calculates the Boltzmann factor for beads inserted in a sequential manner then the chemical potential of the whole chain can be determined following Eqn [11]. Clearly, the implementation of the Widom method to the polymer problem would be relatively easy except for the sampling problems created by the relatively small probability of randomly inserting a test chain, without overlap, into a frozen snapshot of the system at liquid-like densities.

To overcome these problems Frenkel and coworkers[71,72] and dePablo et al.[73] have suggested the use of a biased insertion procedure which favours the growth of low energy conformations. The first bead of a chain is inserted at random into a frozen equilibrium snapshot of the simulation cell, and the energy of interaction of this bead with the rest of the system, $U_{1,N+1}$, tabulated. Subsequently k $(1 \leq k \leq \infty)$ trial positions are generated for the next bead of the chain, with the spatial position of these trials obeying any geometric constraints imposed by the architecture of the chains. The energy of each of these trial positions, $U^l_{2,N+1}$ is then generated, and one position, l, is picked with a weight, w_l,

$$w_l = \frac{\exp(-\beta U^l_{2,N+1})}{\sum_{l=1}^{k} \exp(-\beta U^l_{2,N+1})} \quad [12]$$

Subsequent segments of the chain are picked in a similar manner until the chain of desired length is grown. One now has to take care of the weighting given to low energy states in the biased insertion procedure when one calculates the chemical potential of this inserted chain. This can be done following the umbrella sampling procedure, first proposed by Valleau and Torrie,[43] which suggests that if an arbitrary weighting function w is used to sample from an ensemble, then the ensemble average of a state variable, f, can be obtained from the equation,

$$\langle f \rangle_0 = \frac{\left\langle \frac{f}{w} \right\rangle_w}{\left\langle \frac{1}{w} \right\rangle_w} \quad [13]$$

where the subscripts 0 and w represent properties in the real system, and the weighted ensemble, respectively. Equation [11] can now be simplified with the use of Eqns [12] and [13],

$$\exp(-\beta \mu^r_{\text{chain}}) = \frac{\left\langle \exp(-\beta U_{1,N+1}) \times \prod_{n=2}^{v} \left(\frac{1}{w_n} \times \exp(-\beta U^l_{n,N+1}) \right) \right\rangle_0}{\left\langle \prod_{n=2}^{v} \frac{1}{w_n} \right\rangle_0} \quad [14]$$

where it is to be realized that the first chain segment, which was inserted at random, is not weighted by any factor. Substituting for the weight functions finally yields

$$\exp(-\beta \mu^r_{\text{chain}}) = \frac{\left\langle \exp(-\beta U_{1,N+1}) \times \prod_{n=2}^{v} \left[\sum_{l=1}^{k} \exp(-\beta U^l_{n,N+1}) \right] \right\rangle}{k^{v-1}}$$

$$\equiv \frac{\left\langle \prod_{n=1}^{v} R_{n,N+1} \right\rangle}{k^{v-1}} \quad [15]$$

where $R_{n,N+1}$ represents the Rosenbluth weight for the nth segment of the test chain.[71,72] Equation [15], which is an exact result, represents a relatively simple method for the calculation of the chemical potential of a chain molecule with arbitrary interactions both on a lattice, as well as in free space.[71] Further, it is emphasized that this method will only yield reliable values for the chemical potentials in free space when k tends to ∞, a result which can be illustrated by allowing k to assume the other extreme value of unity. In this situation the method reverts to a random insertion procedure, a technique we have asserted to be of little value when considering a long chain length polymeric system at liquid-like densities. A point to be realized is that Eqn [15] automatically defines the chemical potential of chain molecules in a fashion that is describable as a sequential growth of a macromolecule through a bead-by-bead insertion process. A mathematical expression that describes this procedure is

$$\mu_{\text{chain}}^r(v) = \sum_{n=1}^{v} [\mu_{\text{chain}}^r(n) - \mu_{\text{chain}}^r(n-1)] \equiv \sum_{n=1}^{v} \mu_r(n) \qquad [16]$$

where the chemical potential of a chain of length n can be described by an equation similar to Eqn [15]. This point, which has already been alluded to in Eqn [11], is particularly important since it allows us to establish a direct connection between the Rosenbluth method and the Chain Increment method which we shall describe below.

7.2.2 The chain increment method[78–80,83]

While the Rosenbluth method extends the viability of the original Widom method in the case of polymeric systems, it is not expected to be very accurate for truly macromolecular chains ($v \geqslant$ ca. 100) since the probability of overlap of the chain with itself or one of the other molecules in the system, though reduced in the biased insertion procedure used by the Rosenbluth (or CBMC) method, nevertheless does become significant when one proceeds to longer chain lengths.

The chain increment method,[78–80,83] which calculates the incremental chemical potentials between a homopolymer chain of length v and one of length $v + 1$, $\mu_r(v + 1)$, at any temperature or density, allows one to overcome the disadvantages associated with the Rosenbluth technique for the calculation of chain chemical potentials.

$$\mu_r(v+1) = \mu_{\text{chain}}^r(v+1) - \mu_{\text{chain}}^r(v) \qquad [17]$$

Here $\mu_{\text{chain}}^r(v)$ is the residual chemical potential of a chain of length v ($\equiv \mu_{\text{chain}} - \mu_{\text{chain}}^{\text{ID}}$, where $\mu_{\text{chain}}^{\text{ID}}$ is the chemical potential of the chain in an ideal gas state), and it is emphasized that $\mu_r(v + 1)$, the incremental chemical potential is, in general, a function of v. Before we outline a method for the calculation of this quantity we prove that, in the thermodynamic limit, the chemical potential of a whole chain of length v is formally equal to a sequential sum of the incremental chemical potentials of all chains shorter than v under the same conditions.[80] Consider the definition of $\mu^r_{chain}(v)$

$$-\beta\mu^r_{chain}(v) = \ln Z(N+1, v) - \ln Z(N, v) \qquad [18]$$

Here β is the thermodynamic temperature, N is the number of chains in the system and Z is the configurational part of the canonical partition function. This equation can be rewritten as

$$\exp[-\beta\mu^r_{chain}(v)] = \frac{Z(N+1, v)}{Z(N, v, 1, v-1)} \times \frac{Z(N, v, 1, v-1)}{Z(N, v, 1, v-2)} \cdots$$
$$\times \frac{Z(N, v, 1, 1)}{Z(N, v)} \qquad [19]$$

where $Z(N, v, 1, x)$ corresponds to the configurational partition function of a system comprising N chains of length v, and one chain of length x, where $x < v$. One then defines the incremental chemical potential between a chain of length x and one of length $x - 1$ as

$$\exp[-\beta\mu_r(x)] = \frac{Z(N, v, 1, x)}{Z(N, v, 1, x-1)} \qquad [20]$$

which is consistent with Eqn [8]. With this definition Eqn [19] can be rewritten in a form similar to Eqn [9],

$$\mu^r_{chain}(v) = \mu_r(v) + \mu_r(v-1) + \cdots \mu_r(1) \qquad [21]$$

where $\mu_r(x)$ corresponds to an incremental chemical potential between a chain of length x and one of length $x - 1$ at the appropriate temperature and density. To devise a method to evaluate the incremental chemical potential of a chain of length $x - 1$, $\mu_r(x)$, we note that

$$Z(N, v, 1, x) = \int_V \cdots \int_V d\mathbf{r}_1 \ldots d\mathbf{r}_{v(N)} \, d\mathbf{r}_{v(N)+1} \ldots d\mathbf{r}_{v(N)+x}$$
$$\times \exp[-\beta U(\mathbf{r}_1 \ldots \mathbf{r}_{v(N)+(x-1)}) - \beta U(\mathbf{r}_x)] \qquad [22]$$

Here $U(\mathbf{r}_1, \mathbf{r}_2, \ldots, \mathbf{r}_{v(N)+(x-1)})$ is the potential energy of a system comprising N chains of length v and one chain of length $(x-1)$, and U_x is the energy experienced by the xth bead on the $(N+1)$th chain. From Eqns [20] and [22] it then follows that

$$\exp[-\beta\mu_r(x)]$$
$$= \frac{\int_V \cdots \int_V d\mathbf{r}_1 \ldots d\mathbf{r}_{v(N)+x-1} \exp[-\beta U(\mathbf{r}_1 \ldots \mathbf{r}_{v(N)+(x-1)})] \int_V d\mathbf{r}_{v(N)+x} \beta U[(\mathbf{r}_x)]}{Z(N, v, 1, x-1)} \qquad [23]$$

Recognizing that $\exp[-\beta U(\mathbf{r}_1, \ldots \mathbf{r}_{v(N)+(x-1)})]/Z(N, v, 1, x-1)$ is the probability of the occurrence of a system of N molecules of length v and one molecule of length $x-1$ in state $\mathbf{r}_1 \ldots \mathbf{r}_{v(N)+(x-1)}$ finally yields

$$-\beta \mu_r(x) = \ln \langle \exp[-\beta U(\mathbf{r}_x)] \rangle_{(N)v, x-1} \quad [24]$$

Equation [24] suggests that one method for the calculation of the incremental chemical potential of a chain of length x is to consider a system comprising N chains of length v, and one chain of length $x - 1$, where $0 \leqslant x < v$. One considers a frozen snapshot of this system, inserts a bead onto one of the ends of the chain of length $x - 1$ and evaluates $U(\mathbf{r}_x)$ the test bead energy. The incremental chemical potential, $\mu_r(x)$, is then computed by averaging the appropriate Boltzmann factor over many different realizations of the simulated system.

It is emphasized that the application of the chain increment method in this form to calculate the chemical potential of a chain of length v is only valid for pure homopolymer systems, and that the method needs to be modified to deal with copolymer architectures. Several other salient points associated with this technique are listed below:

1. It should be pointed out that although the chain increment method involves the addition of a bead to an end of a homopolymer chain of length $x - 1$, the property that is calculated is not representative of an end alone, but rather one that characterizes the difference in chemical potential between a chain of length x and one of length $x - 1$. This is an important point, which is illustrated clearly in the derivation presented above.

2. Comparison of the chain increment technique to the Rosenbluth procedure suggests that both methods are similar in that they involve a sequential growth of the chain in the medium. The only difference is that the Rosenbluth method grows the whole chain in a frozen medium, whereas the chain increment method grows it a bead at a time in a series of simulations (see point 4 below). In the absence of sampling problems, therefore, the two methods should yield identical estimates for $\mu_{\text{chain}}(v)$.[81] An advantage of the chain increment method over the Rosenbluth technique is that it is applicable for truly macromolecular systems, since the addition of a single bead to a chain of length 100 or 1000 requires the same effort as adding it to a short chain. Thus the method can be used to estimate the incremental chemical potential of long chain length polymers, and therefore permits us, in principle, to calculate the chemical potentials of chains of any desired length.

3. The chain increment method, in its present form, is valid for the addition of a single monomer to a homopolymer chain. It can be seen

that this is a special case of a generalized method where we can add y ($1 \leq y \leq v - x$) monomers to a chain of length x and obtain the incremental chemical potential between a chain of length x and one of length $x + y$. If we set $x = 0$, and $y = v$ it can be seen that the procedure reduces to the insertion of a whole chain of length v into a frozen medium. This is a particularly interesting result since we can now consider a combination of the Rosenbluth and chain increment methods in ways so that longer sequences (i.e. $y > 1$) can be inserted into a chain thus simplifying the calculation of the chemical potentials for long chains.

4. Finally, it is clear that the chain increment method requires one to conduct v different simulations to determine the chemical potential of a chain of length v. Each simulation corresponds to a system comprising N chains of length v, and one chain of length x (where x is varied systematically from 0 to $v - 1$). Although this method seems inefficient in terms of computer time requirements for the calculation of the chemical potential of whole chains, its real advantages can be realized if the incremental chemical potential has a well defined chain length dependence.

From here it is clearly established that both the Rosenbluth method and the chain increment procedure should yield identical estimates for chain chemical potentials. The chain increment technique enjoys the advantage that it can be applied with relative ease to polymers of any length since the addition of a single bead to a chain of any length is simple: however, this advantage is offset by the fact that one has to conduct v simulations to obtain the chemical potential of a whole chain. Although the Rosenbluth method does not face this computational time problem, it is, however, restricted to relatively short chains. In summary we therefore suggest that a combination of the two methods suggested above would allow for the reliable determination of the free energies of macromolecule systems of arbitrary length.

7.3 Results and discussion

7.3.1 Isolated chains

There has been considerable interest in understanding the chemical potentials of polymeric systems at zero density due to the fact that this quantity is related directly to the partition function for such isolated chains,[46–48] i.e.

$$-\beta\mu^r_{\text{chain}}(v) = \ln Z(1, v) \qquad [25]$$

Several years ago Hammersley[84,85] examined the statistical properties of isolated self-avoiding chains on a lattice and showed that the configurational partition function for such chains could be written in the form,

$$Z(1, v) \equiv Z_v = C\mu^v v^{[\gamma - 1]} \qquad [26]$$

where C and γ are constants, and μ is an attrition constant which reflects the number of empty sites neighbouring an end segment on the chain. Similar forms have been suggested for more complex polymer architectures such as star polymers[86–89] and combs,[89] but the verification of these results from simulation are difficult since the random insertion of these complex chains is associated with an increased probability of overlap, as compared to a linear chain of the same length. It is to be stressed that these results are particularly important since they can aid directly in calculating the dimensions of complex polymer architectures as functions of arm length, number of arms and solvent conditions.

While we shall not illustrate any results for complex chain architectures,[86,87] we would like to illustrate the applicability of the concept of

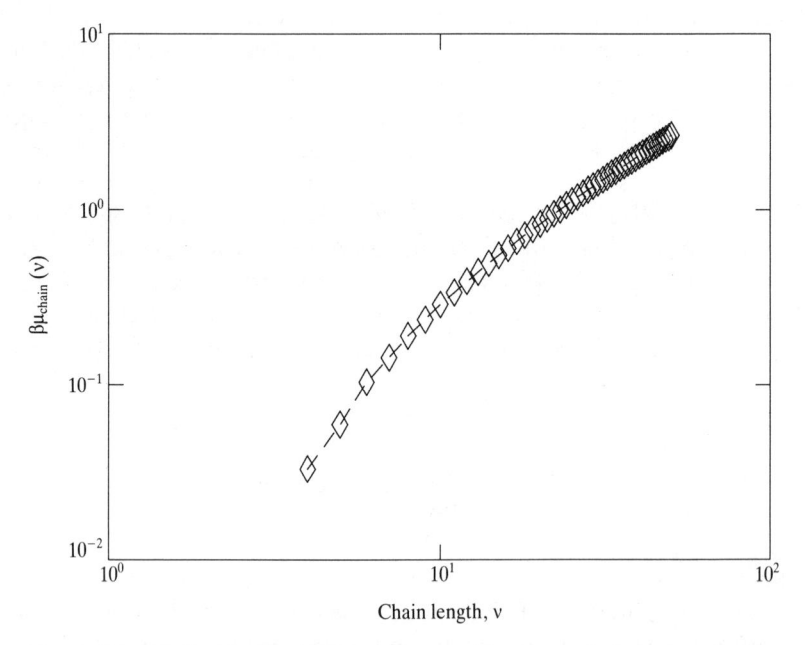

Figure 7.1 Chain chemical potentials as a function of chain length for isolated self-avoiding chains on a lattice. ($\diamond$) are results from the chain increment method, while (-----) are results from the Rosenbluth method. Adapted from Szleifer.[83]

the chain increment and Rosenbluth methods in this context by presenting a few results for linear chains. In Fig. 7.1 we plot the chain chemical potentials for self-avoiding linear chains on a lattice,[83] computed by Rosenbluth as well as the chain increment methods, as a function of chain length on a log–log basis. The lattice simulations were conducted at $T^* = \infty$ (self-avoiding walks). It can be seen clearly that the chemical potentials follow the expected scaling asymptotically for long chain length polymers, although they apparently show strong curvature for polymer chains of relatively short length. In addition, there is quantitative agreement between the two methods for the calculation of chain chemical potentials, in agreement with expectations. Similar findings have been made for polymer chains in free space.[81]

The exponents μ and γ required for detailing the partition function can then be derived using the definition of the incremental chemical potential,

$$\beta\mu_r(v) = -\ln\left[\frac{Z_v}{Z_{v-1}}\right] \approx -\ln\mu - [\gamma - 1]\left[\frac{1}{v}\right] \qquad [27]$$

The right-hand side of this equation, which is valid for $v \gg 1$, suggests that the incremental chemical potential for a self-avoiding chain should be linearly dependent on $(1/v)$, and that the slope of this line yields its chain length dependence. If correct, this is a particularly important result since it shows that the incremental chemical potentials of a few short chains should be sufficient to predict the scaling of chain chemical potentials with their degrees of polymerization. Further, this also allows for an easy determination of the incremental chemical potentials of long chains and therefore the chemical potential of whole chains of any desired length. In Fig. 7.2 we plot incremental chemical potentials for linear chains in free space, $\beta\mu_r(v)$, as a function of $(1/v)^{81}$ at $T^* = 8$ where the chains are swollen. The first fact to be noted is that the incremental chemical potential is a linear function of $1/v$ (for $v \geq 5$) in agreement with the predictions of Eqn [27]. Second, it can be noted that the slope of the line $[\equiv 1 - \gamma]$ is close to zero, in agreement with theoretical predictions.[84] These facts therefore suggest that the chain increment method is an efficient method to obtain the scaling exponents μ and ν necessary to characterize the partition functions for isolated chains. Following a similar methodology, Smit recently has shown that the constants μ and γ in the expression for the partition function for a linear chain on a lattice can be derived accurately.[90]

Similarly, dePablo et al.[73] have examined the chemical potentials of isolated polyethylene-like chains where the methylene groups are modelled

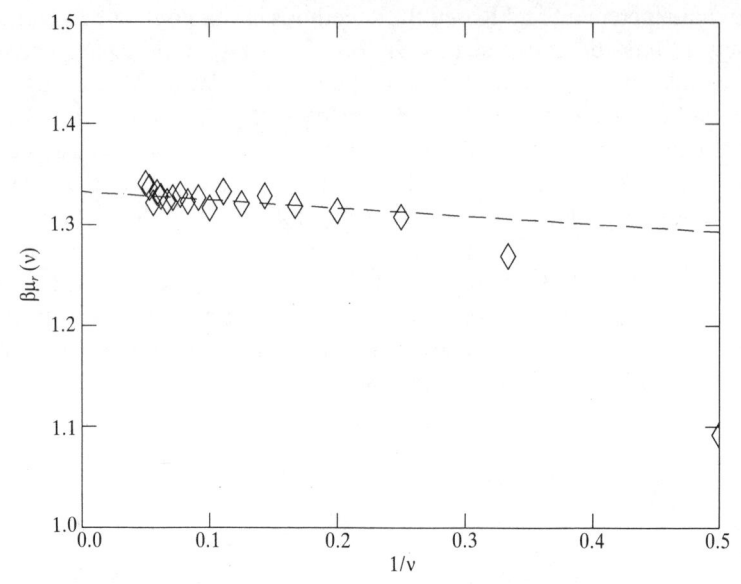

Figure 7.2 Incremental chemical potential, $\beta\mu_r(v)$, as a function of 1/chain length at $T^* = 8$ ($\diamond$) for an isolated single chain in free space using the chain increment method.[81] (-----) represents a best fit straight line.

as united atoms which are connected to adjacent united atoms through a fixed bond angle, and a torsional potential that is characterized by three local minima (corresponding, respectively, to *trans*, *gauche*$^+$, and *gauche*$^-$). It was found that the chemical potentials for these isolated, swollen chains varied as a function of chain length even up to DPs of *ca.* 80, a result that apparently contradicts the findings on the Gaussian, fully flexible chain models employed by Kumar and coworkers.[78] In Fig. 7.3 we replot the data of dePablo *et al.*[73] in a form similar to that suggested by Eqn [27] and show that, although the chemical potentials approach their infinite chain length limit slower than the Gaussian chains utilized by Kumar *et al.*,[78] they nevertheless display a similar scaling relationship with chain length. The cause for this discrepancy in behaviour between the realistic chains and Gaussian models is unclear at this time, but it has been attributed to the fact that the realistic chains, which have more internal constraints, have longer Kuhn segment lengths.

We have also conducted similar studies for collapsed Gaussian chains[81] where the scaling equation of Hammersley is not expected to work in an accurate manner.[84] However, we found that the incremental chemical potentials were linear functions of $(1/v)$ for chains with $v \geqslant 5$ and the slope

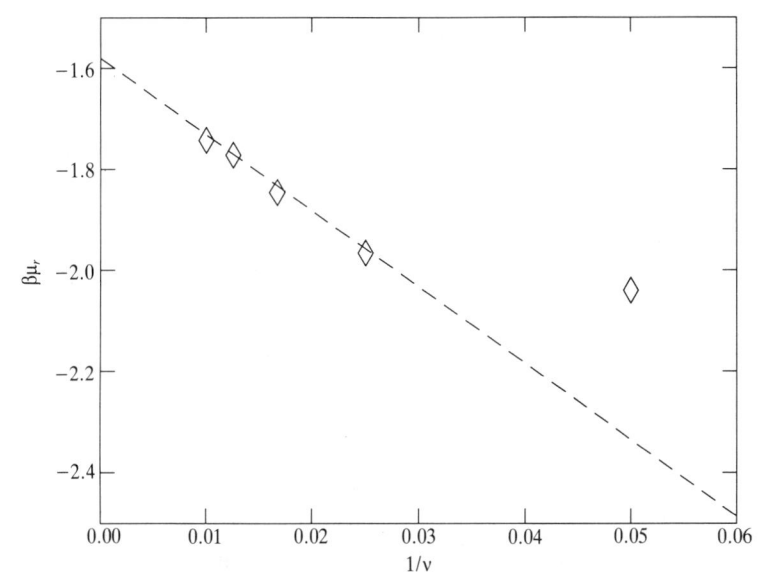

Figure 7.3 Incremental chemical potential, $\beta\mu_r(v)$, as a function of 1/chain length ($\diamond$) for an isolated single realistic polyethylene-like chain in free space using the chain increment method.[78] (-----) represents a best fit straight line. Adapted from dePablo et al.[73]

of the line for collapsed coils ($T^* = 1$) is much larger (≈ 3.0) over the range of chain lengths examined. This apparently illustrates the strong dependence of the chemical potential on chain length in this regime.

The conclusion of these arguments is that the variation of the incremental chemical potential with chain length can be employed to obtain the scaling exponents μ and γ with relative ease. The determination of the corresponding scaling exponents for stars and combs, which are particularly difficult quantities to obtain from traditional simulation procedures, can thus be facilitated if the chain increment method were to be utilized in this context. This work will be the focus of future research in this area.

Further, the incremental chemical potentials for long Gaussian chains that are θ-like or swollen are independent of chain length (to within logarithmic corrections) when one considers chains of length 5 and larger. Under these conditions, therefore, we can estimate the chemical potential of a chain of arbitrary length by multiplying this chain length independent incremental chemical potential (obtained for $v \geqslant$ ca. 5) by the number of beads in the chain, and adding corrections from simulations of short chains for which the incremental chemical potential is a function of chain length. It should be emphasized that such an estimation of the chemical potentials of chain molecules represents a convenient approximation which is only valid for Gaussian chains, and that a procedure following Eqn [9] would always represent the exact route to this quantity.

7.3.2 Polymer melts and blends

The understanding of the chemical potentials and other thermodynamic properties of neat polymer chains has been of considerable interest in the recent past. Sanchez[91] and Kumar[92] have, for example, enumerated the variation of molar volume for polymer melts on a lattice, as well as in free space, as a function of chain length at constant pressure and shown that this quantity obeyed the following relationship:

$$V_i(v) = V_i(\infty) + C_1 \times \frac{1}{v} \qquad [28]$$

Considerable interest has also been devoted to the calculation of chemical potentials for Gaussian chains,[78] realistic chains,[73] as well as for lattice chains[83] in the melt state and it has been found that the chain chemical potentials apparently reach their asymptotic chain length independent values for chains as short as ca. 10. This result is merely a restatement of the Flory theorem on chain dimensions in a melt[93] which states that correlations in a polymer melt are screened for length scales larger than about one monomer size, and suggests that the physics of chains in a melt-like environment are dictated purely by short range effects.

Having explored the relatively uninteresting behaviour of the structure and thermodynamic properties of polymer melts, we now proceed to the more interesting case of polymer blends. Binary polymer blends have also been the focus of intense study in the last few years because they represent systems with controllable physical properties.[94] Although these materials are of considerable practical importance the primary understanding of their thermodynamics has been obtained through the mean-field Flory–Huggins lattice model,[95] which includes several uncontrolled approximations. Recent small angle neutron scattering (SANS) experiments[22] and computer simulations[23] on symmetric polymer blends have shown that the Flory χ parameter must scale as $(1/N)$ at a critical point (specifically the UCST) apparently validating the mean-field predictions.[96]

In spite of the success in predicting the correct chain length dependence of the χ parameter at the critical point, measurements on several miscible off-critical polymer blends have shown that the SANS determined χ parameter (χ_{SANS}) is composition and chain length dependent, apparently in contradiction to one of the primary assumptions of the Flory–Huggins theory.[95] Such results have been obtained experimentally on PS/PVME blends,[17] isotopic blends[15] and blends of ethylene-butene copolymers.[97] In contrast, measurements on isotopic polystyrene blends,[98] as well as some blends of ethylene–butene copolymers, apparently indicate that χ_{SANS} is essentially independent of composition.[99] It is therefore unclear at this time if the Flory–Huggins theory is appropriate to quantitatively model even simple polymer blend systems such as those considered in the SANS experiments.

To resolve this issue we have conducted off-lattice Monte Carlo simulations of symmetric polymer blends under constant pressure, as well as a separate series under constant volume conditions to study the applicability of the Flory theory.[100] These studies also allow for a critical analysis of the effects of equation-of-state (EOS) contributions to the free energy of mixing of simple, symmetric polymer blends. Both blend components were chains of length 25 repeat units, and all bonded beads were held together by harmonic springs of average length σ.[78] All non-bonded beads of these 'Gaussian' chains interacted with a Lennard-Jones potential truncated at 2.5σ.[78] Segments of both chains were assumed to have the same value of σ, the bead diameter, and energy of interaction between like segments (i.e $\varepsilon_{11} = \varepsilon_{22}$, where $\varepsilon_{\alpha\alpha}$ is the Lennard-Jones well depth parameter). However, the energetic interactions between unlike monomers were different and we shall only consider the case where unlike contacts are favoured:

$$\frac{\varepsilon_{12}}{\varepsilon_{11}} = 1.2 \qquad [29]$$

Simulations were conducted at a temperatures, T^* ($\equiv k_B T/\varepsilon_{11}$) of 2 and 2.5, and a pressure P^* [$P\sigma^3/\varepsilon_{11}$] of 0 which corresponds closely to atmospheric pressure, the experimental condition of interest. Since we have shown in separate work using the chain increment method and the Rosenbluth methods[83,101] that the $\mu_r(v)$ for melt chains are essentially independent of chain length [i.e. for $v \geq 3$], we assume that

$$\mu^r_{chain}(v) \approx v \times \mu_r \qquad [30]$$

where μ_r, which represents a chain length independent incremental chemical potential, is the residual segmental chemical potential. This quantity is related to the chemical potential of a chain through the relationship,

$$\mu^{chain}_1 = v_1 \mu^{seg}_1 \approx \ln \phi_1 + v_1 \mu_r \qquad [31]$$

where the first term on the right-hand side represents the ideal gas contribution, and μ^{seg}_1 is the segmental chemical potential. In this context we point to the fact that Flory theory[95] predicts that μ_r for a symmetric

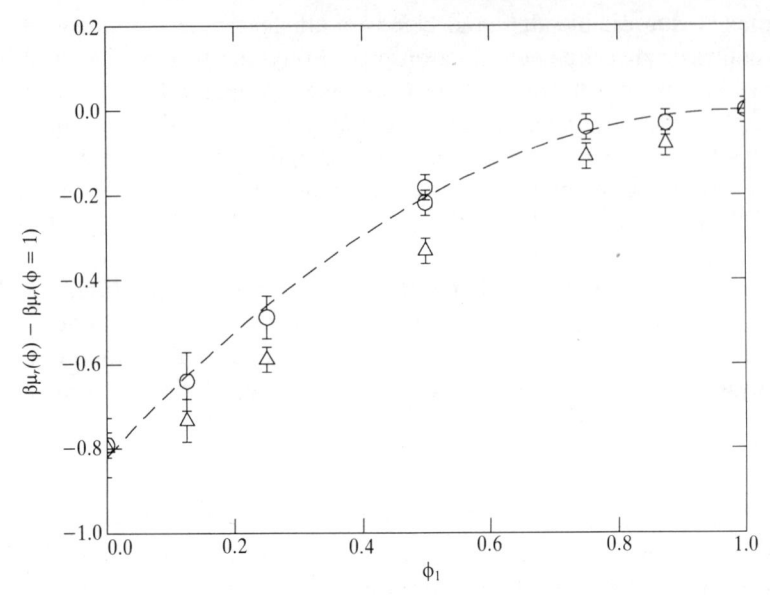

Figure 7.4 Residual chemical potentials for component 1 in a symmetric polymer blend as a function of composition, ϕ_1. ($\bigcirc$) represents results at T^* of 2 and $P^* \approx 0$ in the isothermal–isobaric ensemble, and line is a parabolic fit. ($\triangle$) are results for the same system in the NVT ensemble at $T^* = 2$, and $\rho^* = 0.7$.

blend would assume the functional form

$$\beta\mu_r = \chi'\phi_2^2 \qquad [32]$$

where χ' represents a composition independent interaction parameter. In Fig. 7.4 we show that the blend at $T^* = 2$ and $P^* \approx 0$ apparently obeys this parabolic relationship within simulation uncertainty. This result suggests that the Flory form for the Gibbs energy of mixing with a composition independent χ adequately describes the simulation results at constant pressure.

These results should be contrasted with the findings of Sariban and Binder[26,35] who suggest that the χ parameter, as determined from lattice simulations in the NVT ensemble (where the molar volume of the blend is independent of composition) should be approximately parabolic in composition. To reconcile the findings of the NVT ensemble lattice calculations and our off-lattice simulations in the NPT ensemble, we have separately conducted NVT simulations for our off-lattice symmetric blends where the blend volume was assigned to the pure component molar volume at $T^* = 2$ and $P^* \approx 0$, independent of composition. In Fig. 7.4 it can be seen that these chemical potentials obtained at constant volume cannot be modelled using a Flory form with a composition independent χ. While these simulations apparently are in qualitative agreement with the earlier findings of Sariban and Binder,[35] they surprisingly suggest that the assumptions inherent in the Flory theory make it more appropriate to model systems at constant pressure rather than at constant volume although the initial derivation of the model was actually performed in the canonical ensemble. While the reason for this unusual conclusion is unclear at this time we attribute it to a cancellation of errors which make the predictions of the simple lattice theory appropriate for systems under experimental conditions.

From our simulations it is clearly established that the χ parameters as computed in the constant volume (i.e. independent of composition) and isothermal–isobaric ensembles show different trends, and that the incorporation of EOS effects into the NVT ensemble (to convert it to a constant pressure calculation) actually has the consequence of making the χ parameter composition independent. To understand this effect in more detail we begin by examining the molar volume of the mixture at $T^* = 2$ in the isothermal–isobaric ensemble at $P^* \approx 0$ as a function of composition. As can be seen in Fig. 7.5, the specific volumes of the blend show a negative deviation from ideal mixing. As expected, the specific volume of the blend is also parabolic in composition.[102] The point to be emphasized

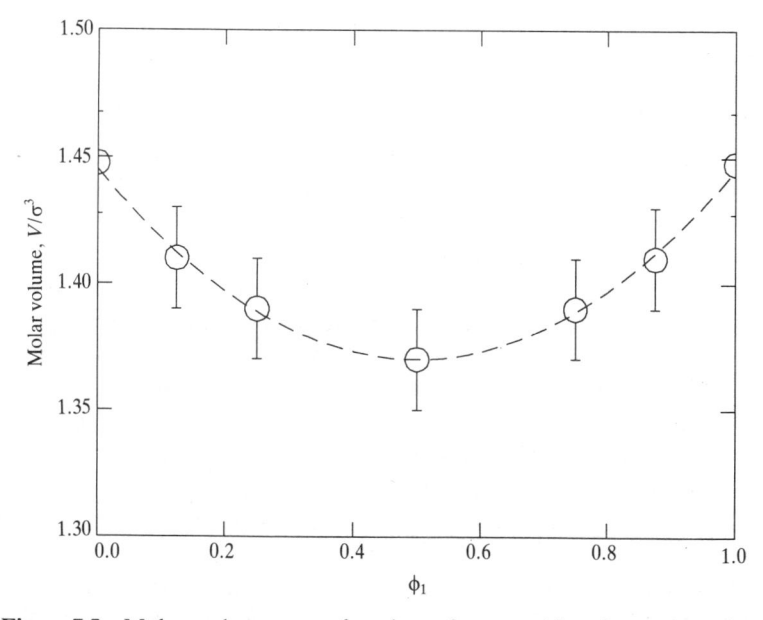

Figure 7.5 Molar volume as a function of composition for symmetric polymer blend shown in Fig. 7.4. (○) represent results at $T^* = 2$ and $P^* \approx 0$

here is that although the Flory theory assumes a zero volume change on mixing, an assumption that is also built into the Sariban and Binder lattice simulations,[35] the blends actually show significant volume changes on mixing when they are considered at constant pressure. We therefore stress that these volume changes on mixing, which are a manifestation of EOS effects, must be directly responsible for the difference in behaviour of the χ parameters determined from the isothermal–isobaric and the canonical ensembles.

We now return to the results in the isothermal–isobaric ensemble and find that χ' assumes values of ca. -0.8 and -0.45 for the blend at the two different temperatures, T^*, of 2 and 2.5, respectively. It should be realized that the numerical values of χ cannot be anticipated a priori since we consider systems in free space. Analysis of the blends shows that the χ' values obtained from the simulations at the two different temperatures cannot be reproduced following the normal definition of the Flory χ parameter, by assuming a constant value of z, the coordination number. While this may represent a general inadequacy of the Flory lattice model,[95] we point to the additional fact that z must change with temperature since the density of the system decreases with increasing temperature. We therefore stress that χ' normally assumes a renormalized value, as has been observed in the critical point simulations of Binder and coworkers,[26] and estimates for this quantity cannot be obtained from the Flory definition of the χ parameter.

In conclusion, we find that the Flory–Huggins form of the Gibbs energy of mixing is adequate in describing these extremely simple polymer blends as long as one considers systems in the isothermal–isobaric ensemble. In contrast, the variation of the Gibbs energy with composition for a system whose molar volume is set to be independent of composition cannot be modelled with a composition independent χ parameter. This represents an unexpected result and suggests that equation-of-state effects, which normally have been ascribed as causes for departure of blend behaviour from Flory theory, result in a composition independent χ parameter in the isothermal–isobaric ensemble. These results also have consequences on the composition dependence of SANS determined χ parameters for polymer blends, and it has been found that χ_{SANS} for this system has a parabolic dependence on composition as shown in Fig. 7.6 due to the fact that the random phase approximation (RPA) which is utilized in data analysis does not incorporate volume changes on mixing.[101]

It should be realized that the simple ideas that are presented here for free energies of symmetric polymer blends, however, are not expected to hold when more realistic blends, which incorporate other factors such as differences in chain stiffnesses[103,104] and segment size and shape disparity,[105,106] are considered. In these situations we expect deviations from Flory theory even in the isothermal–isobaric ensemble.

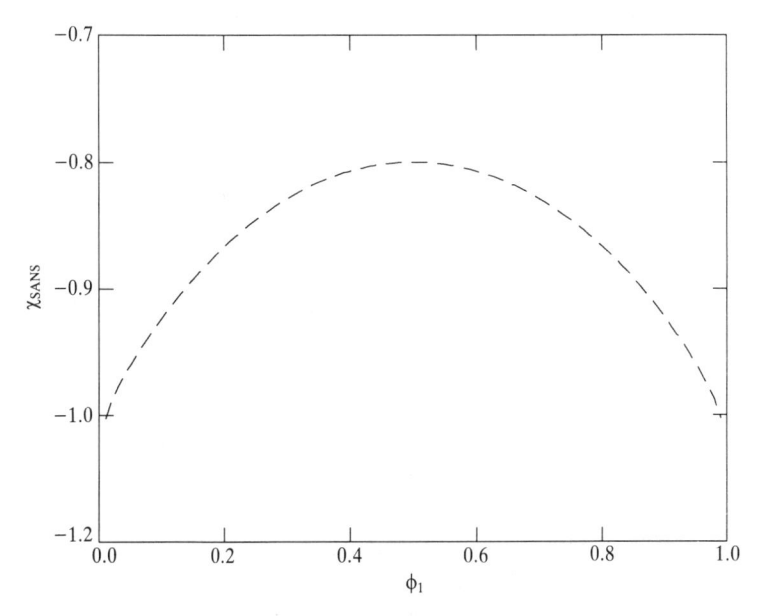

Figure 7.6 Predicted χ_{SANS} as a function of composition for binary blend discussed in Figs 7.4 and 7.5 at $T^* = 2$ and $P^* \approx 0$ (-----).

7.3.3 Chains in solution

The final aspect of the problem of interest relates to the behaviour of chains in solution. This is an area that has received much interest in the past especially in the pioneering work of Okamoto.[65] However, not much interest has been focused on these systems of late from a simulation viewpoint. We begin by considering the case of a bead-spring chain in free space dissolved in its own monomer in the canonical ensemble at $T^* = 2$ (i.e. the density, ρ^*, assumes a value of 0.7 independent of composition). In Fig. 7.7 we plot the incremental chemical potentials of the polymer as computed by the chain increment method, and the chemical potentials of the solvent computed by a standard Widom insertion procedure as a function of composition. It can be seen clearly, in agreement with the results obtained from the polymer blends, that the chemical potentials of both components do not follow a Flory-like form.[12,93] Additionally, it can be seen that the composition dependence of both chemical potentials can essentially be represented by a straight line with the same slope.

To make a connection with the blend results and to elucidate the effects of free volume on the chemical potentials we now consider the variation of the pressure, P^*, with composition for the solutions in question. As shown in Fig. 7.8, the pressure also shows a linear variation with

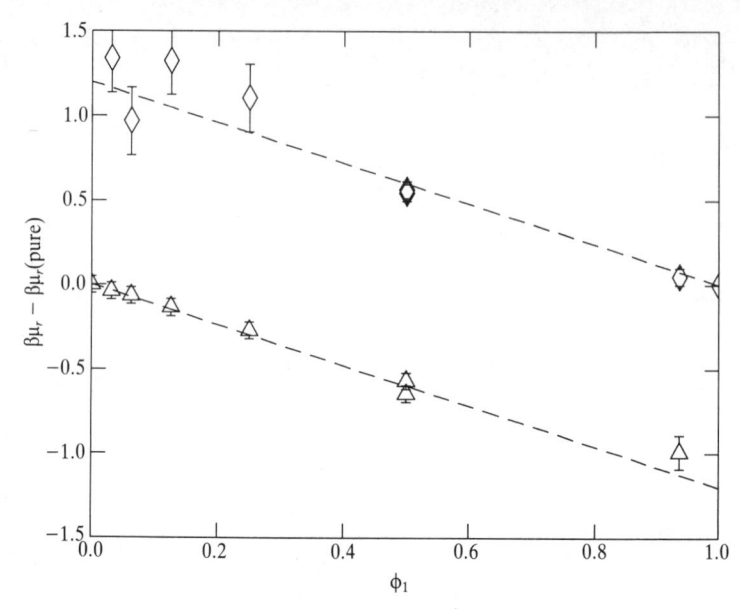

Figure 7.7 Chemical potentials for the components of a polymer solution as a function of composition, where chains of length 20 are dissolved in their own monomer at $T^* = 2$, and $\rho^* = 0.7$. ($\diamond$) are the incremental chemical potentials for these chains computed following the chain increment method,[78] while ($\triangle$) are chemical potentials of the solvent. The lines are linear fits to the data.

composition thus suggesting that a primary reason for the departure of solution behaviour from the Flory theory is the fact that pressure changes with composition in the range of interest. While this result apparently validates the importance of free-volume effects in characterizing the thermodynamics of macromolecular systems, it is to be emphasized that the Flory theory with a composition independent χ parameter will not be adequate to model these simple athermal solutions under constant volume conditions. While these studies therefore emphasize the importance of free-volume effects, we have not conducted simulations at constant pressure to examine the applicability of Flory theory in this context, and this work will be the focus of future research.

A final aspect we would like to focus on are some findings of Smit and Frenkel[71] who suggest that chain chemical potentials, for isolated bead-rod chains dissolved in their own monomer, computed using the Rosenbluth method shows an unexpected kink as a function of chain length for relatively short chains (i.e. for v values as small as 5 for $T^* = 2$ and $\rho^* = 0.6$). All non-bonded beads (including beads on chains, as well

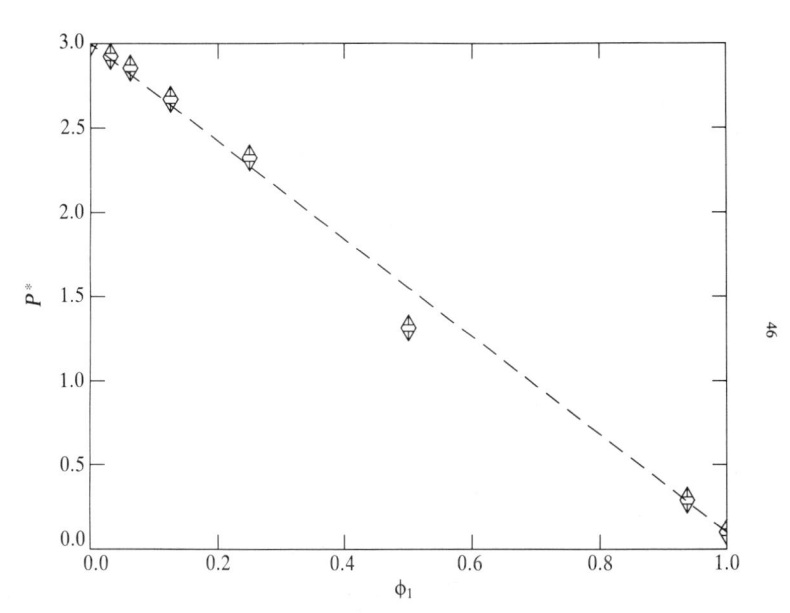

Figure 7.8 Pressure, P^*, as a function of volume fraction for the polymer solutions discussed in Fig. 7.7. ($\diamond$) are simulation results, while (-----) represents a best fit straight line.

as solvent) were assumed to be identical in this case and interacted with a Lennard–Jones potential truncated at its minimum and shifted (the Weeks–Chandler–Anderson potential).[107] Chain chemical potentials computed following the chain increment method (following Eqn [9]), however, do not show this kink (see Fig. 7.9). Recent analytical calculations by Chapman and coworkers[108] also suggest that there should be no kink in the calculated chemical potentials, in agreement with the conclusions of Kumar.[107] Similar results were found for a system where the interactions were assumed to be full Lennard–Jones potentials, rather than the truncated WCA potentials referred to in Fig. 7.9.

To understand the discrepancy between the Rosenbluth method and the chain increment technique, it should be realized that the former method involves the growth of chain in a frozen (or quenched) medium while the latter involves the addition of a bead to a chain that is in equilibrium with its surroundings (an annealed medium). It has been shown for some time that chain conformations are perturbed significantly by the presence of quenched impurities,[109–113] and chains can assume swollen or collapsed conformations, relative to chains in an annealed medium, as a function of the specifics of the interactions between the chain and the medium, as well as chain length. We therefore suggest that chain chemical potentials computed following the Rosenbluth method give rise

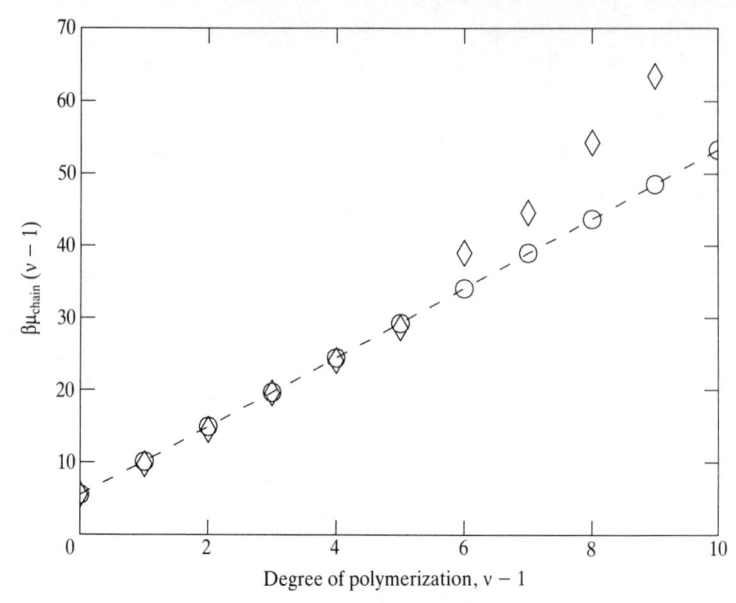

Figure 7.9 Chain chemical potentials as a function of DP for the WCA system at $T^* = 1.2$ and $\rho^* = 0.6$ in (○) annealed and (◇) quenched media. Adapted from Kumar.[107]

to unphysical kinks as a function of chain length due to the fact that chain dimensions in quenched media are perturbed significantly by the presence of the random, frozen solvent molecules.

We illustrate this point by considering the WCA mixture, and in Fig. 7.10 the expansion ratio,[114]

$$\gamma = \frac{R_G^2}{R_{G,\text{ideal}}^2 \times [v-1]^{0.19}} \quad [33]$$

for the quenched and annealed systems are plotted as a function of DP $[\equiv v - 1]$. Here, R_G^2 is the mean-squared radius of gyration of the chains, $R_{G,\text{ideal}}^2$ is the corresponding quantity for an ideal chain of the same length, and the last factor in the denominator accounts for the expansion of chains in a good solvent.[12,93] Here γ should assume a value of unity for chains in a good solvent, independent of chain length. The γ values computed for chains in annealed media apparently obey this relationship for all values of DP examined, in agreement with expectations.[12,93] The expansion ratios for chains in the quenched case are also equal to 1 for chains with DP less than *ca.* 5 at a ρ^* of 0.6. Beyond this value of DP, however, chain dimensions in quenched media are apparently larger than those obtained in annealed media.

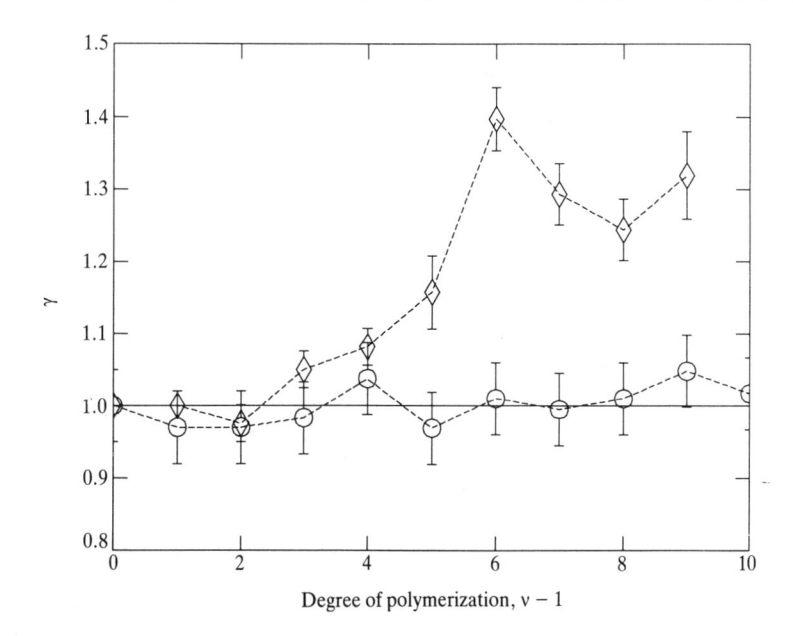

Figure 7.10 Chain expansion ratios as a function of DP for the WCA system at $T^* = 1.2$ and $\rho^* = 0.6$ in (○) annealed and (◇) quenched media. (———) corresponds to the Flory scaling prediction. Adapted from Kumar.[107]

Based on this analysis we therefore stress that thermodynamic properties computed by inserting chains into a quenched medium, such as in the Rosenbluth procedure, do not reflect the properties of chains in a normal condensed phase (which normally represent annealed media). This is particularly important in interpreting the results presented by Smit and Frenkel[71] on the WCA system, where they had found that the chemical potentials as a function of chain length showed an unexpected 'kink'. While these authors have attributed this unusual phenomenon to an unexpected collapse of chains in a good solvent, the analysis presented here suggests that these findings simply represent the shortcomings of the Rosenbluth method, and result because the frozen solvent does not relax to accommodate the chain molecule. This also results in chain dimensions in quenched media which are expanded as compared to a chain under good solvent conditions in an annealed medium. It is thus clearly established that the 'kink' observed by Smit and Frenkel in a plot of $\mu_{chain}(v)$ vs. v, as computed by the Rosenbluth method, does not signal any new physics inherent in the behaviour of polymer chains dissolved in good solvents.

An important question to be reconciled is the difference in behaviour of the off-lattice calculations presented here and the lattice simulations presented by Szleifer and Panagiotopoulos[83] which apparently demonstrate numerically the equivalence of the Rosenbluth and chain increment methods at finite densities and temperatures for polymer–solvent and polymer–polymer systems for chains of DP up to *ca.* 50. We assert based on other work,[115] that a lattice system at a finite density will have properties that are similar to those obtained from a free-space system at a much lower packing fraction. This renormalization occurs due to the finite number of nearest neighbours accessible to a molecule on a lattice, as well as the restrictions placed on the structuring of monomers due to the artificial constraints of a lattice. Due to these facts we suggest that the Rosenbluth method on a lattice will encounter the same problems as the off-lattice system, except that these problems will occur at much higher densities (or longer chain lengths) on a lattice system. As proof of this assertion we point to results of Szleifer and Panagiotopoulos[83] for a polymer melt where it is clear that the results from the chain increment method are systematically somewhat smaller than those from the Rosenbluth method, although the differences are within the uncertainities of the simulations. Although unproven at this time, we therefore suggest that chain properties as computed from the lattice simulation should also yield similar discrepancies between quenched and annealed media, although at much higher densities or chain lengths as compared to a chain in free space.

7.4 Phase equilibrium behaviour

The direct consequence of the calculation of the chemical potentials of molecular systems is in the enumeration of their phase equilibria, a problem which is of much interest from a practical viewpoint. We begin by considering the vapor–liquid equilibrium of short chain molecules, and note that the two most complete pieces of work in this context were performed by Laso *et al.*,[116] who considered realistic models for *n*-alkanes running from pentane to pentadecane using the Gibbs ensemble technique,[117] and by Mooij *et al.*[118] using a similar method for Gaussian chains of lengths up to 15. The Gibbs ensemble technique requires the simulation of two boxes (each representing one equilibrium phase) and requires the exchange of particles between the two cells, in addition to volume changes and equilibration moves for molecules in each box. The novelty of the two approaches referred to above is that the transfer of molecules between two cells is achieved through the use of a Rosenbluth algorithm, so that the probability of chain insertion is enhanced as compared to random insertion. A sample phase diagram for freely jointed

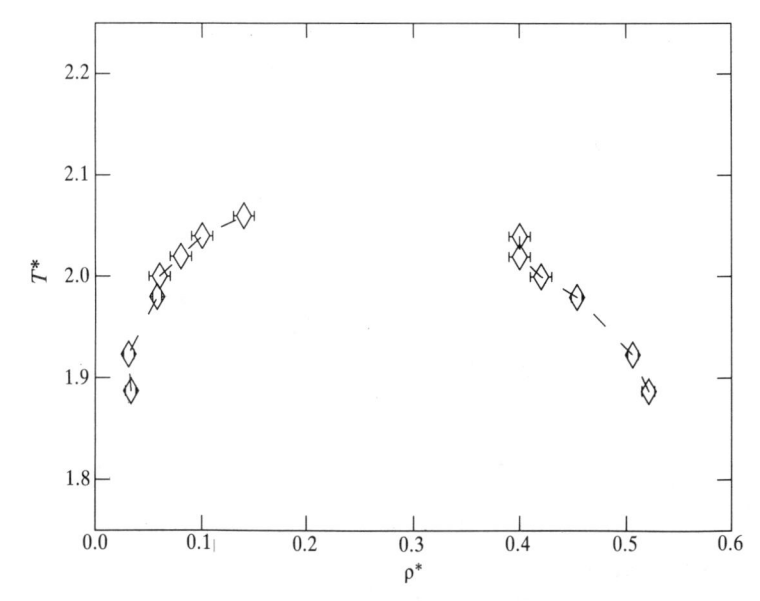

Figure 7.11 Phase diagram from an octamer system calculated using the CBMC procedure coupled to the Gibbs ensemble method. Chains are modelled as beads connected by rods, and all non-bonded beads interact with a Lennard–Jones potential truncated at 2.5σ and shifted. Adapted from Mooij et al.[118]

octamers as adapted from Mooij et al.[118] is then illustrated in Fig. 7.11. As noted by these authors, the critical point for octamers is shifted up by almost a factor of 2 along the temperature axis as compared to a monomer fluid, while the critical composition is moved to lower compositions also in agreement with expectations.

While these results apparently illustrate the viability of the Rosenbluth method for the enumeration of phase equilibria for relatively short chains, it should be realized that such computations cannot be extended to polymer chains of realistic length. This arises since the probability of insertion of a long chain length polymer, though increased through the biased insertion procedure adopted in the Rosenbluth scheme, nevertheless will be extremely small for long polymer chains.

The enumeration of gas–liquid equilibria for longer chains is thus an unsolved problem to date. In this context it is to be emphasized that although such an equilibrium is seldom encountered in practical situations, a knowledge of the critical point for such systems is extremely important from the viewpoint of developing industrial correlations for equation of state constants for use in petroleum industries.[119]

Recently, Panagiotopoulos[120] has utilized the chemical potentials obtained from the chain increment method to obtain an approximate phase diagram for chains of length up to ca. 100. The technique is based on two simplifying assumptions that are listed below. First, it is assumed that the chemical potentials of chains on the gas-phase side of the phase diagram are governed by the ideal gas contribution and are not affected by intermolecular interactions. Further, it has been surmised that the incremental chemical potentials for chains on the liquid side are independent of chain length. It should be realized that this assumption is probably most valid when one considers infinite chain length polymers at the critical point, with any deviations from this limit causing errors associated with this approximation. Under these approximations, however, a phase diagram is obtained as sketched in Fig. 7.12 for Gaussian chains of length 20 and 100, respectively. As with the case of the octamers discussed above, the critical temperature moves to higher values with increasing chain length and to decreasing volume fractions, in agreement with expectations. It is stressed, however, that this phase diagram should only be viewed as a convenient approximation.

Recently Sheng et al.[121] have calculated the chemical potentials of whole chains of length 20 exactly utilizing the chain increment method and Eqn (9) at a variety of densities and temperatures. The 'exact' phase

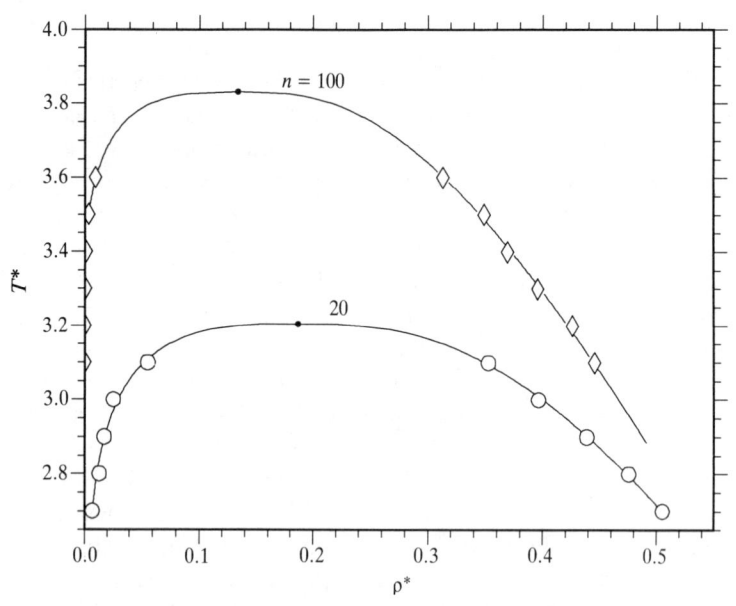

Figure 7.12 Approximate phase diagrams for Gaussian bead-spring chains of length 20 (○) and 100 (◇), respectively. Adapted from Panagiotopoulos.[120]

diagram computed through these chemical potentials is in excellent agreement with the approximate phase diagram for this length computed by Panagiotopoulos.[120] While this represents one isolated example it nevertheless suggests that approximations such as the one utilized above are sufficient to yield phase diagrams of engineering accuracy at an almost negligible computational effort. This area is one of great current interest, and is the field in which progress is expected in the next few years.

A related area of interest would then be phase equilibria between a polymer and a solvent, specifically in the context of solvent sorption by glassy polymers. Very little effort has been devoted in this context except for recent work by dePablo et al.[122] where the Henry's law constant for solvents in polymer matrices, which only relates to the case of an infinite dilution of solvent in the polymer, is enumerated from simulation. The equilibrium solubility of the solvents in the polymers can thus be estimated from this calculation, although it is to be emphasized that little work has been done at finite solvent concentrations. Similar work has also been performed on lattices by Szleifer and Panagiotopoulos.[83]

A final area that is of great interest is the phase equilibria of polymer blend systems. Little work has been done in this area and we point to the pioneering work of Binder and coworkers[26] on the phase diagrams of symmetric polymer blends on a lattice. Corresponding free space results have been obtained recently by Mooij et al.[123] These calculations are based on the simple fact that the particle exchange procedure, which is the difficult part of any of these phase equilibrium calculations, are facilitated in symmetric blends simply by exchanging two polymer chains of different identity between the two boxes. The simulations, alternately, correspond to calculations in the grand canonical ensemble and can be performed with relative ease for chains of length up to ca. 100 under constant volume conditions. Similarly, some results have been obtained for the equilibria blends with asymmetric interactions.[125] However, the general problem of the phase behaviour of polymer blends with chains of arbitrary interactions and sizes (both in length and the monomer size) has not been explored in detail, and remains an unexplored frontier.

In summary, therefore, it is clear that the simulation of the phase equilibrium behaviour of these complex condensed phases has received little interest to date, and the reason for this has been the relative difficulty in performing the insertion move which is necessary for the simulations in the Gibbs ensemble. It is stressed that this field is currently the focus of much research and the use of biased insertion procedures, such as the CBMC move, will greatly aid in these calculations.

Finally, we point to some pioneering work performed by Kofke[125] which allows one, with the knowledge of phase equilibrium at one thermodynamic state, to obtain information regarding equilibria under any other thermodynamic state conditions. This procedure involves the numerical integration of the Gibbs–Duhem equation, and therefore requires one to obtain enthalpies and volumes as a function of temperatures for these systems. The estimation of these quantities is very much simpler than enumerating phase equilibria, and we therefore feel that this method may offer much simplification in the context of polymer phase equilibria.

7.5 Summary

Here we have described recent developments in the computation of the free energies of polymeric systems from computer simulations. We show that this is a particularly difficult problem since the application of methods useful for small molecular systems, such as the Widom test particle technique, are associated with serious sampling difficulties when extended directly to macromolecular systems at liquid-like densities. The two methods discussed, the Rosenbluth technique and the chain increment method, attempt to circumvent the problems associated with the Widom method. We show here that the chain increment method, though computationally far more expensive than the Rosenbluth method, is the only reliable method to determine chain chemical potentials for polymers of arbitrary length and density. The application of these methods to polymeric systems allow us, for example, to calculate the scaling exponents required to describe the partition functions for isolated chains at different solvent conditions, and offers the possibility of determining the corresponding quantities for complicated chain architectures such as stars and combs. They also allow for the systematic evaluation of the effects of free volume (or EOS effects) on the thermodynamics of polymer systems and we show that the inclusion of compressibility effects apparently plays an extremely critical role in determining the applicability of the classic Flory–Huggins theory to polymer blend systems. Finally, we have considered the phase equilibria of polymeric systems and suggest that, although some progress has been achieved, this promises to be an area where significant development will be made in the next few years.

Acknowledgements

The author would like to gratefully acknowledge Prof. K. Binder (Mainz), Dr T. Cagin (Molecular Simulations Inc.), Prof. W. Chapman (Rice University), Prof. J. J. dePablo (University of Wisconsin), Dr D. Frenkel (FOM Institute, Amsterdam), Prof. S. L. Hsu (University of Massachusetts), Prof. D. Kofke (SUNY, Buffalo), Prof. K. E. Gubbins (Cornell University), Prof. H. Meirovitch (Florida State), Dr R. G. Larson (AT&T Bell

Laboratories), Prof. J. M. Prausnitz (University of California, Berkeley), Prof. I. C. Sanchez (University of Texas), Prof. K. S. Schweizer (University of Illinois), and Prof. U. W. Suter (ETH, Zurich), for providing reprints and preprints of their work. The author would also like to acknowledge a long time collaboration and stimulating discussions with Prof. A. Z. Panagiotopoulos (Cornell) and Prof. I. Szleifer (Purdue University). Jeff Weinhold spent much time reading this document and finding even the smallest typographical error. I would like to thank him for his help. The financial assistance of the American Chemical Society, Petroleum Research Foundation is also gratefully acknowledged.

References

1. For example, see: *Monte Carlo Method in Statistical Physics*, Binder K (ed.), Springer-Verlag, 1992.
2. Gubbins K E *Applications of Molecular Theory to Phase Equilibrium Predictions*, Chapter 6 in *Thermodynamic Modelling*, Sandler S (ed.), Dekker, New York, 1992.
3. Flory P J *Principles of Polymer Chemistry*, Cornell University Press, 1953; Freed K F *Renormalization Group Theory of Macromolecules*, Wiley, New York, 1987.
4. deGennes P G *Scaling Concepts in Polymer Physics*, Cornell University Press, Ithaca, NY, 1979.
5. deGennes P G *Phys. Lett A* **38**: 339 (1972).
6. des Cloizeaux J *J. Chem. (Paris)* **36**: 281 (1975).
7. Cherayil B J, Douglas J F, Freed K F *J. Chem. Phys.* **83**: 5293 (1985); Kholodenko A L, Freed K F *J. Chem. Phys.* **78**: 7341 (1983).
8. Hoffman J D, Davis G T, Lauritzen J I *Treatise in Solid State Chemistry*, Vol. 3, Hannay N B (ed.), Plenum Press, New York, 1976.
9. Composto R J, Stein R S, Kramer E J, Jones R A L, Mansour A, Karim A, Felcher G P *Physica B* **156–157**: 434 (1989).
10. Jones, R A L, Norton L J, Kramer E J, Composto R J, Stein R S, Russell T P, Karim A, Mansour A, Felcher G P *Europhys. Lett.* **12**: 41 (1989).
11. Hariharan A, Kumar S K, Russell T P *Macromolecules* **23**: 3584 (1990); *Macromolecules* (1990).
12. Flory P J *J. Chem. Phys.* **9**: 660 (1941); Flory P J *J. Chem. Phys.* **10**: 51 (1942).
13. Huggins M L *J. Phys. Chem.* **46**: 151 (1942).
14. Szleifer I *J. Chem. Phys.* **91**: 6940 (1991).
15. Bates F S, Muthukumar M, Wignall G D, Fetters L J *J. Chem. Phys.* **89**: 535 (1988).
16. Muthukumar M *J. Chem. Phys.* **85**: 4722 (1986).
17. Han C C, Bauer B J, Clark J C, Muroga Y, Matsushita Y, Okada M, Tran-cong Q, Chang T, Sanchez I C *Polymer* **29**: 2002 (1988).
18. Fan C F, Olafson B O, Blanco M, Hsu S L *Macromolecules* **25**: 3667 (1992).
19. Schweizer K S, Curro J G, *J. Chem. Phys.* **89**: 3342 (1988); Schweizer K S, Curro J G *J. Chem. Phys.* **89**: 3350 (1988); Schweizer K S, Curro J G *J. Chem. Phys.* **87**: 1842 (1987); Schweizer K S, Curro J G *Phys. Rev. Lett.* **58**: 246 (1987).

20. Schweizer K S, Curro J G, Grest G S, Kremer K *J. Chem. Phys.* **91**: 1357 (1989).
21. Metropolis N, Rosenbluth A W, Rosenbluth M N, Teller A H, Teller E J *J. Chem. Phys.* **21**: 1987 (1953).
22. Gehlsen M D, Rosedale J H, Bates F S, Wignall G D, Hansen L, Almdal K *Phys. Rev. Lett.* **68**: 2452 (1992).
23. Deutch H-P, Binder K *Europhys. Lett.* **17**: 697 (1992).
24. Sariban A, Binder K *Macromolecules* **24**: 578 (1991).
25. Sariban A, Binder K, Heermann D W *Phys. Rev. B* **35**: 6873 (1987).
26. Kremer K, Binder K *Comput. Phys. Rep.* **7**: 259 (1988).
27. Kremer K, Grest G S, Carmesin I *Phys. Rev. Lett.* **61**: 566 (1988).
28. Kremer K, Grest G S *J. Chem. Phys.* **92**: 5057 (1990).
29. Larson R G, Scriven L E, Davis H T *J. Chem. Phys.* **83**: 2411 (1985).
30. Larson R G *J. Chem. Phys.* **89**: 1642 (1988).
31. Larson R G *J. Chem. Phys.* **91**: 2479 (1989).
32. Larson R G *J. Chem. Phys.* **96**: 7904 (1992).
33. Mackie A D, O'Toole E M, Hammer D A, Panagiotopoulos A Z *Flu. Ph. Eq.*, in press, 1992.
34. Larson R G *J. Chem. Phys.* **96**: 7904 (1992).
35. Minchau B, Duenweg B, Binder K *Polym. Commun.* **31**: 348 (1990).
36. Binder K *Monte Carlo Studies of Collective Phenomena in Dense Polymer Systems*, Chapter 3 in this book.
37. Jaffe R L, Yoon D Y, McLean A D *Computer Simulation of Polymers*, Roe R-J (ed.), Prentice-Hall, 1991.
38. Boyd R H, Pant K *Computer Simulation of Polymers*, Roe R-J (ed.), Prentice-Hall, 1991.
39. Sumpter B G, Noid D W, Wunderlich B, Cheng S Z D *Macromolecules* **23**: 4671 (1990).
40. Winkler R G, Ludovice P J, Yoon D Y, Morawitz H *J. Chem. Phys.* **95**: 4709 (1991).
41. Boone T D, Theodorou D *Preprint*, 1992.
42. Fan C F, Hsu S L *Macromolecules* **24**: 6244 (1991).
43. Valleau J P, Torrie G M *Modern Theoretical Chemistry 5. Statistical Mechanics*, Berne B (ed.), Plenum Press, New York, 1977.
44. Mezei M, Beveridge D L *Ann. New York Acad. Sci.* **482**: 1 (1986).
45. Huang K *Statistical Mechanics*, Wiley, New York, 1987.
46. Mazur J, McCracken F L *J. Chem. Phys.* **49**: 648 (1968).
47. Baumgartner A *J. Chem. Phys.* **72**: 873 (1980).
48. Kremer K, Baumgartner A, Binder K *J. Phys. A: Math. Gen.* **15**: 2879 (1981).
49. Mezei M *Proton Transfer in Hydrogen-Bonded Systems*, Bountis T (ed.), Plenum Press, New York, 1992.
50. Reynolds C A, King P M, Richards W G *Mol. Phys.* **76**: 251 (1992).
51. Watson B S, Chao K-C *J. Chem. Phys.* **96**: 9046 (1992).
52. Bennett C H in *Diffusion in Solids: Recent Developments*, Nowick A S, Burton J J (eds) Academic Press, New York, 1975.
53. Mon K K, Griffiths R B *Phys. Rev. A* **31**: 956 (1985).
54. Swope W C, Anderson H C *J. Phys. Chem.* **88**: 6548 (1984).
55. Widom B *J. Chem. Phys.* **39**: 2808 (1962)
56. Jackson J L, Klein L S *Phys. Flu.* **7**: 228 (1964).

57. Shing K S, Gubbins K E *Mol. Phys.* **43**: 717 (1981).
58. Shing K S, Gubbins K E *Mol. Phys.* **46**: 1109 (1982).
59. Panagiotopoulos A Z, Suter U W, Reid R C *Ind. Eng. Chem. Fundam.* **25**: 525 (1986).
60. Guillot B, Guissani Y *Mol. Phys.* **54**: 255 (1985).
61. Powles J G *Chem. Phys. Lett.* **86**: 335 (1982).
62. Powles J G, Evans W A B, Quirke N *Mol. Phys.* **46**: 1347 (1982).
63. Randelman R E, Grest G S, Radosz M *Mol. Simul.* **2**: 69 (1989).
64. Okamoto H *J. Chem. Phys.* **64**: 2686 (1976); *J. Chem. Phys.* **79**: 3976 (1983); *J. Chem. Phys.* **83**: 2587 (1985).
65. Okamoto H, Itoh K, Araki T *J. Chem. Phys.* **78**: 975 (1983).
66. Dickman R, Hall C K *J. Chem. Phys.* **89**: 3168 (1988).
67. Honnell K G, Hall C K *J. Chem. Phys.* **90**: 1841 (1989).
68. Honnell K G, Dickman R, Hall C K *J. Chem. Phys.* **87**: 664 (1987).
69. Croxton C A *Phys. Lett. A* **70**: 441 (1979).
70. Frenkel D, Mooij G C A M, Smit B *J. Phys. Condens. Matter* **3**: 3053 (1992).
71. Siepmann J I *Mol. Phys.* **70**: 1145 (1990); Mooij G C A M, Frenkel D *Mol. Phys.* **74**: 41 (1991).
72. Frenkel D, Smit B *Mol. Phys.* **75**: 983 (1992).
73. dePablo J J, Laso M, Suter U W *J. Chem. Phys.* **96**: 6157 (1992).
74. Rosenbluth M N, Rosenbluth A W *J. Chem. Phys.* **23**: 356 (1955).
75. Meirovitch H *Phys. Rev. A* **32**: 3699 (1985).
76. Harris J, Rice S A *J. Chem. Phys.* **88**: 1298 (1988).
77. Siepmann J I, Frenkel D *Mol. Phys.* in press, (1992).
78. Kumar S K, Szleifer I, Panagiotopoulos A Z *Phys. Rev. Lett.* **66**: 2935 (1991).
79. Kumar S K *J. Chem. Phys.* **96**: 1490 (1992).
80. Kumar S K, Szleifer I, Panagiotopoulos A Z *Phys. Rev. Lett.* **68**: 3658 (1992).
81. Kumar S K *Fluid Ph. Equil.* **83**: 333 (1993).
82. Smith N C, Fleming R J *J. Phys. A Math. Gen.* **8**: 929 (1975).
83. Szleifer I, Panagiotopoulos A Z *J. Chem. Phys.* **97**: 6666 (1992).
84. Hammersley J M *Quart. J. Math. Oxford* **12**: 250 (1961).
85. An excellent review of this can be found in des Cloizeaux J, Jannink G *Polymers in Solution*, Oxford, 1991.
86. Ohno K, Binder K *J. Stat. Phys.* **64**: 781 (1991).
87. Ohno K, Binder K *J. Chem. Phys.* **95**: 5444 (1991).
88. Grest, G S, Kremer K, Witten T A *Macromolecules* **20**: 1276 (1987); ibid. **22**: 1904 (1989).
89. Lipson J E, Whittington S G, Wilkinson M K, Marin J L, Gaunt D S *J. Phys. A* **18**: 649 (1985).
90. Smit B Personal communication, 1992.
91. Sanchez I C Personal communication, 1992.
92. Kumar S K Unpublished data, 1992.
93. Flory P J *Principles of Polymer Chemistry*, Cornell University Press, Ithaca, NY, 1953.
94. Shibayama M, Yang H, Stein R S, Han C C *Macromolecules* **18**: 2179 (1985).
95. Huggins M L *J. Chem. Phys.* **9**: 440 (1941).
96. Murray C T, Gilmer J W, Stein R S *Macromolecules* **18**: 996 (1985); Hadziioannou G, Stein R S *Macromolecules* **17**: 567 (1984).

97. Balsara N P, Fetters L J, Hadjichristidis N, Lohse D J, Han C C, Graessley W W, Krishnamoorti R *Macromolecules* **25**: 6137 (1992).
98. Schwann D, Hahn K, Streib J, Springer T *J. Chem. Phys.* **93**: 8383 (1990).
99. Krishnamoorti R, Graessley W W Personal communication, 1992.
100. Kumar S K Manuscript in preparation.
101. Kumar S K *Macromolecules*, in press.
102. Prausnitz J M, Lichtenthaler R N, de Azevedo E G *Molecular Thermodynamics of Fluid-Phase Equilibria*, Prentice-Hall, New York, 1986.
103. Liu A, Fredrickson G *Macromolecules* **25**: 5551 (1992).
104. Bates F S, Schulz M F, Rosedale J H, Almdal K *Macromolecules* **25**: 5547 (1992).
105. Schweizer K S, Curro J G *Macromolecules* **23**: 1402 (1990).
106. Frenkel D, Louis A A *Phys. Rev. Lett.* **68**: 3363 (1992).
107. Kumar S K submitted to *J. Chem. Phys.* 1992.
108. Ghonasgi D, Llano-Restrepo M, Chapman W G submitted to *J. Chem. Phys.* 1992.
109. Edwards S F, Muthikumar M *J. Chem. Phys.* **89**: 2435 (1988).
110. Muthukumar M *J. Chem. Phys.* **90**: 4594 (1989).
111. Cates M E, Ball R C *J. Phys. Paris* **49**: 2009 (1988).
112. Baumgartner A, Muthukumar M *J. Chem. Phys.* **87**: 3082 (1987).
113. Kremer K *Z. Phys. B.* **45**: 149 (1981).
114. Baumgartner A *J. Chem. Phys.* **72**: 873 (1980); Kremer K, Baumgartner A, Binder K *J. Phys. A: Math. Gen.* **15**: 2879 (1981).
115. Yethiraj A, Schweizer K S Personal communication, 1992.
116. Laso M, dePablo J J, Suter U W *J. Chem. Phys.* **97**: 2817 (1992).
117. Panagiotopoulos A Z *Mol. Phys.* **61**: 813 (1987).
118. Mooij G C A M, Frenkel D, Smit B *J. Phys. Conden. Matter* **4**: L255 (1992).
119. Tsonopoulos C *Flu. Ph. Equil.*, to appear, 1992.
120. Panagiotopoulos, Personal communication, 1992.
121. Sheng J, Panagiotopoulos A Z, Kumar S K Manuscript in preparation, 1992.
122. dePablo J J, Laso M, Suter U W *Flu. Ph. Eq.*, in press, 1992.
123. Mooij G C A M, Frenkel D, Szleifer I, Panagiotopoulos A Z, Private communication, 1992.
124. Deutsch H-P, Binder K *Makromol. Chemie. Makromol. Symp.*, in press, 1992.
125. Kofke D Submitted to *Mol. Phys.* 1992.

CHAPTER 8

Computer simulation of polymer network formation

B E EICHINGER AND O AKGIRAY

Abstract

Gelation and high elasticity are two manifestations of polymer network formation and structure. The classical Flory–Stockmayer theory of gelation has been of vital importance in understanding the basic features of this important phase transition, but modern techniques relating to renormalization group theory have supplanted classical theories for descriptions of critical phenomena. Percolation simulations have shown that the classical theory is not accurate in the vicinity of the gel point. Off-lattice simulations are advocated here. We describe a simple modelling technique that encompasses significant molecular detail so that real chemical examples can be treated with accuracy. Several validation studies show that the method is accurate, and where there are discrepancies the model can shed light on the chemistry. We end with example results for bimodal networks, RTV adhesives and sulphur cured natural rubber, and show how modelling can be used to aid in both understanding and designing these very complicated network systems.

8.1 Introduction

Three-dimensional random networks of polymer molecules are formed in a variety of circumstances. The first important commercial application of network formation was the vulcanization of natural rubber. The addition of sulphur crosslinkages between the poly(*cis*-isoprene) chains of natural rubber converts the creepy, gummy polymer into a much more useful semi-rigid, non-tacky material. Many other examples of crosslinking reactions may be cited, such as those of paints and coatings, thermosetting adhesives, the clotting of blood, the gelation of agar, and so on. The introduction of bonds between chains, be they covalent bonds as in the case of sulphur vulcanization or physical bonds, e.g. hydrogen bonds or hydrophobic interactions, as in the gelation of agar, inevitably leads to the formation of a random network of chains.

Upon recognizing that network formation is occurring in a polymerization reaction, either by design or inadvertently, it behoves the chemist to understand the conditions leading to the formation of large branched molecules. In those cases where network structure is something to be sought rather than to be avoided, the chemist also desires to apply the time-honoured principle that an elucidation of structure will lead to a better understanding of properties. It is not difficult to understand the essential conditions for network formation: branches can form if and only if the repeat units are capable of forming more than two bonds with their neighbours. The elucidation of structure, however, is a much more difficult problem.

A molecule or aggregate that qualifies as a random network contains many different extended sequences of concatenated bonds (covalent or physical) that proceed from one portion of the structure to another and these sequences of bonds are predictable in statistical terms only. This definition does not require that a random network be large, and it is convenient for our purposes not to draw distinction between large and small networks. Furthermore, the statistical nature of the process of bond formation guarantees that any sample of material consisting of random networks will contain molecules that have different connectivities or topologies. These characteristics of random network formation are the bane of chemistry. The chemist who is used to dealing with pure compounds, or even with clean polymers with molecular weight distributions, may shudder at the thought of systems that can only be described in statistical terms, that do not have isolatable pure components, and which cannot be analysed with step-by-step analytical procedures.

Polymer chemists have long since given up the notion of working with pure polymeric compounds, so this feature of random network structure should not be daunting. More troublesome is the absence of a simple nomenclature that captures the essential structural characteristics of random networks in a way that can be easily conveyed and understood. In contrast, for linear polymers we convey structural information inductively, for example, $-(-CH_2CH\phi-)_n-$ signifies all linear polystyrene molecules. The absence of similar nomenclature for random systems need not impede us, however. Physical chemists are used to dealing with systems that can only be described in statistical terms. Take, for example, the instantaneous configuration of an ideal gas. We do not usually go through the exercise of writing out a typical configuration of coordinates of the molecules, but we all know that such a list would generate a complete description of the system at some instant of time. That one configuration would be no more representative of the system than any other, and it would in fact be misleading to describe the gas in terms of any one configuration.

Given these observations, statistical methods must be applied to the description of an ideal gas, or any other disordered system. In the present instance, these statistical methods are to be applied to random networks. The first real progress in this direction was made in the 1941 gelation theory of Flory.[1-3] By assuming that no intramolecular reactions occur in the condensation of bifunctional and trifunctional units, Flory was able to use elementary statistical methods to show that the critical condition for formation of an infinite molecule, or gel phase, could be simply deduced from the stoichiometry of the reaction. All that was needed for the calculation was the simple, but brilliant, observation that at the gel point there is unit probability that any given branch point will be succeeded by another branch point, and that one by another, and so on to infinity. This work spurred fifty years of research into the theory of gelation and network formation, and the job is still not finished.

Given this success, why should we be further interested in the problem? The reason is that the theory is not without approximations, just as the ideal gas state is only an approximation to real gases. In our case, the limitation that is posed by statistical theories of network formation is that they inevitably discard intramolecular reactions at one or another level of approximation. The necessity for doing so in order to make theory simple is essentially contained in the following trivial observation: every intermolecular bond reduces the number of molecules by one, whereas intramolecular bond formation conserves the number of molecules. Thus, intermolecular reactions add to or create branches, and intramolecular reactions create circuits. By allowing only intermolecular reactions in these polycondensation or crosslinking reactions, there is a strict correspondence between the number of reacted groups and the number of molecules. When intramolecular reactions are allowed, this correspondence is lost, and that makes life very difficult indeed. Theorists have been able to correct their calculations for the intramolecular reactions that lead to the formation of small circuits, called loops, but have been unable to contend with the vast array of larger circuits that are expected to form in typical branching or crosslinking reactions. These several investigations will be briefly reviewed in the next section.

The traditional analytical theories of gelation have been augmented in more recent years by percolation theory, and a great many papers have been published on the connection between gelation, percolation, universality classes, renormalization group theory and related topics. Critical ideas have emerged from renormalization group theory that are essential for an understanding of the problems with classical gelation theory. Considerable effort has also been devoted to computer simulations of percolation as a model of gelation, and this work will be briefly considered under this heading.

Returning now to a more chemical point of view, our intuitive description of random networks is only intelligible to the extent that we understand what is meant by the words connectivity, statistical, topology, etc. Whatever image we might form of the structure of a random net, it is imperative that we be able to encode these structures in a conventional and reproducible fashion. Now, by the definition given above, a mixture of, say, branched alkanes would qualify as a random network. If the alkanes are sufficiently small, the IUPAC conventions for chemical nomenclature may be used effectively to describe their individual structures. However, this scheme quickly becomes excessively burdensome as the molecular weight of the molecules grows, and we must look to other methods to describe their structures. Happily, the mathematicians have a well developed graph theory that provides the tools needed for this task. We will review the parts of graph theory that are useful for our chemical problem in the fourth section.

The problem that one faces in trying to figure out the rules for network formation is that the interesting correlations are of arbitrarily large spatial extent. Whether a circuit containing, say, 23 chains is formed or not depends upon a sequence of bond formation events that link together the 23 chains in a linear sequence with no smaller circuit being formed within them prior to the placement of the last bond in the 23 that closes the circle. The possibilities are immense for all kinds of structures forming before the one of interest, and this makes for a nightmare of interdependent statistics that has not yielded to pencil and paper. However, given a respectable model for the structure of amorphous polymers and some good guesses about the bond formation process, it is possible to come up with computer models that do a very good job of simulating the statistical bond formation for us. The structures that are encoded in the mathematical description are not trivial to decode, but much information of use to the chemist may be retrieved from the computer.

The simulation model that has been developed by us is based on research conducted by coworkers (students and research associates) of the senior author at the University of Washington over a period of about eight years. The basic ideas of the algorithm have been described in some detail in the literature, and they will be reviewed here. Several selected results that serve to validate the method will be discussed. Finally, we end with some new and provocative results that suggest the utility of network modelling as a tool for the rational design of elastomeric materials.

8.2 Analytical theories of gelation

The first theory of gelation was published by Flory[1-3] in 1941, and shortly thereafter Stockmayer[4,5] analysed Flory's treatment in great detail so as to calculate the molecular weight distributions up to the gel point. Stockmayer's significant contribution to the field is recognized in the now standard name: Flory–Stockmayer Gelation Theory. Approximately four other mathematical approaches to the same basic theory have been offered in the ensuing years, and each has certain features that makes it appealing. Our interest in these theories is more than historical. It is crucial to understand the strengths and weaknesses of them to appreciate the need and place for other approaches to the problem.

In his classical gelation theory, Flory approximated real systems by neglecting the effects of cyclic formation. That is, the only branched molecules that were considered in the statistical analysis are those that do not contain circuits. One such molecule is represented abstractly in Fig. 8.1; graph theorists refer to structures such as this as trees. Adding to the mixture of nomenclature from chemists and mathematicians, physicists have come along and called figures of this kind percolation clusters on a Bethe lattice.[6,7] Mathematicians know the Bethe lattice as a Cayley tree.

The branched molecule illustrated in Fig. 8.1 might be formed in the polycondensation reaction of, say, glycerol with adipic acid. Each bond (edge, in the parlance of graph theory) between two junctions represents an adipic diester, and each junction (node or vertex) represents the trifunctional glycerol molecule. The ends may be either unreacted hydroxyl or carboxylic acid groups. To generalize, let there be a maximum of f bonds emanating from each vertex. This figure, and any other similar one that you might like to draw, may be mapped onto an f-valent Bethe

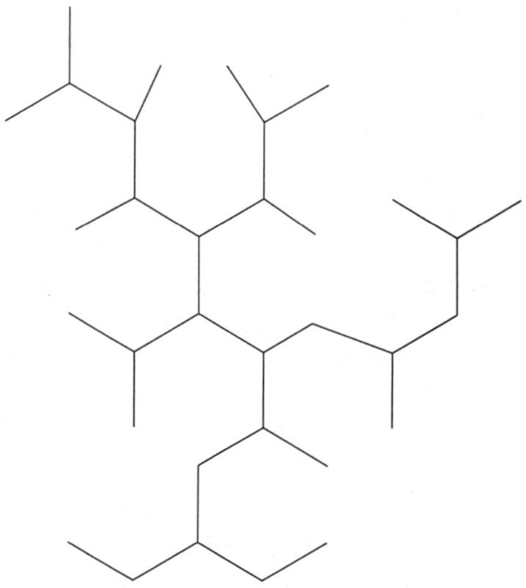

Figure 8.1 A representative trifunctional branched molecule.

lattice that extends to infinity. Every vertex of a Bethe lattice has f neighbours, and the lattice does not contain any circuits. It is constructed by starting with a vertex and attaching f edges to it. Next, connect $f-1$ edges to each end of the first f edges, to get a figure with $f + f(f-1)$ edges. The next layer is again formed by attaching $f-1$ edges to each of the $f(f-1)$ ends in the second layer or generation. This generation has $f(f-1)^2$ edges. By induction, it is easy to see that the ith generation has $f(f-1)^{i-1}$ edges. The cluster in Fig. 8.1 is mapped onto the lattice by selecting a vertex of the lattice to be identified with one of the vertices of the molecules in question, and the bonds of the molecule are placed one-to-one along the edges of the lattice in the obvious way. Dendrimers are today's chemical realization of (perfect?) finite Bethe lattices.

It is interesting to note[6] that the finite Bethe lattice of j generations has

$$N_j = f + f(f-1) + f(f-1)^2 + \cdots + f(f-1)^{j-1}$$
$$= f[(f-1)^j - 1]/(f-2) \qquad [1]$$

edges in all, and the proportion

$$f(f-1)^{j-1}/\{f[(f-1)^j - 1]/(f-2)\} \cong (f-2)/(f-1) \qquad [2]$$

of them are in the last layer. Thus, a large but finite Bethe lattice with $f = 3$ has about half its edges and vertices on the surface. Given this fact, it is important when thinking about mapping acyclic molecules onto the Bethe lattice that one is careful to imagine the lattice to be infinite in extent. It is also clear that this lattice is very peculiar because it cannot be embedded in 3-space. Any reasonable space-filling lattice in 3-space has a surface-to-volume ratio that goes to zero as the radius of the lattice goes to infinity. The fact that the Bethe lattice has so many surface sites is a reflection of the fact that it is only embeddable in an infinite dimensional space.[6]

Why dwell on these seemingly obscure points? The very good reason for doing so is that most of the theoretical work (with the exception of percolation theory) that has been done on gelation is based on approximations that are equivalent to asserting that the molecules that obey the rules can be mapped onto a Bethe lattice. To understand the role that modelling plays in helping to understand network formation, it is important to also understand the classical approach to the problem. When we come to treat the formation of an infinite gel molecule with this theory, the fact that the molecule cannot be embedded in 3-space should be a clue that something is amiss, despite the success of the theory.

8.2.1 Flory–Stockmayer theory[1–5, 8–12]

Flory's treatment of the critical gel point[1] is wonderfully elementary. Let Y_i be the expected number of branches that enter the ith generation of the Bethe lattice, assuming that not all of the branches that could form have formed. From these branches there issue the expected number $(f - 1)\alpha Y_i$ of branches into the $(i + 1)$th generation, where α is the probability that a branch has formed. Thus, $Y_{i+1}/Y_i = (f - 1)\alpha$. Now, if $Y_{i+1}/Y_i < 1$ for large i, the number of branches on the tree will eventually wither away to nothing, and the molecule will be finite. On the other hand, if $Y_{i+1}/Y_i > 1$, then each succeeding generation is expected to have more members than its predecessor, and the molecule will grow to infinity. The point at which $(f - 1)\alpha = 1$ is a critical point, below which the generations are certain to die off, and above which they grow without limit. The value of α at this point is obviously

$$\alpha_c = 1/(f - 1) \qquad [3]$$

This elementary result forms the basis for all subsequent work on the formation of networks.

An illustrative calculation of α for a non-trivial example will provide us with the basic techniques for relating experimentally determined extents of reaction to the branching ratio α. We will use the esterification chemistry mentioned above, which is also one of the examples used by Flory.[1] The reaction of glycerol with a dibasic acid will allow us to see how linear polymerization competes with branching in a particularly simple way. Now, glycerol has two primary and one secondary OH groups. It is probable that the secondary OH reacts more slowly than the primary, and correspondingly we set the probability that a primary hydroxyl group has reacted to be p', and the probability that a secondary group has reacted to be p''. These probabilities are related to the numbers $N'(t)$ and $N''(t)$ of primary and secondary groups, respectively, present in the reaction mixture at time t by $p' = [N'(0) - N'(t)]/N'(0)$ and $p'' = [N''(0) - N''(t)]/N''(0)$. The probability that a carboxyl group has reacted is defined similarly in terms of $N_c(t)$, the number of carboxyl groups present at time t. We will assume that the two carboxyls at the ends of the adipic acid molecule are sufficiently far apart that they have the same reactivity. The stoichiometry is balanced for this calculation, so that $N_c(0) = N'(0) + N''(0) = 2N''(0) + N''(0) = 3N''(0)$. A carboxyl group chooses to react with a primary hydroxyl with probability $2p'/3$ (it has 2 choices for a primary group out of a total of 3 groups, and given the choice, it reacts with probablity p'), while the secondary group is chosen with probability $p''/3$. Thus, $p = \tfrac{2}{3}p' + \tfrac{1}{3}p''$.

The interesting feature of this particular reacting system is that we can grasp the competition between building linear polymer or branches simply by setting $p'' = 0$ or not. If the secondary unit does not react ($p'' = 0$), then only linear chains will be built. We can anticipate that the critical gel point is going to be determined primarily by whether or not p'' has a sufficiently large value to create enough branch points.

Let θ_2 be the probability that chain continuation occurs out from a selected adipic acid moiety; θ_3 will correspondingly be the probability that a chain terminates in a branch point. To calculate α all we need is to assert that α is the probability that a sequence of repeat units in a molecule selected at random is terminated by a branch point. Thus

$$\alpha = \sum_{n=0}^{\infty} \theta_2^n \theta_3$$
$$= \theta_3/(1 - \theta_2) \quad [4]$$

The probability θ_2 for chain extension is calculated as the probability that a carboxyl group is connected to a glycerol unit that has only reacted twice. The probability for esterification by a primary group is $p(2p')/(2p' + p'')$. The first factor is the probability that the carboxyl group has reacted, and the second term, the ratio, is the probability that one of the two primary hydroxyl groups has reacted. Similarly, the probability that the reaction is with a secondary group is $pp''/(2p' + p'')$. The probability that the carboxyl group is connected to a glycerol unit with only two reacted sites is

$$\theta_2 = \frac{2pp'^2(1 - p'') + 2pp'p''(1 - p') + 2pp'p''(1 - p')}{(2p' + p'')}$$

where the first factor stems from the first reaction being with a primary OH group, and the second with the remaining (p') primary OH group; the secondary group is unreacted with probability $(1 - p'')$. The next factor arises from the reaction of a primary group, followed by the secondary, with the remaining primary going unreacted. Finally, the third comes from reaction of the secondary group, followed by the reaction with either (hence the factor of 2) of the primary groups, again with the remaining primary group going unreacted. This equation simplifies to

$$\theta_2 = \frac{2pp'(p' + 2p'' - 3p'p'')}{2p' + p''} \quad [5]$$

Note that if $p = p' = p''$, this result becomes $2p^2(1 - p)$; the factor 2 is a combinatorial factor for the two distinguishable configurations of the twice reacted glycerol (primary–primary and primary–secondary), p^2 is the probability that two sites have reacted, and the factor $(1 - p)$ is the probability that the third site has gone unreacted.

The probability θ_3 that a carboxyl group is attached to a thrice reacted glycerol is

$$\theta_3 = 3pp'^2p''/(2p' + p'') \quad [6]$$

Note that this is just p'^2p'' owing to the definition of p above. Equation [4] is thus simplified, with use of Eqns [5] and [6], to

$$\alpha = \frac{p'^2p''}{1 - \tfrac{2}{3}p'(p' + 2p'' - 3p'p'')} \quad [7]$$

With this equation we may deduce some interesting facts about relative reactivity. In the first instance, suppose that $p'' = 0$. Clearly, $\alpha = 0$, and gelation is impossible. Next, suppose that $p'' = p' = p$. Equation [7] simplifies to

$$\alpha = p^3/[1 - 2p^2(1 - p)] \quad [8]$$

which gives $p^2 = \tfrac{1}{2}$ at $\alpha = \alpha_c = \tfrac{1}{2}$; viewed in another way, p^2 is just the probability that both ends of the adipic acid have reacted. Thirdly, one notes that gel will exist only if $\alpha \geqslant \tfrac{1}{2}$ (see Eqn [3] above). Using this condition to solve Eqn [7] for p'' gives

$$p'' \geqslant 3/(4p') - p'/2$$

or

$$p'' \geqslant \tfrac{1}{4} + \tfrac{5}{4}(1 - p') + \tfrac{3}{4}(1 - p')^2 + \cdots$$

At nearly complete reaction of the primary groups ($p' \approx 1$), at least 25% of the secondary groups have reacted if gel is present. Finally, the formulation of the problem will accommodate $p' < p''$. In this case, gel will form only if

$$p' \geqslant p''[(1 + 3/2p''^2)^{1/2} - 1]$$

If the reaction of the secondary groups is nearly complete, $p' \cong 0.581$ at the gel point.

In his original paper, Flory[1] included experimental data showing that $\alpha_c \cong 0.60$ at the gel point for polymerization of several different dibasic acids with tricarballylic acid and diethylene glycol (not exactly the example above). Thus, theory underestimates the gel point by about 20% (1 part in 5). The discrepancy was ascribed to the presence of intramolecular reactions. (Using the example above for equal reactivity, at a gel point α_c of 0.6, the corresponding value of $p_c \cong 0.83$, whereas if $\alpha_c = 0.5$, the value of p_c is about 0.71. The discrepancy is in a direction that can be blamed on intramolecular reactions, because small circuits do not contribute to the growth of the infinite molecule, and when they form they kill off potential reactive sites, and more reactions are needed to compensate.)

This somewhat intricate example illustrates two important facts about the classical gelation theory: it is not trivial to formulate the probability rules correctly for a realistic problem, and once we accomplish the calculation, the answer cannot be trusted to better than about 20%, at least if Flory's experimental results are the rule (some experiments give better agreement with the theory, others give worse). Nonetheless, we need to thoroughly analyse this problem to establish the base-line for further investigations. Hence, we will now briefly review Stockmayer's derivation[4,5,12] of the molecular weight distribution.

The simple case that will be considered for this purpose is that of the polymerization of a single f-valent monomer. A concrete example is the etherification of pentaerythritol, the first step of which is $C(CH_2OH)_4 \rightarrow (HOCH_2)_3CCH_2OCH_2C(CH_2OH)_3 + H_2O$. The reaction mixture at any time consists of m_n molecules containing n monomer units, given that there were N monomer units at the start of the reaction. At any time during the reaction the total number of molecules is M, and these quantities are related by the sums

$$\sum_{n \geq 1} nm_n = N \qquad [9a]$$

$$\sum_{n \geq 1} m_n = M \qquad [9b]$$

The total number of ways that the N monomer units may be bonded together into m_1 monomers, m_2 dimers, etc., is

$$\Omega = N! \prod_n (\omega_n/n!)^{m_n}/m_n! \qquad [10]$$

where ω_n is the number of different ways that an n-mer can form. This count of configurations is maximized subject to the conditions in Eqns [9a] and [9b], to obtain

$$m_n = A(\omega_n/n!)\xi^n \qquad [11]$$

where A and ξ are Lagrangian multipliers introduced via the constraint equations. The problem now devolves to the calculation of ω_n.

Stockmayer, with the acknowledged help of Maria Goeppert-Mayer,[4] obtained the count of the number of ways of assembling the m_n-mers by the use of an ingenious mechanical analogue (Erector Set). By imagining the monomer units to be mechanical frames with f holes that could accept bolts (bonds), and by counting the number of different ways of taking the bolted frames apart and putting them back together, he obtained the result

$$\omega_n = \frac{f^n(fn-n)!}{(fn-2n+2)!} \quad [12]$$

This non-trivial result must next be inserted into Eqn [11], and the sums in Eqns [9a] and [9b] evaluated so as to determine the values of the Lagrangian multipliers. In fact, the somewhat more general sums

$$T_k = \sum_{n=1}^{\infty} n^k x^n \frac{(fn-n)!}{(fn-2n+2)!} \quad [13a]$$

will also provide average degrees of polymerization (DP). The results are

$$T_0 = \alpha(1 - \alpha f/2)/(1-\alpha)^2 f \quad [13b]$$

$$T_1 = \alpha/(1-\alpha)^2 f \quad [13c]$$

$$T_2 = \frac{\alpha(1+\alpha)}{f(1-\alpha)^2[1-(f-1)\alpha]} \quad [13d]$$

where α is a parameter defined for the convenience of the mathematical analysis by $\xi f = \alpha(1-\alpha)^{f-2}$. However, α is exactly the branching probability defined in Flory's work. It is seen that, for $\alpha < 1$, T_0 and T_1 are convergent. Thus, the number average DP (DP_n), is finite for all $\alpha < 1$. On the contrary, T_2, which determines the weight average DP_ω, diverges for $\alpha = 1/(f-1)$, which is the critical gel point defined above.

This recapitulation of the Flory–Stockmayer theory demonstrates how the conditions for the critical extent of reaction are formulated, and shows briefly how the molecular weight distribution can be calculated. The interested reader might like to consult the second[2] of Flory's early papers, or his first book, to see a derivation of the molecular weight distribution (MWD) with use of recursion relations.

8.2.2 Gordon–Dusek cascade theory[13-20]

In the early 1960s, Gordon[13] applied the techniques of cascade theory to the gelation problem, and in several ensuing papers[14-16] the theory has been extended and modified, primarily by the group in Prague headed by Dusek.[15-20] The method makes use of generating functions to evaluate the molecular weight distribution.

A generating function is a 'simple' analytic function of one or more variables whose series expansion in powers of the variable(s) generates coefficients that 'count' something. For example, the function $(1+t)^n$ has the power series expansion

$$(1+t)^n = \sum_{k=0}^{n} \frac{n!}{k!(n-k)!} t^n$$

which qualifies it as a generating function. The coefficient of t^n is the binomial coefficient, which counts the number of ways of selecting k objects from a pool of n objects.

We will now formulate the generating function for all trees, such as the one depicted in Fig. 8.1. The chemistry will be as simple as possible, e.g. the etherification of pentaerythritol. Let $w_n(\alpha)$ be the weight fraction of n-mer in the polymerizing mixture. We want to construct the generating function

$$W(\alpha, t) = \sum_{n \geq 1} w_n(\alpha) t^n \qquad [14]$$

that will provide us with the distribution that is sought, and which furthermore will generate averages by elementary operations such as $\langle n \rangle_w = \partial W(\alpha, t)/\partial t|_{t=1}$.

Imagine the set of all molecules that can be mapped onto the f-valent Bethe lattice. Pick a vertex of the lattice to be the 0th generation of the set of trees. From this vertex there issues a maximum of f branches, and the generating function that counts the number of distinct ways that $0, 1, 2, \ldots, f$ branches can form will be denoted by $F_0(t)$ (the parameter α is implicit, and is omitted from the equations for clarity). The variable t will be used to count the number of monomer units in the molecules, so that if only the 0th and 1st generations exist, $W(t) = tF_0(t)$. But, each branch from the first generation serves to seed the generation of a set of $f - 1$ branches, the ends of those can be succeeded by another $f - 1$ branches, and so on. Hence, the generating function for all possible trees is defined recursively by

$$W(t) = tF_0(tF_1(tF_2(tF_3(t \ldots tF_k(tF_{k+1}(t \ldots)))))) \qquad [15]$$

where $F_0 \neq F_1 = F_2 = \cdots = F_k = \cdots$, and this because the root of the tree grows up to f branches, whereas all succeeding vertices grow $f - 1$ branches at most. This equation, while it looks formidable as it stands, is not all that difficult. Define $r = tF_1(r)$; clearly this is the recursion relation for the inner term in Eqn [15]. We thus have the two equations:

$$W(t) = tF_0(r) \qquad [16a]$$

$$r = tF_1(r) \qquad [16b]$$

to solve to obtain the generating function, once F_0 and F_1 are specified. But these are just the generating functions for single generations, and they are easily seen to be given by

$$F_0(t) = (1 - \alpha + \alpha t)^f \quad [17a]$$

and

$$F_1(t) = (1 - \alpha + \alpha t)^{f-1} \quad [17b]$$

(Expand Eqn [17a] in powers of t. The coefficient of t^k is a binomial coefficient for the number of distinct ways of choosing k functional groups for reaction out of the total of f functional groups, multiplied by $(1 - \alpha)^{f-k}\alpha^k$, which is the probability for reaction of k groups and non-reaction of $f - k$ groups.)

The solution of Eqns [16a, b] and [17a, b] is[13]

$$r = t^{1/f} W^{1-1/f} \quad [18a]$$

and

$$\alpha = \frac{1 - (W(t)/t)^{1/f}}{1 - W(t)[W(t)/t]^{-1/f}} \quad [18b]$$

The functions of most interest are $W(1) = S$, which is just the weight fraction of sol, and the derivatives $W'(1)$, $W''(1)$, etc. When $t = 1$, Eqn [18b] becomes

$$\alpha = \frac{1 - S^{1/f}}{1 - S^{1-1/f}} \quad [19]$$

from which a trivial factor of $1 - S^{1/f}$ may be divided out to give

$$\alpha = \left[\sum_{k=0}^{f-2} S^{k/f}\right]^{-1} \quad [20]$$

When $\alpha = 1$, the solution of Eqn [20] is $S = 0$, corresponding to the absence of a sol fraction. That is, there is only a gel molecule present. For $\alpha < \alpha_c$, the trivial solution of Eqn [19] is $S = 1$; only sol is present in this case. The non-trivial solution of Eqn [20] at $S = 1$ is $\alpha = 1/(f - 1)$, which is the Flory critical point. Inverting Eqn [20], one finds that for $f = 3$, $S = (1 - \alpha)^3/\alpha^3$, $\alpha > \alpha_c = 1/2$. Expressions for S for larger f are complex, as one might easily guess either from Eqn [20] or from Stockmayer's result, Eqn [12]. Solution of the equations with use of a computer algebra program would be an efficient way of generating explicit results for the power series expansion, but numerical solutions of the equations are the norm in the literature.

Dusek and co-workers[17–20] have extended cascade theory to account for the formation of loops. A loop is, by definition, a small cycle that can form during a three-dimensional polymerization. In endlinking reactions, such as the etherification of the longer chain analogue of pentaerythritol, $C[(CH_2O)_nH]_4$, it is possible to form a ring, or loop, by the reaction of

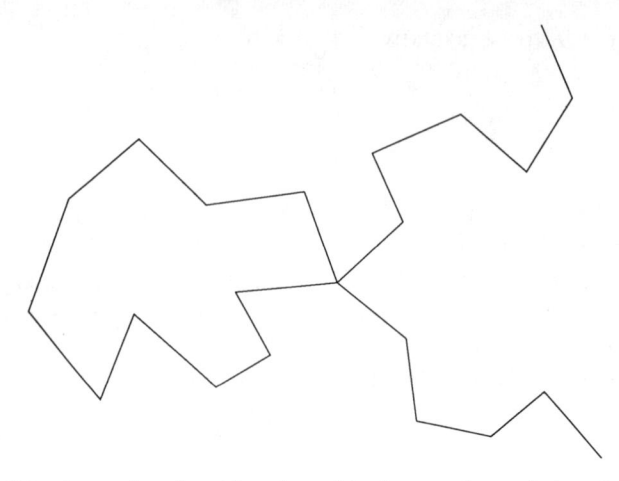

Figure 8.2 A tetrafunctional junction with a loop and two chains attached.

OH groups on the same molecule, to generate a structure as shown in Fig. 8.2. This is only the smallest of the multitude of cycles that can form in reactions of this type, but in many cases it is the most important. In other types of chemistries than the simple one used for illustrative purposes, similar structures are usually unavoidable. For reactions in somewhat dilute solutions, and well below the gel point, the statistics will favour the formation of loops rather than larger cycles. There is an advantage in being able to analyse this problem so as to correct the equations for loop formation. Cascade theory is not able to cope with any but the cycles that are formed within a single generation, simply because the recursion structure of the theory treats one generation at a time, and there seems not to be a way to bend the equations to incorporate more complex interactions between different generations.

Let q be the probability for loop formation that results from bond formation between the 'brothers' within a given generation.[15] (This discussion, in conformity with much of the literature, omits the possibility for more than one loop forming at a given vertex.) The generating function for the first generation, the analogue of Eqn [17a], is thereby modified to the form

$$F_0(t) = q(1 - \alpha + \alpha t)^{f-2} + (1 - q)(1 - \alpha + \alpha t)^f \qquad [21a]$$

and that for subsequent generations is

$$F_1(t) = q(1 - \alpha + \alpha t)^{f-3} + (1 - q)(1 - \alpha + \alpha t)^{f-1} \qquad [21b]$$

The first term of either equation is for a vertex configuration having one loop and $f - 2$, respectively $f - 3$, vertices as offspring in the next generation. The second terms are obviously the same as Eqn [17a, b], as modified by a factor for the probability that a loop does not form.

The probability for loop formation, the factor q above, is determined by the relative concentrations of intramolecular and intermolecular partners for a given functional end-group of the parent molecule. The former concentration is related to the probability that there will be a null, or very small, end-to-end vector for the two arms of the primary molecule. This factor varies as $n^{-3/2}$ for molecules that obey Gaussian chain statistics. The intermolecular factor is obtained simply from the overall concentration of reactive groups, assuming a homogeneous, random mixture of reactive groups. The loop probability factor thus depends upon chain length and extent of reaction α. At this point the mathematics becomes sufficiently intricate that further progress with it would require most of the pages remaining in this chapter, and so we depart for another topic. However, what we have learned from this section is that it is not difficult to write down the generating functions that will give us the information that we need for the branching problem, and even to correct the theory for the first order formation of cycles, but that the equations are not easy to solve except in the simplest cases.

While this discussion has focused on the prediction of the gel point, as will much of the following review, it is also of importance to use network theory to calculate the modulus of elasticity. Here one enters murky territory, because there is not yet universal agreement on what constitutes a good theory of elasticity. Most, but not all, of the theories make some use of a count of elastically active chains as defined by Scanlan[21] and Case.[22] The definition is somewhat round-about. An elastically effective junction (crosslink) has at least three paths (sequences of chains and junctions) leading to infinity. An elastically effective chain is bonded to elastically effective junctions. This definition discounts dangling chain ends and divalent junctions. The former are presumed to be incapable of supporting an equilibrium stress, while the latter may or may not be parts of chains that are elastically effective. In an f-functional polycondensation reaction, junctions that have anywhere from 3 to f reacted sites are elastically effective, and it is not difficult to show that these junctions are counted with appropriate derivatives of generating functions. It is found that the number of elastically effective chains, N_e, in trifunctional reactions is given by

$$N_e = (1 - q)\alpha^3(1 - v)^3[3 - 2\alpha(1 - v)]$$

where v is a respectably complicated algebraic function (not reproduced here) that solves a recursion relation.

8.2.3 Miller–Macosko probability theory[23–27]

The probabilistic scheme that was formulated by Miller and Macoski[23,24] has enjoyed considerable popularity because it is somewhat clearer than the other methods, easier to modify for particular types of reactions, and the equations are easier to solve. It is most closely related to Flory's methods, but is more systematic in the development of the rules for putting together the probability factors. The discussion begins with the law for total probability:

$$E(Y) = E(Y|A)P(A) + E(Y|\tilde{A})P(\tilde{A}) \qquad [22]$$

Here $E(Y)$ is the expected value associated with the occurrence of event Y, $P(A)$ is the probability that event A has occurred, the complement of A is $\tilde{A}$, and the conditional expectation that Y occurs given that event X has occurred is $E(Y|X)$. Event A is bond formation in our case, where $\tilde{A}$ is the absence of bond formation. (The reason for using expectation values rather than probabilities for the present discussion is that we are going to be dealing with the expected value of the molar mass, which is *not* a normalized probability *per se*, but which could be normalized, and hence reduced to a probability.)

The trick that Miller and Macosko[23] use to make the formation of the recursion probability rules transparent is 'looking in' and 'looking out'. This is best explained by example – the etherification of pentaerythritol again. Pick an A group (the molecule is RA_4) at random, and label it A'. The weight of the remainder of the molecule, $W_{A'}^{out}$, 'looking out' from A' is given by

$$W_{A'}^{out} = \begin{cases} 0 & \text{if } A' \text{ does not react} \\ W_{A''}^{in} & \text{if } A' \text{ does react (with } A'') \end{cases} \qquad [23]$$

where $W_{A''}^{in}$ is the weight 'looking in' to A''. By Eqn [22] we have

$$E(W_{A'}^{out}) = E(W_{A'}^{out}|A)P(A) + E(W_{A'}^{out}|\tilde{A})P(\tilde{A})$$
$$= E(W_A^{in})\alpha + 0(1-\alpha) = \alpha E(W_A^{in}) \qquad [24a]$$

The term $E(W_A^{in})$ is just the mass, M, of RA_4 (neglecting the loss of mass that might accompany a condensation reaction), plus the expected weight of the molecule 'looking out' from the arms attached to the same R group of which A'' is a member. That is,

$$E(W_A^{in}) = M + (f-1)E(W_A^{out}) \qquad [24b]$$

The total weight of the molecule containing the selected A' group is $W = W_A^{in} + W_A^{out}$, and the weight average molecular weight, M_w, of the whole molecule is simply $M_w = E(W_A^{in}) + E(W_A^{out})$. Solution of Eqns

[24a, b] gives this weight as

$$M_w = \frac{(1 + \alpha)M}{1 - \alpha(f - 1)} \quad [25]$$

which is the same as that obtained by dividing Eqn [13d] by Eqn [13c] in Stockmayer's theory.

Miller and Macosko[24] also formulated equations for the weight average molecular weight of the sol in the post-gelation regime, and obtained results equivalent in value to those of the other treatments given above. The mathematical structure of the algebraic equations in this treatment bears some semblance to the results of cascade theory, even though generating functions are not used. In fact, the equations of the two theories look very much the same, with probabilities in the Miller–Macosko treatment replacing generating functions in the Gordon–Dusek theory. More recent work by this group includes the crosslinking of primary chains and simultaneous linear polymerization with termination and crosslinking.[25–27]

8.2.4 Stepto kinetic theory[28–30]

The gelation theory that has been used by the Stepto group is more directly derived from the spirit of the Flory–Stockmayer methods than are the Miller–Macosko and cascade theory methods. The main emphasis of this work in recent years has been on the correction of the statistics for loop formation with use of the theory formulated by Ahmad and Stepto.[28] The method gives a first order correction for the formation of independent rings only, but it goes beyond the two methods just described in that rings of all sizes, not just loops, are encompassed.

For an $RA_2 + RB_f$ polymerization, where the only allowed reaction is of A with B, the Flory theory gives $\alpha = p_A p_B$. A correction for ring formation modifies each of the extent-of-reactions parameters, p_A and p_B, by the factor $(1 - \lambda_{AB})$, where λ_{AB} is a ring-forming parameter, i.e. the probability for forming a ring. The gelation condition becomes

$$\alpha_c(f - 1)(1 - \lambda_{AB})^2 = 1 \quad [26a]$$

or equivalently, with $\lambda'_{AB} = \lambda_{AB}/(1 - \lambda_{AB})$

$$\lambda'_{AB} = [\alpha_c(f - 1)]^{1/2} - 1 \quad [26b]$$

The ring forming parameter, λ'_{AB}, is given by the mass action expression,

$$\lambda'_{AB} = c_{int}/c_{ext} \quad [27]$$

where c_{int} and c_{ext}, are, respectively, the concentrations of active ends on the same molecule and on different molecules. The former factor is calculated on the assumption that the chains are fully relaxed, regardless of how many smaller chains they might be built up from. Thus, the chain ends have a Gaussian distribution, and the c_{int} factor entails the probability for a zero end-to-end vector as calculated from the Jacobson–Stockmayer theory,[31] to wit

$$c_{int} = (f-2) \sum_{i=1}^{\infty} (3/2\pi vib^2)^{3/2} N^{-1} \qquad [28]$$

where N is Avogadro's number, v is the number of statistical bonds in one chain (one chain runs from the unreacted A group to an unreacted B group in $ARABRB_{f-1}$), b^2 is the mean square length of one statistical segment in this chain, with v and b^2 defined such that $\langle r^2 \rangle = vb^2$. The sum over i in Eqn [28] captures the probability for formation of all rings consisting of an integral number i of chains. This sum results in the Riemann zeta function, $\zeta(3/2)$. Thus, for a given chain structure, c_{int} is a constant that can be computed from chain configuration statistics.

The factor $(f-2)$ in Eqn [28] is an approximation to the number of opportunities for reaction with B groups that are available for intramolecular reaction. Stepto's picture[30] is that rings form after some stage of strictly intermolecular polymerization in which approximately two of the B groups on each RB_f have been used to build linear and branched molecules containing dangling unreacted A and B groups. The term c_{ext} in Eqn [27] is the instantaneous concentration of unreacted groups at the gel point. This is approximated by replacing the instantaneous total concentration of groups, As and Bs, by the initial total concentration.

Stepto and co-workers[29, 30, 32–34] have conducted extensive tests of this treatment of ring formation in a series of papers. The results are generally in fairly good agreement with experiment, indicating that the first order correction that has been applied can account reasonably well for ring formation that occurs up to the gel point. However, it must be admitted that in the post-gelation region interdependent cycles must be formed, and the theory cannot be expected to be accurate. In fact, there are only a few cases where strict linearity between λ'_{AB} and $1/c_{ext}$ is found.

8.2.5 Other theories, including percolation theory

The relation between graph theory and gelation theory was first emphasized by Gordon,[15] and several other research groups have exploited this connection to gain alternative views of the branching process. Notable amongst these is the treatment of Durand and Bruneau,[35–37] who used combinatorial techniques[35] to count the trees that obey the Flory–Stockmayer branching rules, giving results equivalent to those obtained by Stockmayer for the molecular weight distributions. They also developed simplified methods[36,37] for handling cases where distributions of functionality are of interest, and obtained generalizations of the theory for these cases. While graph theoretical methods provide a general and formal mathematical framework for discussing the problems of gelation, they have not yet provided tools that enable the theory to be extended beyond the restrictions imposed by the assumptions (most particularly, the absence of circuits) that enter into the Flory–Stockmayer theory.

Percolation theory,[6,7] within the larger setting of renormalization group (RG) theory,[38] has attempted to do just that. The many interesting results that resulted from two decades or so of analytical and simulation research into percolation have done much to clarify the relation between classical branching theory and the real world. In particular, it is now quite well established that the classical theory, with its attendant predictions on how various properties of the gelling system depend on distance from the critical gel point, is correct only in spaces of dimensionality six and higher.[6,39–41] This is a remarkable result in view of what was said above about the Bethe lattice only being embeddable in an infinite dimensional space; one might think that the classical theory would only be correct for $d = \infty$. The observation that the critical dimensionality is 6 is a consequence of relationships between the critical exponents that emerge from RG.

One emphasis of percolation theory has been to model gelation, to evaluate the critical exponents in the vicinity of the gel point, and to characterize the universality class into which gelation falls. In percolation theory one usually, but not always,[42] begins with a lattice in a d-dimensional space, and uses a Monte Carlo technique to form bonds at random between the lattice sites. Sites that are joined to one another by bonds are clusters, and at any stage of the bond-forming process there will be a distribution of clusters having different sizes and connectivity (topology). The critical percolation threshold is reached when the clusters of bonded sites have grown sufficiently large so that there exists at least one cluster that spans the dimensions of the simulation container. For large simulations it is found that there is inevitably just one large cluster that satisfies this condition, together with many smaller clusters. It is clear that the largest cluster corresponds to the gel component, and the smaller ones to the sol. Both site percolation models, in which the bonds always exist but the sites are only occupied when selected by the random process, and bond percolation models, in which the sites are always occupied but the bonds do not exist until selected, have been used to model gelation.[7]

The classical theory, Eqn [25] above, gives

$$DP_w \propto (\alpha_c - \alpha)^{-1}, \quad \alpha \to \alpha_c^- \qquad [29]$$

while modern critical theory[6] leads one to write

$$\mathrm{DP}_w = C(\alpha_c - \alpha)^{-\gamma}, \quad \alpha \to \alpha_c^- \qquad [30a]$$

and

$$\mathrm{DP}'_w = C'(\alpha - \alpha_c)^{-\gamma'}, \quad \alpha \to \alpha_c^+ \qquad [30b]$$

where γ and γ' are critical exponents (not necessarily identical on the two sides of the critical point, and certainly not the same as the classical exponent of 1), and C and C' are called critical amplitudes. The other critical exponent that is most interesting with respect to the classical theory (and this brief discussion) is defined by the gel fraction G, for which it is postulated that

$$G = B(\alpha - \alpha_c)^\beta, \quad \alpha \to \alpha_c^+ \qquad [31]$$

The classical theory gives $\beta = \gamma = \gamma' = 1$. Percolation simulations give $\beta \approx 0.40$–0.45, and $\gamma = \gamma' \approx 1.74$–$1.78$ in three dimensions,[6] with other results for lattices of other dimensionality.[40,41] However, it is found that the exponents do not depend on the details of the lattice, as lattices with space groups different from the simple cubic give the same exponents, and this in turn demonstrates that percolation follows the principles of universality.[6,7]

The percolation exponents are obviously different from the classical in the vicinity of the gel point. Furthermore, the critical points are shifted. For example, bond percolation on a triangular lattice[6] has a critical point at $\alpha_c \approx 0.35$, whereas for the comparable $f = 3$ case, Flory–Stockmayer theory gives $\alpha_c = 0.50$. From a practical chemical point of view, the shift in the location of the critical point is probably of greater importance than whether or not universality is obeyed. Assuming for the moment that percolation is a more accurate description of the physical phenomenon than is the classical theory, we conclude that the percolation simulations have much to recommend themselves over the classical theory.

There are many other critical exponents that can be discussed, and the important relations between the exponents, including hyperscaling results, all need to be considered for a thorough understanding of the subject.[6,7] A similar comment applies to the classical theories, and the interested reader is well advised to explore the interesting mathematical techniques that have been developed to cope with the problems that the various approaches to this theory have generated.

8.3 Simulation techniques

So, where do we stand? On the one side, we have a classical theory, in a variety of disguises, that can be applied to quite complex chemical problems of interest, including mixtures of functionality and chain length.

The predictions will not, however, distinguish between different chemical species of the same functionality unless one of the ring correction methods is applied. Even so, the equations are sufficiently complicated that one would want to use a computer to solve them. Indeed, much of the published work has been aided by small programs that solve the equations. The results will not be accurate, because the theory is simply not valid for three-dimensional space, nor do the ring correction factors go far enough to encompass all of the cycles that are formed. On the other side, percolation includes circuit formation, and on that account it is believed to capture the essential cyclization reactions that the classical theory omits. Yet percolation is generic; there is no obvious way to develop percolation models that incorporate the chemical details that are of interest to chemists. The physicists[38] who work in this area want to know why diverse phenomena, such as magnetism, gelation, and the gas–liquid transition, have something in common, whereas chemists are more interested in finding out why polyurethane networks are different from epoxy thermosets. Unfortunately, neither the classical theory nor percolation can provide the latter explanation. There is a clear need for a modelling method that can address these problems.

Given the shortcomings of these approaches to network problems, one would like to devise a modelling method that circumvents the use of a lattice so that realistic chemical structures can be handled. This means allowance for molecules of different sizes, stiffnesses, functionalities, rates of reactions and dilutions. The model should be capable of yielding gel points, molecular weight distributions, elastically effective and ineffective chains, distributions of extents of reactions for each type of functional group and any other quantity that the chemist might like to understand so as to better manipulate formulations so as to tailor properties.

The simulation model for polymer network formation that has been developed by the coworkers[43–55] of the senior author has been effectively used to answer a number of these questions, and is even now being further extended to answer all of them. The model and its results will occupy the remainder of this chapter.

Graph theory provides the proper home for all discussions of networks. Regardless of how a class of network structures might be formed, whether it be with use of probability rules, combinatorial methods, generating functions, Monte Carlo techniques, or by drawing a sketch on a blackboard, the structures are ultimately graphs consisting of vertices and edges (atoms and bonds). That is saying no more than that the molecular structures that chemists draw, with atomic symbols and strokes representing atoms and their bonds, are recognized by mathematicians as graphs. So what do we gain by calling them graphs instead of molecules? The answer is that graph theory does not have the built-in cumbersome size limitations that much of our chemical nomenclature possesses. It can be anticipated that an algorithm is going to have to encode chemical structure information with generic graph theoretical methods if the structures are large. That is the first advantage; the second is that mathematical techniques that might have arisen in other contexts, problems such as the routing of telephone calls on a communications network, might be useful for chemical problems. (The underlying similarity between several different problems involving random networks is reason enough to work in this area. We mention inorganic glasses, amorphous metals and semiconductors, and neural networks as other physical systems requiring similar techniques.)

While we can hope that someone else will solve our problems for us, that has not happened more than once. (That important exception was provided by McKay,[56] who solved the problem of the eigenvalue spectrum of random regular graphs.) The basic reason for the paucity of solutions to related problems is that graph theory by itself cannot handle metrical relations between vertices and edges. The only concept of, say, diameter in graph theory is in terms of the number of edges. The length of an edge does not enter. But that is vital chemical information that cannot be ignored. We chemists are thus reduced to solving our own problems, precisely because we believe that bond lengths are important. But we are not proud, and we take what we can from other disciplines.

8.3.1 The model and algorithm

The crucial element that is absent from classical theory and graph theory, and that is only crudely approximated by percolation theory, is the distribution of molecules in space. Certainly molecules are distributed in space, and they must move with respect to one another to come together to form bonds. Whether or not their opportunities for doing so are free or restricted to a lesser or greater extent depends on details of structure, density, temperature, relative rates of diffusion and reactivity, and perhaps other factors. However, if one pursues this line of reasoning it does not take too long to conclude that a detailed description of all these factors will require a molecular dynamics method, and hence will be restricted to a few thousand atoms at most (with present hardware). The typical networks that we are interested in modelling have chains containing a hundred or more backbone atoms, and therefore such an approach is restricted to a handful of crosslinks at most. Yet percolation simulations have shown that the gel point can only be comprehended if we can handle thousands of crosslinks. This forces us to dispense with detail, and to consider a model that captures the essence of the spatial distribution and the other factors mentioned above, but which will be nonetheless sufficiently accurate so that the chemistries of interest can be modelled.

There is another reason for wanting to handle large systems, and this one is somewhat more subtle. Every small system is subject to fluctuations. In simple fluids and solids, the fluctuations of primary interest are thermally induced. For example, the number of near neighbours to a given noble gas atom in the liquid state is a topological quantity that fluctuates about a mean value because the atoms are in constant motion with respect to one another. In the case of random networks, the topological fluctuations are all important, but they are permanently installed once the bonds are formed. By topological fluctuations we mean that the local connectivity patterns have a wide variation, and one must expect that different regions have properties that are likewise subject to wide variations. Several lines of evidence suggest that the correlation length for these fluctuations is extremely large, with the consequence that large systems are required to get good statistics. At this point universality enters the picture again. If we are to believe in universality, i.e. that molecular details at distances much less than the correlation length of the fluctuations are unimportant to critical behaviour, we will not need a highly accurate model for the local liquid-state structure. However, the location of the critical point is dependent on structural details, for certainly the functionality of the crosslinks, loop formation, chain lengths, and other factors, are known to shift the gel points. Thus we are left with a somewhat schizophrenic situation in which detailed chemical structure is at once important and unimportant. The model we have devised seems to contend with this disease in a productive way.

Given chains and crosslinkers with defined structure, the basic modelling technique is to replace detailed molecular structure by vectorial information that locates the functional groups with respect to one another within the molecule, and dispenses with all other structural information, save for a few essential parameters that characterize the reactions of the functional groups, the masses of the molecules, etc. For example, if the crosslinker is f-functional, one of the functional groups in the molecule is chosen as the root or origin of a local coordinate frame, and $f - 1$ vectors locate the other groups relative to the root. The roots are thrown into a simulation container with use of a random number generator with a uniform distribution, and the other groups are located by giving the molecule a random orientation. Polymer molecules are handled similarly, but with vectorial displacements between the groups drawn from either a Gaussian distribution or from a Rotational Isomeric State[57] Monte Carlo (RIS–MC) sample, as one wishes. Overall system stoichiometry, density, homogeneous dilution, simulation size and number of samples are selected by the modeller.

A few words are in order here about the validity of this procedure. We approximate the liquid state by a random distribution of molecules, thus ignoring intermolecular correlations while treating intramolecular correlations with considerable accuracy. The overall density is presumably well approximated by this method, but there is no guarantee that there will not be local holes or highly dense regions. Whether this works or not is ultimately to be decided by the accuracy of results, and we will soon demonstrate that the results are quite accurate. At this stage of the discussion, we simply remark that the RG postulates[6,7] suggest that we might be able to get away with the approximation. In any event, ignoring intermolecular correlations should be all the better approximation the more dilute are the functional groups (this either means long chains in dense polymers or added diluent or both). We do not claim that the method will work for densely crosslinked systems, such as thermosets, but at the same time we have not yet attempted to apply the model to these systems so we do not know how good or bad it is.

Bonds are formed between reactive functional groups by a method that mimics a diffusion controlled reaction.[44] At each iteration of the bonding algorithm, pair distances between unreacted groups are computed and sorted from smallest to largest. If two reactive groups are within an incrementally specified distance of one another, reaction takes place, provided neither of the groups has reacted with something else that happened to be closer. The distance parameter is the radius of a 'capture sphere', which represents the average volume within which the crosslinkers diffuse in some interval of time. In most cases to be cited, the molecules are not actually moved with respect to one another to form the bonds; the algorithm merely records the bond formation event in a connectivity table, and goes on to the next pair of groups.

Much coding effort went into improving this model by using various movement schemes; for example, moving the smaller molecule while the larger one remained fixed. After much work we found that the movements simply did not matter.[48] The reason for this is probably the same one that explains universality. The systems that we are treating consist of large molecules, and details at the level of a few Ångstroms are unimportant. The movements that real molecules undergo in order to form bonds are ultimately dictated by structurally unbiased thermal motions. Movements are as likely to destroy a bias in the long-range connectivity as they are to create one. Moving the molecules in the simulations is likewise equally likely to make or break a bias. The net result is the same as not moving anything at all. The exception, where a bias does count, occurs when the chains are relatively short and the crosslinkers large. In this case the probability for loop formation is influenced by the addition of the crosslinker to a chain end, and we find better results by moving the

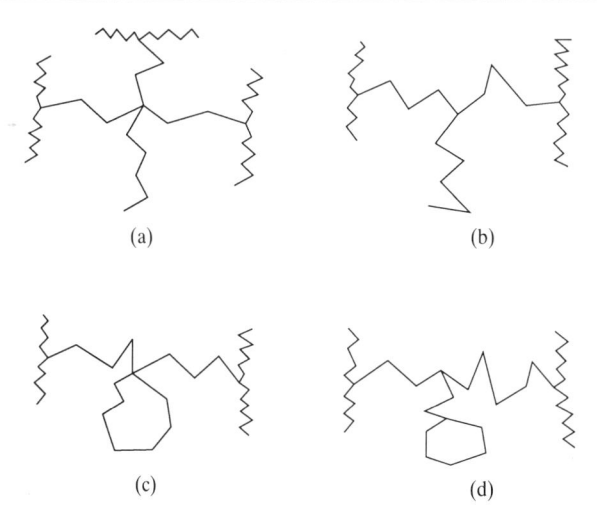

Figure 8.3 Defect structures in trivalent and tetravalent networks. The jagged lines represent connections to the gel. (a) and (b) illustrate dangling ends in tetravalent and trivalent networks, respectively. Loops are depicted in (c) and (d).

coordinates of the crosslinker onto the chain end as the crosslinker makes its first bond.[53] Stepto urged this improvement on us, and it has proven to be important.

Having produced a connectivity table by means of the bonding algorithm, we use a breadth-first search algorithm[58] to sort the connected components from one another. The components are then interrogated to determine their masses, small cyclic structures, dangling ends, extents of reactions, etc. Gel points are usually computed as simple averages[46] of the extents of reaction at the extremum in the Reduced DP_w (RDP_w) and the inflection point in the DP_w. The RDP_w is the weight average DP with the exclusion of the largest connected component in the mixture. Up to the gel point the RDP_w is virtually identical with the DP_w, and beyond the gel point it is just the DP_w of the sol fraction. An exception to the simple averaging was used for work on critical exponents, where more sensitive methods were used to locate the gel points.[46]

Elastically effective chains are determined by correcting the total count of chains in the gel component for the number that are dangling ends, loops and multiple edges. Figure 8.3 shows four such structures (in these sketches the jagged lines indicate the remainder of the gel). In Fig. 8.3(a) the dangling end is seen to reduce the number of chains by one, but in Fig. 8.3(b) the reduction factor is 2 because the two chains that are joined to the trivalent junction are not both fully effective. Together they form a chain that is twice as long as one of them (assuming the chains have the same length, but then effective chains all contribute the same, regardless of their length), and hence they constitute but a single chain. Similarly, a loop in a tetravalent network (Fig. 8.3c) decreases the number of chains by 2 (the loop itself and one of the chains of the two that are linked at the effectively divalent junction), but in a trivalent network (Fig. 8.3d), the reduction factor is seen to be three. We have developed code to walk down into several layers of vertices from a starting end or loop, but higher order corrections are scarcely worth the effort since more complex structures have a small probability of survival at high extents of reaction.

Another quantity of interest in the theory of elasticity is the cycle rank. In Flory's later work on elasticity,[59,60] much of it in collaboration with Erman,[61,62] he proposed that the so-called 'phantom' modulus should be determined by the cycle rank density.[59] The cycle rank, ξ, is a measure of the number of cycles in a graph, and is defined by

$$\xi = v - \mu + 1 \qquad [32]$$

where v is the number of chains in the graph and μ is the number of junctions. The number of edges in a tree is one fewer than the number of vertices. Any edges that are added to a tree with a fixed number of vertices must form circuits, and the difference between the number of edges and vertices counts the circuits, to within a negligible factor of 1 for large systems. The connection with elasticity is as follows: just up to the gel point there are no circuits according to Flory–Stockmayer theory, and because the system is not a gel it does not have an equilibrium modulus. After this the gel point circuits are formed, and the modulus becomes positive. Therefore, only circuits contribute to the modulus. This definition still requires correction for loops and dangling ends, but that is not difficult to accomplish.[60] An alternative explanation of how the cycle rank enters into phantom network theory was presented by one of us.[63] For reviews of rubber elasticity theory, see Refs 16 and 64.

8.3.2 Review of previous results

The systems that were studied in the 1980s were selected to develop and test the simulation method. Not all of the results will be quoted here, but those that are most pertinent to validation will be. Our work began with endlinked silicone systems because there exists a substantial body of experimental stress–strain work on these 'model' networks. In Leung's pioneering work[43–45] the basic algorithmic design was established, and with it he studied the difference between the classical theory, in particular the Miller–Macosko theory, and simulations for trifunctional[44] and tetrafunctional[45] systems. This effort established that very significant differences do exist, which encouraged further work.

The algorithm was modified to model radiation curved systems. Both polydimethylsiloxane[49] and polyethylene[50] were studied, and for the latter system it was shown that a crosslinking to scission ratio of 6 to 1 gives a virtually perfect account of the experimental data. The modelling method used in this work was quite simple. Backbone atoms on chains were chosen with a geometric or Flory distribution of chain length between them. These sites were then located in the simulation container with the rules described above, and the sites were subsequently selected randomly to be either scission or crosslinking sites. Systematic adjustment of the ratio of the two types of sites gave different results for gel fractions as a function of the extent of reaction, and the ratio that optimized the fit to the experimental data is the value mentioned above.

As much as one would like to test the algorithm on a wide range of systems with various chemistries, there is precious little comprehensive experimental data available for this purpose. Experimentalists with interests in rubber elasticity do not usually measure gel points, and those interested in gelation rarely measure the modulus. Extents of reaction at the gel point, measures of the extent of intramolecular reaction, sol fractions beyond the gel point, molecular weight distributions in the sol, equilibrium modulus of elasticity, and similar quantities are all predictable with the algorithm, but there are no reports of a single comprehensive experimental data set for any system. The work that comes closest to providing a testing ground for the model is that on urethanes by the groups in Manchester, Prague and Strasbourg; indeed, some of the more revealing work that we have done has been on these systems. The hydroxylic molecules used in the experiments were telechelic polyoxyethylene (POE) and star-branched polyoxypropylenes (POP). The isocyanates were hexamethylene diisocyanate (HDI), its urea condensation product obtained by partial hydrolysis and condensation (Desmodur N 75, Bayer), and methylene diphenylene diisocyanate (MDI).

Table 8.1 shows a comparison of simulations with the experimental results of Gnanou et al.[65] on networks prepared from POE and Desmodur N 75. The first column is the number average molecular weight of the POE, the second is the volume fraction of polymer at the time of cure, and the third is the weight percentage of sol fraction that was extracted from the gel at 'complete' reaction. (We generally attempt to determine the extent of reaction from sol fractions if no other data are available.) The fourth and fifth columns contain the values of the modulus calculated from the cycle rank via Miller–Macosko theory and from the simulations, respectively. The final column gives the experimental results. (Gnanou et al. reported more experimental data than is presented here. Our previous work[53] encompassed only a subset of their results.) The branching theory predictions differ from the measured values by an average of 31.5%. The

Table 8.1 Comparison of calculated, simulated and experimental elastic moduli for POE + plurifunctional isocyante.

M_n	v_{2c}	w_s, %	$10^{-4}E_G{}^a$	$10^{-4}E_G{}^b$	$10^{-4}E_G{}^c$
1810	0.160	1.00	6.00	8.01	7.05
	0.356	1.30	10.7	15.9	11.6
	0.525	0.50	18.4	27.4	26.5
	0.612	1.30	22.8	29.3	32.1
	1.00	0.30	31.0	48.5	45.0
5600	0.190	5.20	1.20	1.61	1.101
	0.320	2.20	2.76	3.81	4.32
	0.480	1.82	4.24	5.59	6.14
	0.570	0.97	5.94	7.63	14.1
	1.00	0.20	12.4	15.4	29.6

[a] Modulus (in N m^{-2}) calculated from Miller–Macosko theory.[23,65]
[b] Modulus calculated from the cycle rank obtained from simulations.[53]
[c] Measured modulus.[65]

comparable measure of error for the simulations is 24.5%. From this point of view the simulations are about 30% more reliable than the classical theory. However, that is not the whole story. Except for one line of Table 8.1, classical theory consistently underestimates the experimental values, whereas for the simulations, one half are above and the other half are below the experiments. The absence of a systematic error in the simulations is encouraging. (While the Miller–Macosko cycle rank fairly consistently underestimates the experimental moduli, the 'affine' model based on the number of elastically effective network chains seriously overestimates the measured values.[65])

The results of Stanford and Stepto[66] were more challenging to model. Cryoscopic determinations of M_n and titrametric measurements of unreacted NCO groups were reported in the form of the fraction of intramolecular reactions in the endlinking of polyoxypropylene triols with HDI. Measurements from the onset of reaction up to the gel point for a variety of dilutions in benzene provide a very complete set of unique data with which to compare the results of simulations. The results (not shown here, but extensively described in Ref. 47) of the simulations are quite remarkable. Several different treatments of the arm-length distribution of the triols all yielded results that account very nicely for the data on the neat and highly concentrated systems (without any adjustable parameters, an important point), but with high percentages of diluent the simulations consistently underestimated the extent of intramolecular reaction. In other words, changing the probability for loop formation (so as to get more intramolecular reactions) by skewing the arm length distribution to favour short arms while keeping the overall M_n fixed did not improve the agreement with experiment on the highly dilute systems, but at the same time did not destroy the agreement with the bulk systems (loops are not very important in bulk). A considerable amount of L. Y. Shy's work went into exploration of improvements of the model, but in the end the discrepancy persisted. We were led to postulate[47] that the hydroxyl ends of the triols are 'sticky' in benzene, thereby rendering them more highly correlated and more prone to cyclization than even the Jacobson–Stockmayer theory based on Gaussian statistics for short chains is able to capture. Thus, the model that gave a good account of the data in some cases and not in others provided a clue into an unanticipated complexity of the chemistry. Whether or not the conjectured association of the ends in benzene is real or not remains to be verified with experiment. On the other hand, those early results were obtained in the Gaussian approximation, before we had explored the use of RIS–MC chains. Given the insensitivity of the modelling results to large changes in the MWD of the arms of the triol, it is unlikely that a higher approximation to the chain-end vector distribution would offer an improvement.

The next set of results that we would like to discuss is again for urethane systems. In two papers Lee[54,55] described results obtained on POP tetrols condensed with HDI[30,33,54] and on POP triols reacted with MDI.[55,67,68] The results for the first system are summarized in Table 8.2. The HDI molecule was treated as a rigid unit of length 1.26 nm between the NCO groups, and the POP tetrol was very carefully represented with RIS–MC. Simulations were performed on two different tetrols with the number average molecular weights given in the first column, at the experimental volume fractions of polymer (tetrol + HDI) given in the second column. The third column contains the measured branching probability at the gel

Table 8.2 Comparison of simulated and experimental gel points and elastic moduli for POP tetrols + HDI.

M_n	v_{2c}	α_c[a]	α_c[b]	$10^{-5}G$[c]	$10^{-5}G$[d]
1200	1.000	0.445	0.456	10.5 ± 1.9	9.8
	0.851	0.462	0.462	10.9 ± 2.9	9.4
	0.700	0.492	0.480	11.1 ± 2.3	8.6
	0.511	0.515	0.522	10.1 ± 1.5	7.8
	0.301	0.545	0.577	8.5 ± 1.7	5.7
2100	1.000	0.445	0.432	8.1 ± 1.7	6.8
	0.796	0.471	0.446	9.5 ± 1.2	6.4
	0.603	0.502	0.498	7.7 ± 1.3	5.9
	0.500	0.512	0.507	7.2 ± 0.6	4.7
	0.401	0.535	0.528	7.7 ± 1.3	4.0

[a] Critical branching probability $\alpha = (p_A p_B)_c$ at gel point from experiment.[30,33]
[b] Critical branching probability from simulations.[54]
[c] Experimental reduced modulus (N m^{-2}).
[d] Reduced modulus from simulations calculated from the cycle rank.

point determined from the unreacted end groups, using the classical definition $\alpha = (p_A p_B)$ (see the above discussion of classical gelation theory). Column four gives the simulation results. The average percentage difference between columns 3 and 4 is 3.8%, and the average difference, $|\Delta\alpha_c|$, is 0.0096. An error of 3.8% in α_c is a 1.9% error in p_A, which is perhaps comparable to the experimental uncertainty. Since there were no adjustable parameters in the simulations whatsoever, the agreement is spectacularly good. The percolation threshold for the square planar lattice[6] is $\alpha_c = 0.5$, which is quite close to the mean, 0.492, of the experimental results. We know of no percolation simulations that include the influence of dilution that are directly comparable with the experimental data. The classical prediction for this system is $\alpha_c = 1/3$, which is quite far from the experimental values.

A comparison of the experimental reduced moduli in column 5 and the results of simulation in the last column of Table 8.2 reveals greater uncertainties, both in the experimental values and in the simulation results. Given the unsettled state of elasticity theory, the differences between the columns are not disturbing. At least the simulations give all the right trends. The surprising feature of the comparison is that moduli calculated on the basis of the cycle rank are as close to the experimental values as they are. The accepted wisdom in elasticity theory is that networks such as these should more nearly obey the affine model. However, the affine moduli would be about twice as large as the cycle rank moduli, and that is clearly not the interpretation of choice. One of our motivations for doing simulation work was to explore the need for an entanglement contribution to the modulus. These results for the moduli, as well as those immediately above, argue persuasively that a correction is not required. In any event, the experimental moduli are somewhere between the phantom and affine moduli, and thus are encompassed by Flory–Erman theory without the need for an entanglement contribution.

The second paper[55] in this series was devoted to the POP triol + MDI systems studied by Ilavsky and Dusek.[67,68] The results are not quoted in detail here. Briefly stated, the cycle rank moduli from the simulations fail miserably to account for the data. Departures from the measurements are systematic, with the simulations falling anywhere from 5% to 77% below the measured values. In this case the moduli calculated from the number of elastically effective chains gave somewhat better results. However the results might be interpreted, there is a dilemma. How can it be that changing from tetrols + HDI to triols + MDI, while keeping the basic modelling method intact, causes the accuracy of the simulations to change from excellent to very poor? All of our previous work with 3-functional systems made perfectly good sense relative to 4-functionals.

We concluded that there is nothing wrong with the modelling except that we did not put in something special to account for the association of MDI units. Therefore, the systematic errors must be due to the experimental system being reinforced by aggregation of MDI, which is well known to happen in segmented urethanes. Stepto[33] has arrived at the same conclusion by a completely different argument. This phenomenon was not anticipated by the experimentalists.[67,68] Here again, modelling is telling us something about the chemistry that is not intuitively obvious, and which should not be ignored in the total analysis of these systems.

The final set of results that we want to discuss in any detail is that on critical exponents. Leung and Shy[46] ran a wide series of simulations on polydimethylsiloxane and various molecular weights that are endlinked with 3-functional and 4-functional crosslinkers at a few different dilutions. They found $\beta \approx 0.300 \pm 0.024$ and $\gamma \approx 1.77 \pm 0.16$. The value of β for off-lattice gelation appears to be 25% lower than the percolation result cited above, while γ has the same value within the uncertainty. Furthermore, the critical amplitudes were found to be quite different from percolation. This is an unsettling result, and one that certainly needs to be further investigated. If off-lattice gelation is not in the same universality class as percolation, there are going to be several surprised physicists.

There are four open questions that need to be answered before this discrepancy can be resolved.

1. Are the off-lattice simulations inaccurate because they were done on smaller systems than the percolation work, and therefore need corrections for finite size scaling?

Percolation on a lattice generates cycles that are an incomplete set; for example, a square lattice cannot accommodate a cycle with an odd number of edges.

2. Is percolation an inadequate model for gelation because the underlying lattice omits a significant fraction (approximately 1/2 in rectilinear lattices) of all the cycles that can form in an off-lattice gelation model?

The last two-part question is best introduced with the following observation[7]: '... close to the critical point the system averages over a length L much larger than the size of a single molecule or the range of its interaction. Therefore, molecular details become unimportant when L has become sufficiently large. This argument not only makes the universality principle plausible but also indicates that exceptions exist if the range of interaction is infinite'. One of us[63] proved that chains in 'perfect phantom tetravalent networks' are contracted by a factor of two owing to the presence of cycles. Thus, there is at least some evidence to suggest

that there is some aspect of an infinitely long ranged potential in this problem.
3. Since cycles can be infinitely large, and since the dimensions of the chains in a cycle differ from those in a tree, do cycles effectively create an infinitely long range potential?
4. Does gelation not obey universality because random polymeric networks have an effective potential of infinite range, or is there some screening principle that renders large cycles ineffective in exerting their influence?

We believe that answers to all three questions need to be given before we can feel comfortable. We hope to answer 1 in the near future with simulation on large systems; we leave it to others to ponder 2, 3 and 4.

The other intriguing aspect of this work on critical exponents was that the classical theory gives a reasonably good prediction for the gel points that are found in the simulations. The reason that this is so seems to be the following: molecules that obey the classical rules can only be embedded in spaces of dimension 6 and higher. Given such trees, one would have to prune them to embed them in 3-space, which lowers the extent of reaction. However, the unreacted groups that are left from the pruning are now available for bonding to the ends of remaining branches, thereby creating circuits and increasing the extent of reaction. There seem to be two compensating effects that make, in many cases, the classical theory more accurate than one might expect.

8.4 New results

We would like to conclude with three modelling studies that have not previously appeared in the literature. These are (1) endlinked bimodal networks; (2) RTV adhesives with different end group functionalities; and (3) the influence of dilution on random crosslinking. In all cases the simulation results are not dependent upon any adjustable parameters, and they are all done in the Gaussian approximation. We do not claim that the results are of the utmost accuracy that can be obtained from the modelling procedure because of the use of the gaussian. That disclaimer aside, we hope that the reader will find the examples illustrative and interesting.

8.4.1 Bimodal siloxanes

Mark and coworkers have published extensively[69] on the properties of bimodal networks prepared from various ratios of short and long chain siloxanes that are endlinked with tetraethylorthosilicate (TEOS). The

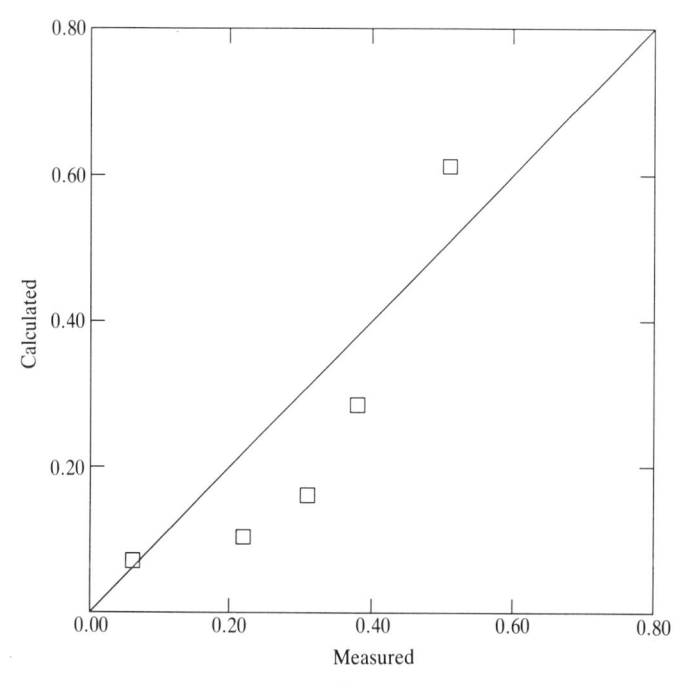

Figure 8.4 Comparison of measured[69] and calculated moduli from simulations for bimodal networks having different proportions of short chains.

recent work of Xu and Mark[69] provides the data to be analysed. These systems were modelled with a log-normal distribution of molecular weights for the two polymeric components, using values of M_w/M_n of 3.00 and 2.75 for the components with M_n of 500 and 18 000, respectively.[70] Xu and Mark reported graphical data for the reduced modulus of networks prepared from 0 to 96 weight per cent of short chains, and we read values for the zero strain reduced moduli from their graph. Figure 8.4 presents these data in the form of the calculated modulus versus the measured modulus. The best fit to the data is obtained with use of the number of elastically effective chains, in contrast to the cycle ranks that fit the urethanes described above. (We still do not understand why the phantom model gives better results in some cases and the affine is more accurate in others.) The number of elastically effective chains, v, in this case is related to the cycle rank, ξ, by the equation $\xi = (1 - 2/f_{av})v$, where f_{av} is the average effective junction functionality.[60] We find that f_{av} is about 3.15 overall for these four-functional systems, which gives an impression of the extent of the defects, most of which are loops. The simulations give quite good agreement with experiments, although there is a systematic curvature in the results that merits further study. It could be the result of the Gaussian approximation, which is not expected to be accurate for the very short chains used in the experiments.

In this case, simulations show that a significant enhancement of the small strain modulus will result from the admixture of short chains into the recipe. Unfortunately, the model will not predict toughness, which is the desirable property that stimulated Mark's work on these systems. To do that would require the use of rubber elasticity and possibly fracture theory, which is a development for the future.

8.4.2 RTV adhesives

A common room temperature vulcanizing (RTV) adhesive is based on acetate end-capped siloxanes. Upon absorbing moisture from the air, the acetate end groups are hydrolysed, and Si–O–Si bonds are formed between the chain ends, and two molecules of acetic acid are released for every Si–O–Si bridge that is formed. Our selection of this system for a modelling study resulted from discussions with J M O'Reilly (Eastman Kodak). We began the project with the casual expectation that placing the $f > 2$ functionality at the chain ends rather than on the crosslinks (which in this case have $f = 2$) would give roughly the same results as the usual endlinked siloxanes, where the ends have $f = 1$ and the crosslinks have $f = 3$ or 4. Nothing could be further from the truth. Polydimethylsiloxanes with $f = 3$ give the results shown in Fig. 8.5. The cycle rank increases to a maximum and then decreases as the reaction proceeds to completion! The modulus does the same. Upon reflecting on this result, it makes perfect sense.

When two trivalent ends condense, one forms a chain of twice the length, with a four functional group in the middle (two trivalent ends minus two functions used for the bond formation). The next intermolecular reaction at this centre leaves a five functional node; the next generates a six functional node, and so on. The functionality of the evolving junction point increases as the reaction proceeds. This is quite different from the usual situation, where the functionality of the junction points decreases monotonically with increasing extent of reaction. As the junction functionality increases, the probability for loop formation increases, until eventually loop formation takes over and converts elastically effective chains into elastically ineffective loops. As this happens, the modulus must decrease.

In the absence of information to the contrary, a chemist who observed an RTV preparation building up a modulus and then losing it would be inclined to first look for reversibility of the reaction, since this could happen if the Si–O–Si bonds were susceptible to hydrolysis resulting from the continuing absorption of moisture. However, the model provides the correct interpretation; the modulus falls because of loop formation. To

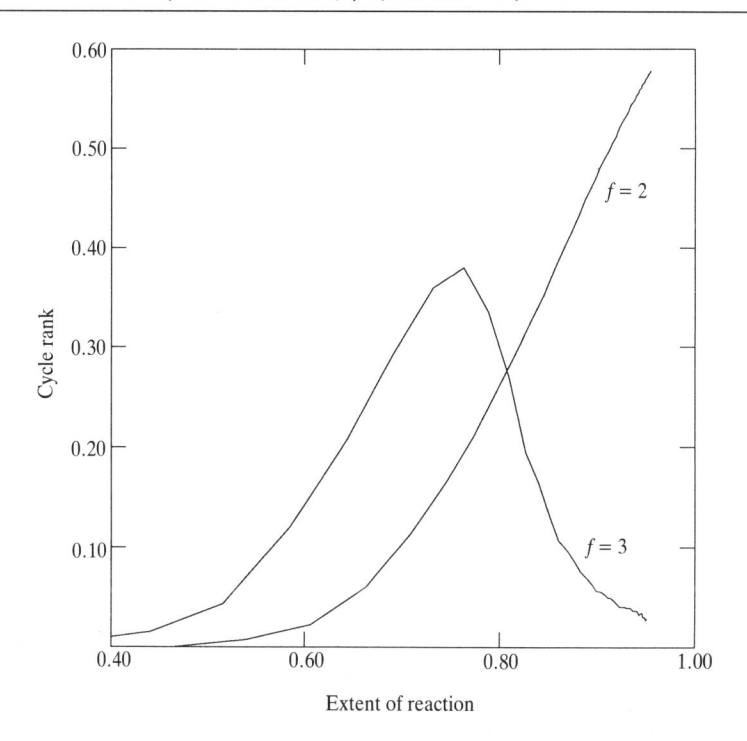

Figure 8.5 Predicted cycle ranks for RTV trivalent and divalent telechelic polydimethysiloxane cured with water. The maximum in the cycle rank for the trivalent materials results from enhanced loop formation as the reaction proceeds.

avoid the problem, the junction functionality must be reduced to minimize loop formation. We thus set the functionality of the ends to two, with the result shown in Fig. 8.5. The cycle rank rises smoothly up to its maximum value at complete reaction and the modulus, for the same overall chain length, achieves a larger value for difunctional ends than it does for trifunctional ends. Silicone RTV adhesives are made with difunctional ends.[71]

8.4.3 Sulphur-cured polyisoprene

The final example is the sulphur cure of polyisoprene (PIP). In this case the crosslinking sites occur at random along the backbones of the chains. The chain sites are selected with probability p, where p is determined by the overall stoichiometry of the system, thus generating a Flory distribution of chain lengths between sites. The Gaussian distribution was again used for the site-to-site vector distribution, with the variance of the Gaussian calculated with use of the characteristic ratio for PIP from RIS.

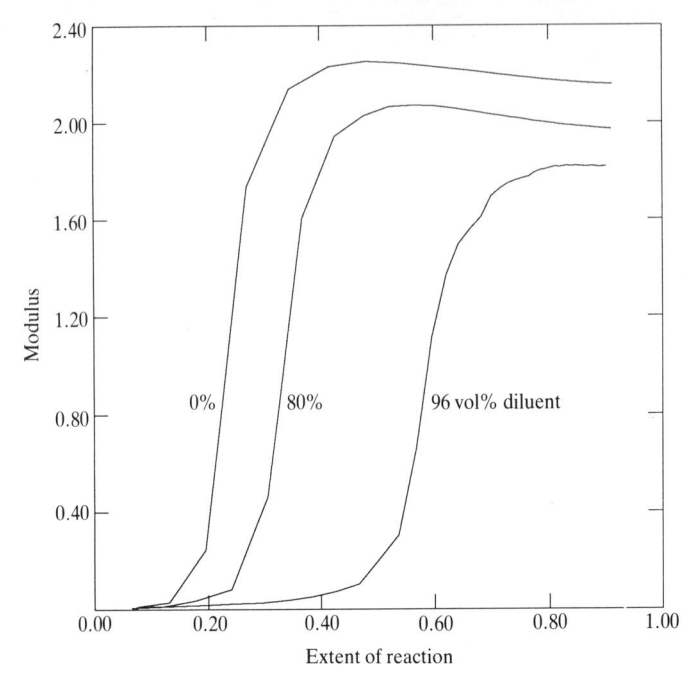

Figure 8.6 Predicted moduli for sulphur cured polyisoprene at various dilutions indicated. The crosslinks are an equimolar mixture of monosulphide and disulphide bridges, present in constant proportion relative to the amount of polymer (2 volume per cent sulphur for the undiluted system). The extent of reaction is the proportion of the total sulphur that has reacted.

The PIP chains are monodisperse with a MW of 10 000, and the volume per cent of monosulphide and disulphide crosslinkers are each set to 1%. The extent of reaction shown in Fig. 8.6 is with respect to the total of 2% divalent sulphur crosslinkers. Our interest here is in exploring the influence of homogeneous dilution on the modulus. Figure 8.6 shows how the modulus of the networks depends upon both extent of reaction and dilution. Clearly, to get the maximum modulus one wants to crosslink in the bulk with about 0.8% by volume of sulphur, assuming that monosulphide and disulphide bridges (in equal proportion) are the dominant crosslink species. Turning the problem around, modelling could be used to infer the average polysulphide bridge length, given the experimentally determined optimal sulphur concentration.

Figure 8.7 shows how the MW of the sol fraction varies with dilution. As expected, the more dilute the curing system, the larger are the sol molecules once the systems have gelled. This figure also illustrates the point that was made earlier about the RDP_w; this measure of molecular

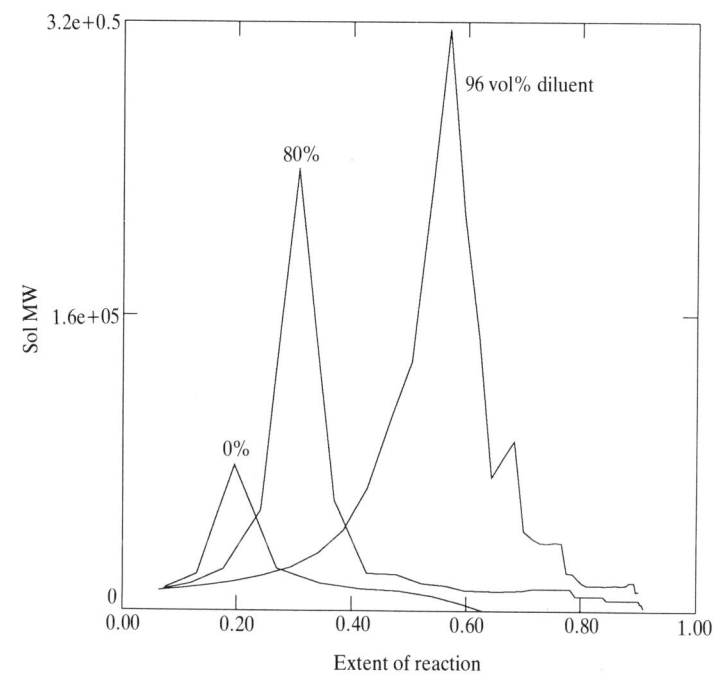

[Graph with y-axis "Sol MW" ranging from 0 to 3.2e+0.5, x-axis "Extent of reaction" from 0.00 to 1.00, showing three curves labeled "0%", "80%", and "96 vol% diluent"]

Figure 8.7 Reduced weight average molecular weights for the same conditions (2 volume per cent sulphur relative to polymer) as in Fig. 8.6. A lower bound for the gel point occurs at the maximum in each of these curves. The gel point for the undiluted system occurs with about 0.4% (0.2 × 2%S) sulphur in the cure system, for example.

size goes through a maximum as the reaction proceeds (remember that the location of the maximum provides a lower bound to the gel point). It is seen that significant shifts in the gel point require quite high dilutions for random crosslinking. The gel points for endlinked systems are more sensitive to dilution than are these systems.

8.5 Conclusions

In this chapter we have explained the fundamentals of the classical theory of gelation, and have shown that the theory is neither trivial to formulate nor accurate in its predictions. The classical theory does not have the flexibility to distinguish between large and small crosslinkers nor between stiff and flexible chains, nor does it provide a fundamentally correct description of the all-important circuits that are formed in polymer networks. While percolation theory is appealing in its simplicity, it only provides a generic solution to a somewhat artificial lattice model for the polymerizing systems of interest. We argue that off-lattice simulations based on realistic molecular structures can improve upon both the classical and percolation approaches to gelation and network structure. Several studies have been cited that demonstrate the accuracy of the predictions that can be made with an algorithm that is not overly CPU intensive.

Now that we are able to model networks with considerable reliability, several questions regarding rubber elasticity can be addressed. We have noted that in some cases the phantom model is nearest to experiments, while in others the affine model works best. It will be interesting to see if Flory–Erman theory alone can account for these observations. In any event, we have not found it necessary to invoke chain entanglements to account for the data. The experimental results generally fall between the phantom and affine models when interpreted with simulations. The exceptional MDI-polyol system is a case where this statement is not true, but there are forceful arguments for believing that the discrepancy is due to aggregation of the MDI units.

The new results that are presented here serve to show the range of behaviour that can be modelled. We hope that they will stimulate further work in network theory and experiment. There are still surprises and counter-intuitive examples emerging from simulations of network formation.

Acknowledgements

The authors wish to thank their coworkers at the University of Washington and at Biosym Technologies, Inc. for critical contributions to the work described here.

References

1. Flory P J *J. Am. Chem. Soc.* **63**: 3083 (1941).
2. Flory P J *J. Am. Chem. Soc.* **63**: 3091 (1941).
3. Flory P J *J. Am. Chem. Soc.* **63**: 3096 (1941).
4. Stockmayer W H *J. Chem. Phys.* **11**: 45 (1943).
5. Stockmayer W H *J. Chem. Phys.* **12**: 125 (1944).
6. Stauffer D *Introduction to Percolation Theory*, Taylor & Francis, London, 1985.
7. Stauffer D, Coniglio A, Adam M *Adv. Polym. Sci.* **44**: 103 (1982).
8. Flory P J *Ind. Eng. Chem.* **38**: 417 (1946).
9. Flory P J *Chem. Rev.* **39**: 137 (1946).
10. Flory P J *J. Am. Chem. Soc.* **69**: 30 (1947).
11. Flory P J *J. Am. Chem. Soc.* **74**: 2718 (1952).

12. Flory P J *Principles of Polymer Chemistry*, Cornell, Ithaca, NY, 1953.
13. Gordon M *Proc. Roy. Soc. A* **268**: 240 (1962).
14. Dobson G R and Gordon M *J. Chem. Phys.* **43**: 705 (1965).
15. Dusek K, Gordon M, Ross-Murphy S B *Macromolecules* **11**: 236 (1978).
16. Dusek K, Prins W *Adv. Polym. Sci.* **6**: 1 (1969).
17. Dusek K, and Vojta V *Br. Polym.* **9**: 164 (1977).
18. Mikes J, Dusek K *Macromolecules* **15**: 93 (1982).
19. Dusek K *Macromolecules* **17**: 716 (1984).
20. Dusek K, Spirkova M, Havlicek I *Macromolecules* **23**: 1775 (1990).
21. Scanlan J *J. Polym. Sci.* **43**: 501 (1960).
22. Case C *J. Polym. Sci.* **45**: 397 (1960).
23. Macosko C W, Miller D R *Macromolecules* **9**: 199 (1976).
24. Miller D R, Macosko C W *Macromolecules* **9**: 206 (1976).
25. Miller D R, Macosko C W *J. Polym. Sci., Polym. Phys.* **25**: 2441 (1987).
26. Miller D R, Macosko C W *J. Polym. Sci., Polym. Phys.* **26**: 1 (1988).
27. Miller D R *Makromol. Chem., Macromol. Symp.* **30**: 57 (1989).
28. Ahmad Z, Stepto R F T *Colloid Polym. Sci.* **258**: 663 (1980).
29. Stepto R F T in *Advances in Elastomers and Rubber Elasticity*, Lal J and Mark J E (eds), Plenum, New York, 1986, pp. 329–345.
30. Stepto R F T in *Biological and Synthetic Polymer Networks*, Kramer O (ed.), Elsevier, London, 1988. Ch. 10, pp. 153–183.
31. Jacobson H, Stockmayer W H *J. Chem. Phys.* **18**: 1600 (1950).
32. Stanford J L, Stepto R F T *Br. Polym. J.* **9**: 124 (1977).
33. Stepto R F T *Acta Polym.* **39**: 61 (1988).
34. Stepto R F T, Eichinger B E in *Elastomeric Polymer Networks*, Mark J E and Erman B (eds), Prentice-Hall, Englewood Cliffs, NJ 1992. Ch. 18, pp. 256–279.
35. Durand D, Bruneau C M *Macromolecules* **12**: 1216 (1979).
36. Durand D, Bruneau C M *Eur. Polym. J.* **21**: 527 (1985).
37. Durand D, Bruneau C M *Eur. Polym. J.* **21**: 611 (1985).
38. Stanley H E *Introduction to Phase Transitions and Critical Phenomena*, Oxford, New York, 1971.
39. Essam J W, Gaunt D S, Guttmann A J *J. Phys.* **A11**: 1983 (1978).
40. Nakanishi H, Stanley H E *Phys. Rev.* **B22**: 2466 (1980).
41. Nakanishi H, Stanley H E *J. Phys. A* **14**: 93 (1981).
42. Gawlinski E T, Stanley H E *J. Phys. A* **14**: L291 (1981).
43. Leung Y K, Eichinger B E, in *Characterization of Highly Cross-Linked Polymers*, Labana S S and Dickie R A (eds), ACS Symposium Series No. 243, 1984. Ch. 2, pp. 21–32.
44. Leung Y K, Eichinger B E *J. Chem. Phys.* **80**: 3877 (1984).
45. Leung Y K, Eichinger B E *J. Chem. Phys.* **80**: 3885 (1984).
46. Shy L Y, Leung Y K, Eichinger B E *Macromolecules* **18**: 983 (1985).
47. Shy L Y, Eichinger B E *Brit. Polym. J.* **17**: 200 (1985).
48. Neuburger N A, Eichinger B E *J. Chem. Phys.* **83**: 884 (1985).
49. Shy L Y, Eichinger B E *Macromolecules* **19**: 2787 (1987).
50. Galiatsatos V, Eichinger B E *J. Polym. Sci., Polym. Phys. Ed.* **26**: 595 (1988).
51. Shy L Y, Eichinger B E *J. Chem. Phys.* **90**: 5179 (1989).
52. Galiatsatos V, Eichinger B E *Rubber Chem. Tech.* **61**: 205 (1988).
53. Lee K J, Eichinger B E *Macromolecules* **22**: 1441 (1989).

54. Lee K J, Eichinger B E *Polymer* **31**: 406 (1990).
55. Lee K J, Eichinger B E *Polymer* **31**: 414 (1990).
56. McKay B D *Linear Algebra Appl.* **40**: 203 (1981).
57. Flory P J *Statistical Mechanics of Chain Molecules*, Interscience, New York, 1969.
58. Nijenhuis A, Wilf H S *Combinatorial Algorithms*, Academic Press, New York, 1975. Ch. 18.
59. Flory P J *Proc. Roy. Soc. London A* **351**: 351 (1976).
60. Flory P J *Macromolecules* **15**: 99 (1982).
61. Erman B, Flory P J *J. Chem. Phys.* **68**: 5363 (1978).
62. Erman B, Flory P J *J. Polym. Sci., Polym. Phys. Ed.* **16**: 1115 (1978).
63. Eichinger B E *Macromolecules* **5**: 496 (1972).
64. Eichinger B E *Ann. Rev. Phys. Chem.* **34**: 359 (1983).
65. Gnanou Y, Hild G, Rempp P *Macromolecules* **20**: 1662 (1987).
66. Stanford J L, Stepto R F T *Br. Polym. J.* **9**: 124 (1977).
67. Ilavsky M, Dusek K *Polymer* **24**: 981 (1983).
68. Ilavsky M, Dusek K *Macromolecules* **19**: 2139 (1986).
69. Xu P, Mark J E *J. Polym. Sci., Polym. Phys. Ed.* **29**: 355 (1991).
70. Mark J E, Xu P Private communication.
71. Hartman B, Torkelson A in *Concise Encyclopedia of Polymer Science and Engineering*, Kroschwitz J I (ed.), Wiley, New York, 1990, pp. 1056–1057.

CHAPTER 9

Computer simulations of biopolymers

D J OSGUTHORPE AND P DAUBER-OSGUTHORPE

9.1 Introduction

Biopolymers form an important class of polymers which are crucial to many physiological phenomena. There are three types of natural polymers – proteins, nucleic acids and carbohydrates. The monomeric units of peptides and proteins are the amino acids. Peptides consist of up to 50 amino acid residues and regulate a variety of biological functions by acting as hormones, neuro-transmitters or inhibitors. Proteins are longer polymers of amino acid residues and are concerned with biological events in cells as well as across cell membranes. Proteins serve as enzymes, catalysing biochemical reactions, antibodies, fighting foreign bodies such as viruses, transport devices, enabling important materials such as oxygen or ions to move to their place of utilization, and structural elements, defining or controlling architectural properties. Nucleic acids are built from nucleotides and play an essential role in the storage and transmission of the genetic code of living organisms. Carbohydrates consist of saccharides and are mainly important for energy storage and metabolism.

These polymers bear many similarities to synthetic polymers, but also have significant differences. For all polymers the chemical composition of the polymer is the major, if not sole, determinant of the properties or function of the polymer. The overall objective of computer simulations of polymers in general, as well as of biopolymers, is to gain an understanding of how these properties are determined, to predict the effects of changes in composition on polymer properties, and ultimately to be able to design polymers with required properties.

The fundamental techniques for investigating biopolymers and industrial polymers are laid on the same foundations. The observed structural and dynamic properties of the polymers are determined by the underlying interatomic forces, and statistical mechanics is the basis for most simulations of polymer properties. However, the application of these techniques is significantly different due to the different nature of the two polymer types. Some of the major differences are summarized in Table 9.1.

Table 9.1 Distinguishing features of synthetic and biopolymers.

	Synthetic polymers	Biopolymers
Polymer length	Variable	Precise
Molecular shape (conformation)	Amorphous	Unique
Monomer types	One or two	4 (DNA/RNA), 20 (peptides/proteins)
Solvent effect	Resistant	Important structural effects
Inter-chain links	Many	Rare
Experimental structural data	Minimal	X-ray, NMR

The chemical composition of industrial polymers is not always fully known. Although the chemical structure of the monomers is known, the exact length of individual molecules, and, for copolymers, the sequence of monomers may vary. Both proteins and nucleic acids have an exact chemical composition, in which the sequence of the monomers and the length of each polymer molecule are known precisely. The only type of biopolymer which is similar to industrial polymers in this respect is the polysaccharides, which have unspecified numbers of sugar molecules and crosslinks.

Most polymer molecules do not adopt a single conformation – different molecules have different torsion angles. Hence the polymer chains are 'tangled' together in a random manner, giving rise to an 'amorphous' structure. On the other hand, for proteins and nucleic acids the shape of the molecules is very highly specified. Proteins have a very well defined shape, and it is this shape which determines the function of the molecule. For example, if the protein is an enzyme, the shape of the active site cavity which binds the substrate of the reaction, defines what molecules will be substrates of the enzyme. Enzymes work either on their own or in multi-enzyme complexes, but still only a few molecules are involved in such complexes. Only the fibrous proteins, such as keratin and collagen, have extended structures involving many protein molecules. However, even in this case the structure is well defined both in terms of the individual molecules and with regard to the way they pack together. Nucleic acids have well-determined conformations, as seen in the double helix structure of DNA. However, in this case this structure does not affect their primary role of carrying the base sequence information but is important in, for example, allowing a metre length of DNA to be stored in a cell ≈ 5 micrometres in diameter. Some nucleic acids, such as transfer RNA (tRNA) and ribosomal RNA, do have a specific conformation which is important for their function. Further, the conformation of nucleic acids is thought to have a significant role in gene regulation, as the adoption of a particular conformation may prevent parts of the nucleic acid molecule from being copied, thus preventing the gene from being active.

A polymer molecule is composed from only a few chemically different monomer units, typically 1 or 2. Biopolymers incorporate a large number of monomer units; proteins are composed of the 20 naturally occurring amino acids and nucleic acids are composed of 4 monomers. The larger number of monomer units used, particularly for proteins, means that many more different functional groups can be introduced into a particular molecule to add a specific feature required for function.

For most polymers solvents do not play a major role in determining the properties of the polymer. (Some polymers swell in the presence of solvent, which affects their mechanical properties, a feature which has been used in packaging applications.) Many biopolymers exist and function in the aqueous environment inside the cells. Recently, work has begun to focus on proteins that exist in the cell membranes, which is an oily environment created by the long aliphatic chains of the molecules in the lipid bilayer. The conformation of all biological polymers is highly dependent on the solvent. If this solvent is changed even slightly, e.g. a different pH or ionic strength, then the structure of the molecule changes, which generally leads to a total loss of biological activity.

Some polymers have crosslinking bonds between polymer chains, formed essentially randomly, thus it becomes almost impossible to talk about a 'molecule' of the polymer. Most biological polymers do not have extensive crosslinking. Crosslinking is found to a large extent in polysaccharides, which act as storage blocks of monosaccharides and have no particular biological function. The cell walls of bacteria are formed from a polymer of modified saccharides and amino acids with extensive crosslinking, which makes the cell wall a highly rigid structure as its function is to protect the fragile lipid bilayer membrane surrounding the cell. Crosslinking is also found in some fibrous proteins, the major one being keratin, the protein component of hair. In these proteins a disulphide bond is formed between cysteine residues in different peptide chains, thus linking the chains together. However, in most proteins, disulphide bonds are intramolecular links which have a role of stabilizing the structure of a single protein molecule.

The major technique for understanding structure at the atomic level, crystallography, can only be used on materials which form crystals. Polymer crystallography is a challenging task since the polymers are only partly crystalline, interspersed with amorphous regions. Crystal structures of a number of polymers have been determined. (See Chapters 4 and 5.) Biopolymers, particularly globular proteins, can form true crystals. The detailed atomic structure of hundreds of proteins has been determined by X-ray crystallography. The number of protein structures determined is increasing rapidly and their quality constantly improving. Recently, progress has been made in determining the structure of biopolymers in solutions as well using NMR techniques. This extensive structural data has been the basis for much of the understanding of how biopolymers function, and a 'launching pad' for computer simulations.

These major differences in the nature of industrial polymers and biopolymers have led to a very different focus in calculations on biopolymers, and in particular on proteins. Whereas for synthetic polymers the composition of the monomer is only one component in defining the properties of the polymer, for biopolymers the type and sequence of monomers define the structural properties of the molecule (in a particular solvent), and the structure is closely related to the function or biological activity of the molecule. Thus, most of the central questions in the investigations of biopolymers are concerned with understanding and predicting the structure and function of the polymer. Some of these central questions are given below.

What are the biologically active conformation(s) of a biopolymer? The protein folding problem, i.e. determining the three-dimensional structure of the protein from the sequence of residues, is one of the active fields of research since the 1970s. Approaches to this problem include energy-based systematic search of the conformational space available to the polymers (particularly for short polymers such as peptides or segments of a large polymer),[1-3] or 'knowledge based' statistical methods[4-7] which utilize data from experimentally known structures.

Many biological events involve 'molecular recognition' phenomena such as the action of an enzyme on a substrate, eliciting a receptor response by the binding of a peptide hormone or a small ligand, DNA–protein interaction, etc. Understanding the underlying factors responsible for molecular recognition is important for rational drug design, e.g. agonists or antagonists to peptide hormones, inhibitors of enzymes, anti-cancer or anti-bacterial DNA binding drugs. Similarly, modified enzymes are designed for biotechnology purposes such as increased thermostability, modified substrate specificity, etc. The computational techniques used to carry out investigations of this type range from mimicking the original binding ligand in terms of shape or electrostatic distribution, through docking algorithms to fit new ligands into the active site of known enzymes, to detailed energy calculations of molecular assemblies.

Although we have emphasized the importance of the well defined structure of biopolymers for biological action, their dynamic behaviour is emerging lately as an important ingredient as well. Collective motion of large parts of molecules occur in important phenomena such as allosteric regulation, domain closure on ligand binding to enzymes and partial DNA helix unwinding during drug intercalation. The main computational methods for elucidating intramolecular motion are molecular dynamics and normal mode analysis.

In the following sections we will describe first the methods and techniques most commonly used in the study of biopolymers. We will then present a few examples to demonstrate how these techniques are applied to specific problems. To conclude we will also mention a few new approaches which are currently being applied to the study of biopolymers.

9.2 Methods and techniques

9.2.1 Potential functions

The most fundamental approach for the theoretical determination of molecular properties is quantum mechanics, which relates these properties to the motions and interactions of electrons and nuclei. By solving Schrödinger's differential equations chemical properties can be calculated from first principles, without experimental data. Thus these methods are termed *ab initio* methods.[8] However, solving Schrödinger's equations is a very demanding mathematical and computational problem and is therefore only practical for small systems. Due to the large size of biopolymers, most theoretical investigations of these systems involve molecular mechanics calculations, based on an empirical potential energy function, rather than quantum mechanics calculations. In molecular mechanics calculations the energy surface of the molecular system is represented by a mathematical function. The form of the mathematical representation as well as the constants are chosen so that experimental data are reproduced. The analytical function, usually expressed in terms of internal coordinates and interatomic distances, and the appropriate constants are known as a force field. The concept of a force field emerged from normal mode analysis techniques employed to interpret vibrational spectra, although originally these force fields did not include interactions between non-bonded atoms.[9] In general, force fields include terms which describe how much energy is required to stretch each bond from its equilibrium value, the energy stored in each angle when bent from its standard value and the intrinsic energy required to twist the molecule about a bond (i.e. change a torsion angle). Finally, the non-bond interactions between atoms are described by Lennard–Jones and electrostatic terms:

$$V = \sum \{D_b[1 - e^{-\alpha(b-b_0)}]^2 - D_b\} + \tfrac{1}{2}\sum H_\theta(\theta - \theta_0)^2$$
$$+ \tfrac{1}{2}\sum H_\phi(1 + s\cos n\phi) + \tfrac{1}{2}\sum H_\chi \chi^2$$
$$+ \sum\sum F_{bb'}(b - b_0)(b' - b'_0)$$

$$+ \sum \sum F_{\theta\theta'}(\theta - \theta_0)(\theta' - \theta'_0) + \sum \sum F_{b\theta}(b - b_0)(\theta - \theta_0)$$
$$+ \sum F_{\phi\theta\theta'} \cos \phi (\theta - \theta_0)(\theta' - \theta'_0) + \sum \sum F_{\chi\chi'} \chi\chi'$$
$$+ \sum \varepsilon[(r^*/r)^{12} - 2(r^*/r)^6] + \sum q_i q_j / r \qquad [1]$$

A typical force field is that used by Hagler et al. and Osguthorpe et al. with a Lennard–Jones 6-12 potential and partial atomic charges for the non-bonded interactions.[10] Similar force fields are the basis of all current programs that perform molecular mechanics calculations, such as MM2,[11] CHARMM,[12] AMBER,[13] GROMOS,[14] and DISCOVER.[15] The potentials used in these programs differ somewhat in the exact terms used in the potential function. For example, a harmonic term is sometimes used to represent bond stretching, some or all cross terms may be omitted and in some cases explicit terms are added to describe hydrogen bonds.

There are two aspects of the force field which determine its ability to reproduce experimental data, the functional form and the constants (potential parameters). The functional form determines the shape of the energy surface and a specific functional form may not be able to reproduce all deformations correctly. For example, it was found necessary to add a bond-torsion cross-term to the above force field in order to be able to reproduce the abnormal bond lengths in anomeric systems.[16].

The second aspect is the values of the potential parameters, which is where the above force fields differ most. In general, the potential parameters are determined by varying the potential parameters until calculated properties reproduce experimental data. This procedure was developed by Lifson and coworkers[17] to produce a consistent force field (CFF), in which the parameters are fitted using a least squares procedure such that they are the optimum parameters for the molecules considered.

The parameters we are using for the valence force field were determined by fitting a wide range of experimental data including crystal structures (unit cell vectors and orientation of the asymmetric unit), sublimation energies, molecular dipole moments, vibrational spectra and strain energies of small organic compounds. Ab initio molecular orbital calculations have also been used in conjunction with experimental data to give information on charge distributions (used to derive partial atomic charges), energy barriers and coupling terms.[10, 18–20]

The accuracy and reliability of the force field also depend on the experimental data used in fitting the potential parameters. For example, most available experimental data are for molecules that are not strained and so a set of potential parameters from such data cannot be expected to reproduce highly strained conformations with the same degree of accuracy. Another example: since the cross terms affect the shape of the energy surface at regions far from the equilibrium structure, force fields with no cross terms may yield correct structural properties but will not reproduce the dynamic behaviour of the molecule (e.g. vibrational frequencies) to the same degree of accuracy.

9.2.2 Simulation methods

One of the basic (and oldest) techniques of molecular modelling is energy minimization, which yields a low energy equilibrium structure. (The corresponding experimental data would be a structure determined at 0 K.) However, the potential energy surface of a biopolymer is very complex. In particular, each of the torsional terms in the energy expression has more than one minimum. Since the torsions are the variables which define the conformation of the molecule, many conformations with local minimum energy exist for large biopolymers. This means that given an arbitrary starting conformation of a protein or nucleic acid no minimization algorithm can guarantee that the lowest energy conformation it finds is the lowest energy conformation of the whole energy surface. This problem is known as the global minimum problem. In order to find the conformation corresponding to the global minimum energy of a molecule, or a set of low energy 'accessible conformations' one can either search the conformational space by changing all degrees of freedom systematically, or apply 'directed' searches such as molecular dynamics (MD)[21] or Monte Carlo (MC).[22] In addition, at physiological temperatures molecules are not frozen in a particular structure. Rather, fluctuations around one equilibrium conformation as well as conformational changes occur. Any measurable property will therefore reflect the existence of many structures. In particular, systems which are inherently disordered, such as solutions, are not amenable to investigation by minimization techniques. To deal with this aspect of molecular behaviour the statistical mechanics techniques, such as MC or molecular dynamics, are required. MC techniques introduce random changes into the structure of the system. These changes can be changes in the torsion angles defining the conformation of a molecule, or the position and orientation of molecules in the system. Molecular dynamics is the solution of Newton's equations of motion for the atoms of the system. The motion of the molecules is determined by the forces acting on the atoms as defined by the potential function. Both of these techniques are iterative and average values of properties calculated over a large number of iterations (10000–100000) are, by statistical mechanics, equivalent to what the experimentalist can measure. Their main drawback is the large amount of computer time required, as now the energy calculation must be iterated many times. These methods were first introduced for protein systems over 10 years ago[23–28] and are coming into wider use as computer power increases have allowed both larger and longer simulations to be performed.

As mentioned before, the dynamic properties of biopolymers play an important role in many biological events. Obviously, methods other than straight minimizations are required in order to study such phenomena. Molecular dynamics simulations and normal mode (NM) analysis are most suited for this purpose.

9.2.2.1 Minimization

Given a potential function as described above, minimization, i.e. the solution of the following equation:

$$\partial V/\partial x_i = 0 \quad i = 1, 3N \qquad [2]$$

finds the structure with the lowest potential energy. The numerical techniques employed to minimize the energy of a molecular system include first order techniques, which make use of the first derivatives of the energy with respect to the coordinates, such as steepest descent and conjugate gradient, or second order methods, such as Newton–Raphson or quasi-Newton, in which second derivatives of the energy are utilized as well.[29] Minimizations can be used to reproduce (or predict) observed properties such as structure or strain energy. In addition, the calculations can provide a tool for revealing the origins of any strain in the molecule or locating favourable interactions, by partitioning the overall energy into its basic components, i.e. bond strain, angle distortion or non-bond atom–atom interactions.

9.2.2.2 Systematic conformational searches

The most complete information about the energy surface of a molecule can be obtained by varying each of the conformational degrees of freedom in a systematic way and calculating the corresponding energy. A classic example of such energy mappings are the Ramachandran maps for amino acids. These maps depict the energy of an amino acid residue as a function of its two major degrees of freedom, the backbone torsion angles ϕ and ψ. The original maps treated the molecules as rigid bodies with fixed bond lengths and valence angles. More recent applications allow relaxation of the molecular geometry and are referred to as flexible geometry maps. In this case, two harmonic terms are added to the energy function defined in Eqn [1], which constrain the torsions ϕ and ψ to adopt the values ϕ_0 and ψ_0:

$$V = K_\phi(\phi - \phi_0)^2$$
$$V = K_\psi(\psi - \psi_0)^2 \qquad [3]$$

By varying ϕ_0 and ψ_0 and performing an energy minimization for each different pair of values, a complete map of the potential energy of the molecule can be computed. (Note that without the constraint the molecule would simply adopt its nearest local energy conformation upon energy minimization.) Such a map reveals all the stable conformations (local minima), their relative energies and the transition barriers berween them. It is a useful tool for studying the conformational preferences of a single residue and is particularly important in designing 'novel' residues.[30]

The brute force approach, i.e. the systematic search of all conformational space, can only be used for simple systems. Even for a simple peptide such as luteinizing hormone releasing hormone (LHRH), such a search is prohibitive. If we limit the search to the two major torsion angles ϕ and ψ, then a 30-degree search of this 10 residue peptide would require the generation and energy calculation of $(360/30)^{10 \times 2}$ or 3.8×10^{21} separate conformations. Thus, when this type of approach is applied to molecules (or molecular fragments) of the size of peptides, the search has to be performed in a rational and more efficient manner. First, these studies are only undertaken for peptides which are partially constrained, such as cyclic peptides or peptide loops connecting structural elements of a protein, and have known 'anchoring' points. Extensive 'pruning' tests are carried out while constructing the structures to rule out highly strained structures, or structures incompatible with the cyclization or anchoring constraints. In addition, only a set of pre-selected favourable residue conformations are utilized instead of changing each torsion of each residue in fixed intervals.

9.2.2.3 Monte Carlo

The MC procedure consists of generating a large number of states of the system and calculating the value of properties of interest for each state. The macroscopic property is given by the Boltzmann statistical average of the value of the property for all states,

$$\langle X \rangle = 1/N \sum_i \frac{X_i \, e^{-E_i/kT}}{e^{-E_i/kT}} \quad [4]$$

where $\langle X \rangle$ is the calculated value of the thermodynamic property, which corresponds to the experimentally measured property. X_i and E_i are the value of the property X and the energy of the system at configuration i and N is the number of states.

In practice configurations are generated using the Metropolis[31] method. This method generates configurations with a Boltzmann weighting, $e^{-E_i/kT}$, as opposed to generating configurations at random and then weighting them by $e^{-E_i/kT}$. This leads to faster convergence and the thermodynamic properties are given simply by the average over the converged configurations. The procedure of generating configurations is as follows. One molecule is chosen at random and translated and rotated at random. (In conformational MC one torsion angle is chosen at random and rotated at random.) The change in energy from the previous configuration, δE, is computed and a random number, R, between 0 and 1 is generated. The new configuration is accepted if δE is negative (the new configuration has a lower energy than the previous one), or if $e^{-\delta E/kT}$ is greater than R. Otherwise the previous configuration is taken to be the new configuration and is added again into the averages. This procedure is reiterated a large number of times to produce an ensemble of configurations for which the average properties are calculated.

9.2.2.4 Molecular dynamics and normal mode analysis

The motion of an assembly of particles is described in classical mechanics by the Lagrange equations of motion:

$$\frac{d}{dt}\frac{\partial K}{\partial \dot{q}_i} + \frac{\partial V}{\partial q_i} = 0 \quad i = 1, 2, 3, \ldots, 3n \qquad [5]$$

where K and V are the kinetic and potential energy of the system, and $\dot{q}_i$ and q_i are the velocities and coordinates.

One approach to the solution of the Lagrange equation was taken in normal-mode analysis.[9] By the introduction of approximations into the representation of the potential energy an analytical solution to the equations was obtained. The conditions necessary to achieve these solutions are: only small fluctuations around a minimum energy conformation are considered and the potential energy is approximated by a harmonic (multi-dimensional quadratic) function of the coordinates. In a coordinate system of mass weighted Cartesian displacements, $q_i = \sqrt{m_i}(x_i - x_i^0)$, the potential energy is expanded in a Taylor series which is truncated after the second-order terms, to give the set of equations:

$$\ddot{q}_i + \sum_{j=1}^{3n} f_{ij}q_i = 0 \quad i = 1, 2, 3, \ldots, 3n \qquad [6]$$

where f_{ij} are the second derivatives of the energy with respect to the coordinates. By diagonalizing the matrix of second derivatives f_{ij}, a set of $3n$ solutions is obtained:

$$q_i = l_{ik}\cos(2\pi v_k t + \varepsilon_k) \quad k = 1, 2, 3, \ldots, 3n \qquad [7]$$

where the v_k are the set of characteristic frequencies, and ε_k are the corresponding phases. The coefficients, l_{ik}, define the normal modes of motion. The frequencies are obtained from the eigenvalues, $v = \lambda^{1/2}/2\pi$, and the coefficients, l_{ik}, are the normalized eigenvectors. Six of these solutions have frequencies of zero since they correspond to translational and rotational motion of the system as a whole.

An alternative approach to the solution of the Lagrange equations was adopted in MD simulations. A full representation of the potential energy was retained and the equations were solved numerically instead of analytically. In a system of Cartesian coordinates, x_i, the Lagrange equations correspond to Newton's equations of motion:

$$F_i = m_i \ddot{x}_i \quad i = 1, 2, 3, \ldots, 3n \qquad [8]$$

In MD simulations the forces on each atom, F_i, and hence the acceleration, $\ddot{x}_i$, are calculated from the gradient of the potential energy, then a small time step ($\Delta t \approx 10^{-15}$ s) is applied to generate the next set of coordinates and velocities. The process is repeated to produce a set of atomic trajectories in which the position of each atom and its corresponding velocity are defined as a function of time.

Structure determination. Many MD simulations are carried out in order to find one or a set of low energy conformations for a biopolymer. Since the main objective of these simulations is the exploration of conformational space, many of them are carried out at elevated temperatures to facilitate conformational transitions. In order to optimize the search for local minima 'simulated annealing' or 'quenched dynamics' simulations have been developed. These consist of high temperature simulations, which enable the overcoming of high conformational barriers, and thus a large area of the conformational space is scanned. At regular intervals along this high temperature trajectory, minimizations or additional MD simulations in which the temperature is lowered gradually are carried out. This process allows the system to relax into low energy regions of the surface.

Two recent developments have made MD a very important tool in the generation of experimental 3-D structures from X-ray and NMR data. These are MD with NOE distance constraints[32-34] and MD with structure factor constraints.[35]

Dynamic behaviour. Although the laws governing atomic motion are very simple, the resultant trajectories can appear very complex, almost noise like. The most common methods for analysing MD simulations are by a dynamic display on a molecular graphics system, and by monitoring and plotting structural properties such as torsion angles, interatomic distances, radius of gyration, etc. Fourier transforms of specific properties (or of their correlation functions) have been used in order to obtain the underlying frequencies of oscillations of these properties. The problem of understanding the dynamic behaviour of proteins has recently been addressed in several studies. One method for identifying collective motions in MD simulations made use of the equal time covariances and cross correlations of atomic fluctuations, and revealed that regions of secondary structure move in a correlated manner.[36] Another method projected the atomic trajectories in MD simulations onto previously calculated axes of normal modes,[37] or onto axes obtained by principal component analysis.[38]

We have used digital signal processing techniques to characterize the motion in MD simulations.[39] First, Fourier transforming all the atomic trajectories yields the frequency distribution.

$$\left. \begin{array}{l} g(v) = (1/kT)v^2 \sum H_i^2(v) \\ H_i(v) = \displaystyle\int_{-\infty}^{\infty} q_i(t)\, e^{-j2\pi vt}\, dt \end{array} \right\} \quad [9]$$

where $q_i(t)$ are the mass weighted displacement coordinates and $H_i(v)$ are the corresponding Fourier transforms. Filtering techniques enable focusing on frequency ranges corresponding to motions of interest and eliminating the rest. This involves three steps: Fourier transforming the coordinates, applying the filter function and inverse Fourier transforming back to the time domain:

$$\left. \begin{array}{l} H_i(v) = \displaystyle\int_{-\infty}^{\infty} x_i(t)\, e^{-jvt}\, dt \\ F(v) = 1 \quad v_{min} < v < v_{max} \\ F(v) = 0 \quad v < v_{min};\, v > v_{max} \\ H'_i(v) = H_i(v) \cdot F(v) \\ x'_j(t) = \displaystyle\int_{-\infty}^{\infty} H'_i(v)\, e^{jvt}\, dv \end{array} \right\} \quad [10]$$

Thus it is possible to remove high frequency bond stretches and valence angle bending and focus on the low frequency conformational motion.[40–42] This approach was extended to enable extraction of vectors defining the characteristic motion for each frequency in the MD simulation. These vectors are analogous to those obtained from normal mode analysis and provide a pictorial description of the motion as well as means for comparing the results of the two methods.[43]

9.2.3 Entropy and free energy

Entropy and other thermodynamic properties such as free energy cannot be obtained easily from MD or MC simulations. Meaningful determination of entropy requires sampling of the complete conformational space available to the system, which for a large biopolymer is prohibitively expensive in computer resources. Various approximations and techniques have been developed to circumvent this problem. One approach is based on the approximation of small harmonic oscillations around an equilibrium conformation. Rigid geometries were assumed in early calculations.[44,45] Einstein's harmonic oscillator approximation[46] was introduced to compare the entropies of different conformations using flexible geometry calculations.[26,47] This method involves calculating the vibrational frequencies of the molecule, and using this discrete frequency distribution to calculate the thermodynamic properties. This method was extended for estimating the conformational free energy from an MD trajectory.[40] A continuous frequency distribution $g(v)$ is obtained by Fourier transforming the trajectories of coordinates using Eqn [9]. From the continuous frequency distribution, the thermodynamic properties can be obtained by

$$\left. \begin{array}{l} E_0 = 1/2 \int hvg(v)\,dv \\[1em] E = E_0 + \int [g(v)hv/(e^{hv/kT} - 1)]\,dv \\[1em] A = E_0 + \int g(v)kT \ln(1 - e^{-hv/kT})\,dv \\[1em] S = (E - A)/T \end{array} \right\} \quad [11]$$

Another method for determining free energy is based on a quasi-harmonic approximation. A set of 'effective' force constants is obtained from MD or MC simulations. These force constants are used in a similar manner to normal mode analysis to determine characteristic frequencies and hence the thermodynamic properties.[48]

In the last few years thermodynamic integration and perturbation techniques have been used frequently in order to evaluate relative free energies.[49-52] These techniques refer to hypothetical thermodynamic cycles such as

$$\begin{array}{ccc} A + B & \xrightarrow{4} & AB \\ {\scriptstyle 1}\downarrow & & \uparrow{\scriptstyle 3} \\ A + B' & \xrightarrow{2} & AB' \end{array} \quad [12]$$

$$\Delta(\Delta G_{AB}) = \Delta G_4 - \Delta G_2 = \Delta G_1 + \Delta G_3$$

Thus the relative free energy of binding A and B ($\Delta(\Delta G_{AB})$), is obtained by converting molecule B to molecule B' and converting the complex AB to complex AB'. This method can be applied, for example, to calculating relative free energy of binding different ligands (B) to the same enzyme or receptor (A), or of binding the same ligand (A) to different mutants of the enzyme (B). In practice, this free energy perturbation method involves changing the molecular mechanics parameters during an MD simulation and determining the free energy change during the process. The essential part of the calculation is to define a Hamiltonian (the potential energy) for two related states of the system, B and B' which are linked by a coupling parameter, λ:

$$H_\lambda = \lambda H_B + (1 - \lambda)H_{B'} \quad 0 \leqslant \lambda \leqslant 1 \qquad [13]$$

The change between B and B' can be done gradually using the slow growth technique or in integration steps or 'windows'. For recent applications to biopolymers, see McCammon[53] and Lee.[54] Although this is a very powerful method its application to large biological systems has to be done with great caution. In particular, the convergence of the calculations needs to be verified by a series of long simulations, and the reversibility of any structural changes checked.

Another method for estimating the free energy of a system from a set of conformations generated by MD or MC simulations was developed recently. With this method the entropy, as defined by the sampling probability, can be approximated from the frequency of occurrence of 'local states', the set of neighbouring torsion and valence angles defining a local conformation of the chain.[55]

9.2.4 Environmental effects

As mentioned before, the environment can have a significant effect on the structure (and hence on the function) of the biopolymers. Since most biopolymers function in an aqueous solution, it is important to take into account the effects of the surrounding water molecules. Similarly, when comparing structures obtained by theoretical calculations and experimental crystal structures, it is important to understand the effect of crystal forces on the single molecule. Initial attempts to incorporate solvent effects in calculations centred on mimicking the solvent's screening effect on the electrostatic interactions. This was achieved by including a dielectric constant (D) in the term representing the electrostatic interactions (Eqn [1]).

$$V_{\text{elect.}} = \sum q_i q_j / Dr \quad \text{or} \quad V_{\text{elect.}} = \sum q_i q_j / Dr^2 \qquad [14]$$

The first form assumes a constant dielectric whereas the second represents a 'distance dependent dielectric'. The latter is aimed to account for the fact that the screening decreases when the charges are closer.

This approximate representation of the solvent is of limited value in the regions of interface between the solvent and solute. Water molecules interact with biopolymers in a very specific manner, forming hydrogen bonds with polar groups on the polymer, and hydrogen bond networks among themselves. This behaviour cannot be accounted for by a macroscopic property such as a dielectric. Consequently, it is necessary to include the water molecules explicitly, at least in crucial regions. The most accurate method for taking environmental effects into account is by including all molecules explicitly. This is accomplished by utilizing 'periodic boundary conditions'. A basic system is usually a cube with a biopolymer molecule surrounded by solvent molecules. This cube is embedded in a cubic lattice of images of the original system. The interactions of the atoms in the original cube with all atoms (original and images) within a cutoff distance are taken into account. Any changes in position of atoms in the original cube are repeated in the image cubes.[21] Periodic boundary conditions can be used to simulate solutions, in which case the original cell has to be big enough so that the solute interacts with solvent molecules only. Alternatively, the lattice can reflect the observed crystal structure of a biopolymer, with the original cube being the asymmetric unit of the crystal.

9.3 Examples of applications to peptides and proteins

9.3.1 Design of bioactive peptide analogues

Our first example looks at the problem of how to determine the conformational requirements for activity in systems where the structure of the receptor is not known. This is a very common situation in the pharmaceutical industry and much time and effort has been devoted to this problem. Here we concentrate on the application of molecular dynamics to this problem and look at the specific example of peptide hormones, systems which are very difficult to handle because of the highly flexible nature of peptides.

Luteinizing hormone releasing hormone (LHRH) is a decapeptide with the sequence:

p-Glu—His—Trp—Ser—Tyr—Gly—Leu—Arg—Pro—Gly—NH$_2$

This hormone is of pharmaceutical interest since an antagonist of this hormone can be used to control fertility. Since the receptor of this hormone is not known, the 'active conformation' has to be deduced from data about the hormone and putative agonists and antagonists. The working hypothesis is that the hormone, its agonists and antagonists all have common conformational features which are necessary for binding to the receptor. The hormone and its agonist (but not antagonists) have additional features which elicit the receptor's biological response.

Early conformational studies of the parent peptide and some agonists suggested that LHRH had conformations characterized by a β turn at residues 5–8.[56–58] A common practice in designing peptide analogues is to design analogues with restricted conformational space. Cyclization is one of the ways to achieve this restriction.[59] A set of LHRH analogues, cyclized through the side chains of residues 5 and 8, was found to be potent agonists.[60] The most highly potent was

Ac—D—Phe(p-Cl)—D—Phe—(p-Cl)—D—Trp—Ser—
Glu—D—Arg—Leu—Lys—Pro—D—Ala—NH$_2$

An exhaustive study of the conformational preferences of this analogue was carried out using conformational search MD and minimization techniques.[61] A conformational search of the cyclic part of the analogue was carried out first, followed by minimizations of all the generated conformations that were consistent with cyclization. Among the 22 conformations obtained after minimization, only two had a β-turn conformation at positions 6–7, and these were ≈ 10 kcal/mol higher in energy than the most stable conformation. In all other generated conformations the central two residues were in extended or γ-turn conformations. An MD study of this analogue was performed next. The initial structure was based on the lowest energy conformation found for the cyclic part. Additional investigations of the conformational preferences of the N and C terminal of the peptide defined a plausible conformation for three more residues. The others were set to an extended conformation. After about 24 ps the peptide adopted a β-sheet conformation, with a β-turn between residues 3–6. (See Fig. 9.1.) Molecular dynamics simulations starting from the conformation with a β-turn at residues 5–8 resulted in higher energy conformations. Similar studies were carried out on other antagonists cyclized through the side chains of residues 5 and 8. A total of 1.5 ns of MD simulations were carried out, starting from different initial conformations as suggested from conformational searches and previous studies. A master list of accessible conformations for the peptide and its cyclized analogue was compiled. Each of the compounds was template forced onto all conformations, thus investigating the ability of each analogue to adopt these conformations. A consistent trend emerged indicating that, for this series of analogues, a β-sheet structure with a turn at residues 3–6 is most favourable. All structures with a 5–8 turn were of higher energies.[62]

In the preferred structure that emerged from these simulations the N and C terminals are in close proximity. This suggested that the next step

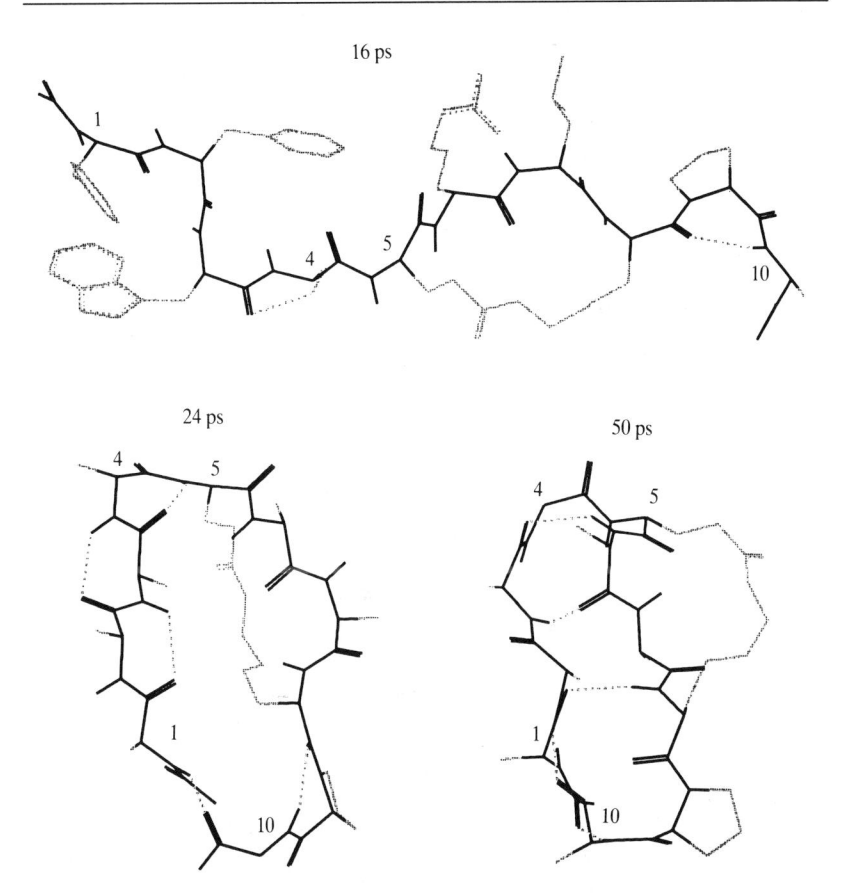

Figure 9.1 Minimized conformations of the cyclic antagonist of LHRH, accessed during the MD simulation. Backbone and side chain atoms are shown in black and grey, respectively. Only heavy atoms and the amide hydrogens are shown. For the last two conformations some of the side chain atoms are not shown. Hydrogen bonds are shown by dotted lines.

towards a more active antagonist might be to join the two ends by an amide bond. Calculations showed that the bicyclic analogue maintained the preference for a 3–6 rather than a 5–8 turn. In addition, any other possible patterns of hydrogen bonding consistent with a β-sheet structure were ruled out by using an N-methyl residue at position 10. The stable conformation for this analogue is depicted in Fig. 9.2. The two bicyclic analogues, with D–Ala[10] and with D–MeAla[10] were synthesized. The first analogue was not active, however, the methylated analogue showed considerable activity.[63]

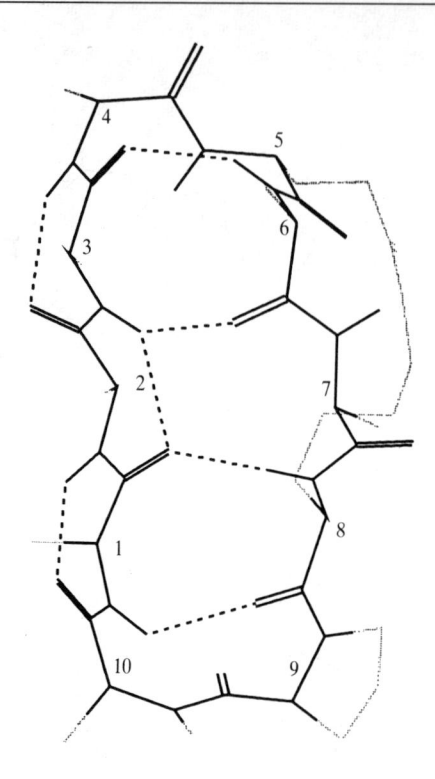

Figure 9.2 N-methyl-D-Ala^{10} bicyclic LHRH antagonist. Backbone and side chain atoms are shown in black and grey, respectively. Only heavy atoms and the amide hydrogens are shown. Some of the side chain atoms are omitted for clarity.

9.3.2 Enzyme–ligand interactions – structure, energetics and dynamics

The rapidly growing number of high resolution structures of biological systems provides an excellent starting point for simulations which aim to understand in detail the energetic and dynamics of the system. Not only does this increase our scientific understanding of biological systems but again this information is very valuable to the pharmaceutical or biotechnology industry. The example we have chosen to demonstrate this type of study is the enzyme phospholipase A$_2$, PLA$_2$.

PLA$_2$ catalyses the hydrolysis of the *sn*-2 acyl side chains of phosphoglycerides. Novel inhibitors of this enzyme are of interest as control of its activity could provide therapeutic treatment of diseases such as rheumatoid arthritis, atherosclerosis and asthma. This section describes a study aimed at understanding the mode of binding and the hydrolysis mechanism in terms of structure, energy and dynamics, at the molecular level. This understanding should facilitate a rational design of substrate analogues or inhibitors.

The highest resolution X-ray structure of bovine pancreatic PLA_2[64] provided initial positions for the heavy atoms of the protein, the calcium ion and water oxygens. Water molecules were generated to solvate the active site and charged residues. Hydrogen atoms were added to the protein and the water. The minimization was carried out in stages, allowing first relaxation of water molecules and hydrogen atoms, and then allowing all atoms to move. Protein ligand complexes were generated by docking the ligand into the minimized apo-enzyme structure and reminimizing. MD simulations started from the corresponding minimized structures.

9.3.2.1 Enzyme–substrate complexes

In order to dock the substrate in the active site, the side chains of some of the residues had to be moved. A few alternative positions for the side chain of one of these residues, Tyr^{69}, were examined in detail. The energetics of the apoenzyme and the complexes are given in Table 9.2. A comparison of the experimental and minimized apoenzyme structures and the complex is given in Fig. 9.3. The active site structure of the best model (substrate C) is shown in Fig. 9.4. The hydrolysis reaction centre maintained all the features consistent with the suggested mechanism: the catalytic couple His^{48}—Asp^{99} interacts through a —N^ε—O^γ hydrogen bond. A catalytic water is hydrogen bonded to the N^δ of $His,^{48}$ and is 3.18 Å from the carbonyl carbon of the ester. The carbonyl oxygen of the ester function is 2.7 Å from the calcium ion and 1.9 Å from the N—H of Gly^{30}. Both of these interactions polarize the carbonyl and thus activate it towards nucleophilic attack. The phosphate group of the lipid is stabilized

Table 9.2 Intermolecular energies of ligand–enzyme complexes[a]

Complex	Protein–lipid	Protein–calcium	Lipid–calcium
Apoenzyme		−424.1	
Substrate-A	−65.7	−325.3	−240.6
Substrate-B	−75.1	−325.5	−238.8
Substrate-C	−109.7	−393.7	−165.1
D enantiomer	−67.8	−335.6	−215.4
Proton relay	−102.0	−333.1	−428.6
Charge separation	−127.9	−302.4	−443.5
Phosphonate inhibitor	−99.9	−341.1	−380.2
Amide inhibitor	−189.7	−405.2	−91.0

[a] Energies are in kcal/mol

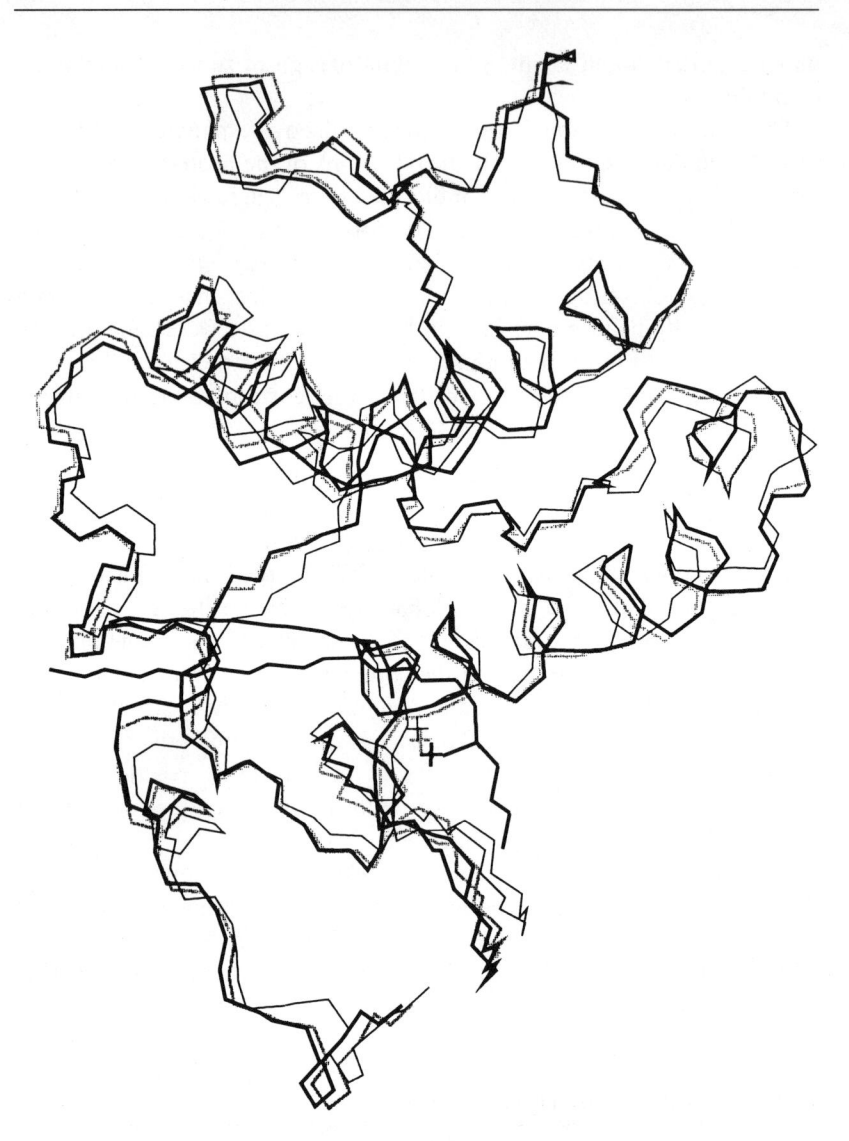

Figure 9.3 The structure of PLA$_2$. The experimental and minimized structures of the apoenzyme are shown by thin black and thick grey lines, respectively. The minimized structure of the enzyme–substrate complex is shown in thick black lines. Only one atom, C$_\alpha$, is shown for each protein residue, and heavy atoms are shown for the substrate. The position of the calcium ion is shown by +.

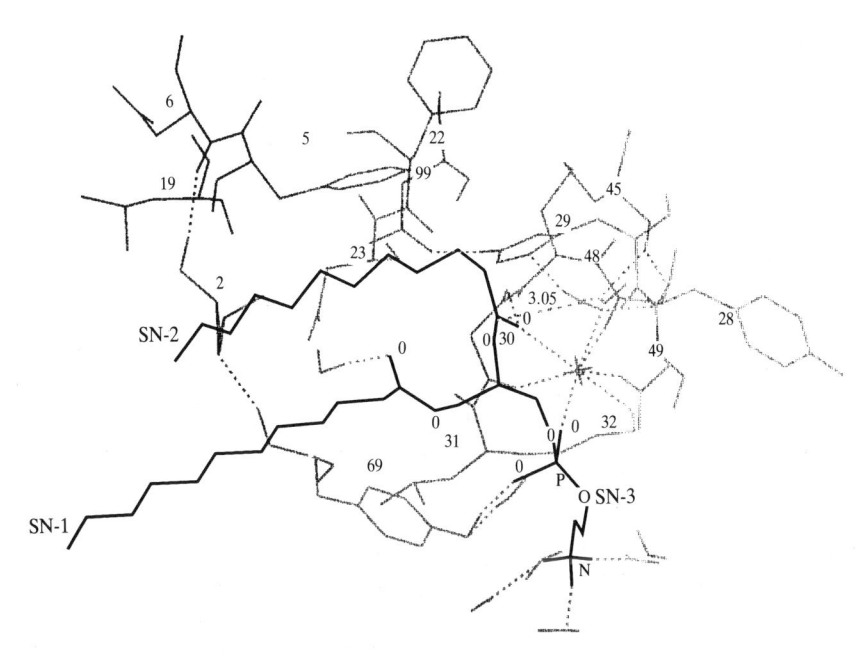

Figure 9.4 The active site region of the enzyme–substrate region. The protein and calcium ion are shown in grey and the substrate in black. Only heavy atoms and amide hydrogens are shown. Hydrogen bonds are indicated by dotted lines.

by interactions with the calcium ion, and hydrogen bonding to the NH of Gly[32], and OH of Tyr.[69] The interactions of the calcium with the four residues observed in the X-ray structure (three carbonyls of alternate residues in the loop 28–32 and the side chain of Asp[49]) were maintained, and in addition the ligand's phosphate oxygen is pointing towards the calcium.

PLA_2 is stereoselective and hydrolyses only the phospholipid with L stereochemistry. Although the D enantiomer cannot be hydrolyzed, it binds to the enzyme just as strongly and acts as a competitive inhibitor. After changing some of the torsion angles of the *sn*-3 chain the D enantiomer was docked into the active site of the enzyme. The position of the three chains of the two enantiomers in the complexes is similar, and the phosphate group is coordinated to the calcium ion and Tyr[69]. The energy of the two complexes is very similar, in agreement with their similar experimental binding constants. However, the *sn*-2 carbonyl of the D enantiomer is pointing in a different direction and thus is not polarized by the calcium ion and the NH of Gly[30], and is not located in a suitable position for an attack by the 'active water'.[65]

The Tyr[69] residue fixes the position and orientation of the phosphate of the substrate so that both calcium binding and correct reaction centre orientation can occur only for the L enantiomer and for the R_p sulphur substrate analogue. When Tyr is replaced by Phe the orientation of the phosphate can be changed to accommodate the 'wrong' enantiomers.[66]

9.3.2.2 Conformationally restricted ligands

Two analogues with restricted conformational freedom were examined in order to test the plausibility of the conformation of the docked lipid. One is an inhibitor in which the *sn*-1 and *sn*-2 chains are linked by a five-membered lactone ring, and the other a substrate in which the *sn*-1 and *sn*-3 positions are linked to form a cyclopentane ring. The lipid 3-arachidonyl-4(O-phosphoethanolamino)-methyltetrahydrofuran-2-one was found to be a potent inhibitor of PLA_2.[67] This molecule could adopt eight chiral isomers. Each of the eight possible chiral isomers was built and their conformations were adjusted to mimic the docked substrate by constrained minimizations. A good fit to the model substrate was obtained for the (3S, 4R) structure. All other structures have large deviations or an unfavourable orientation of the lactone ring. Conformationally restricted phospholipid analogues based on 1,2,3-cyclopentane-triol have been investigated.[68–71] It was observed that of the six possible positional/stereo-isomers only one enantiomer (laevorotatory) of the *all trans* isomer showed a measurable rate of reaction. The ability to overlay the various isomers on the substrate model in the proposed active site conformation was examined. Only for the laevorotatory *all trans* isomer, the pucker that is required to get the appropriate orientation of the two carbonyls results in positioning the phosphate in a conformation corresponding to the one in the minimized complex.

9.3.2.3 Substrate and transition state surrogate inhibitors

A phospholipid *sn*-2 amide analogue was found to be a potent inhibitor.[72] The *sn*-2 ester group in the substrate–enzyme complex was replaced with an amide and reminimized. The geometry of the active site of the complex is similar to that of the substrate complex: the calcium–protein interactions were maintained; the ligand's phosphate is coordinated to the calcium and to Tyr[69]; the ligand's amide carbonyl is coordinated to the calcium and the NH of Gly[30]; the ligand's amide NH points towards the N^δ of His[48]. The inhibitor–calcium interactions are less favourable than the substrate–calcium interactions, but this is more than compensated for by better inhibitor–protein interactions. A subsequent X-ray structure of a similar amide inhibitor[73] reveals considerable similarity to this model.

The most potent transition state surrogate inhibitors of PLA_2 is a phosphonate analogue.[74,75] The binding of this inhibitor to PLA_2 was modelled by replacing the reactive centre of the tetrahedral intermediate in the charge separation complex with a phosphonate group. The minimized complex retains the general features of the charge separation complex. A hydrogen bond links an oxygen of the phosphonate and N^δ of His^{48}; the other phosphonate oxygen is hydrogen bonded to the NH of Gly^{30} and coordinated to the calcium. The phosphate group of the ligand is also coordinated to the calcium. The energetics of the inhibitor–enzyme complex are somewhat less favourable than those of the corresponding tetrahedral intermediate complex, but significantly more favourable than those for the substrate complex. Thus, by mimicking the transition state, in terms of internal structure and charge distribution, it is possible to achieve a tighter binding than by mimicking the substrate's structure and charges. An X-ray structure of a phosphonate–PLA_2 complex[76] verified the characteristic features of our model.

9.3.2.4 Molecular dynamics

Molecular dynamics simulations revealed that the major features of the complex are maintained, although some local rearrangements are observed. The secondary structure elements (helices and sheets), as well as the active site and calcium environment were preserved. The ligand itself undergoes quite large fluctuations. However, atoms at the vicinity of the reaction centre and the *sn*-3 chain have very small fluctuations. The nature of the low frequency collective motions of the apoenzyme was studied using novel filtering techniques to reveal active site breathing and helix accordion motions.[42]

The magnitude and phases of the Fourier transform of the atomic fluctuations define the typical modes of motion in the MD trajectory.[77] This enabled the comparison of the modes corresponding to low frequency motions of the isolated and complexed ligand. The lowest frequency, and highest amplitude mode of motion in the MD simulations of the isolated and complexed lipid are shown in Fig. 9.5, and are compared to the low frequency mode obtained by NM analysis. The motion of the isolated ligand in the NM and MD calculations is very similar, with a scissoring motion of the *sn*-1 and *sn*-2 chains. In the complex, the mode of motion of the *sn*-2 chain is similar to that of the isolated ligand, but the other two chains (which interact with the protein or calcium) have very reduced amplitudes. In this case the *sn*-1 and *sn*-2 chains move in the same direction.

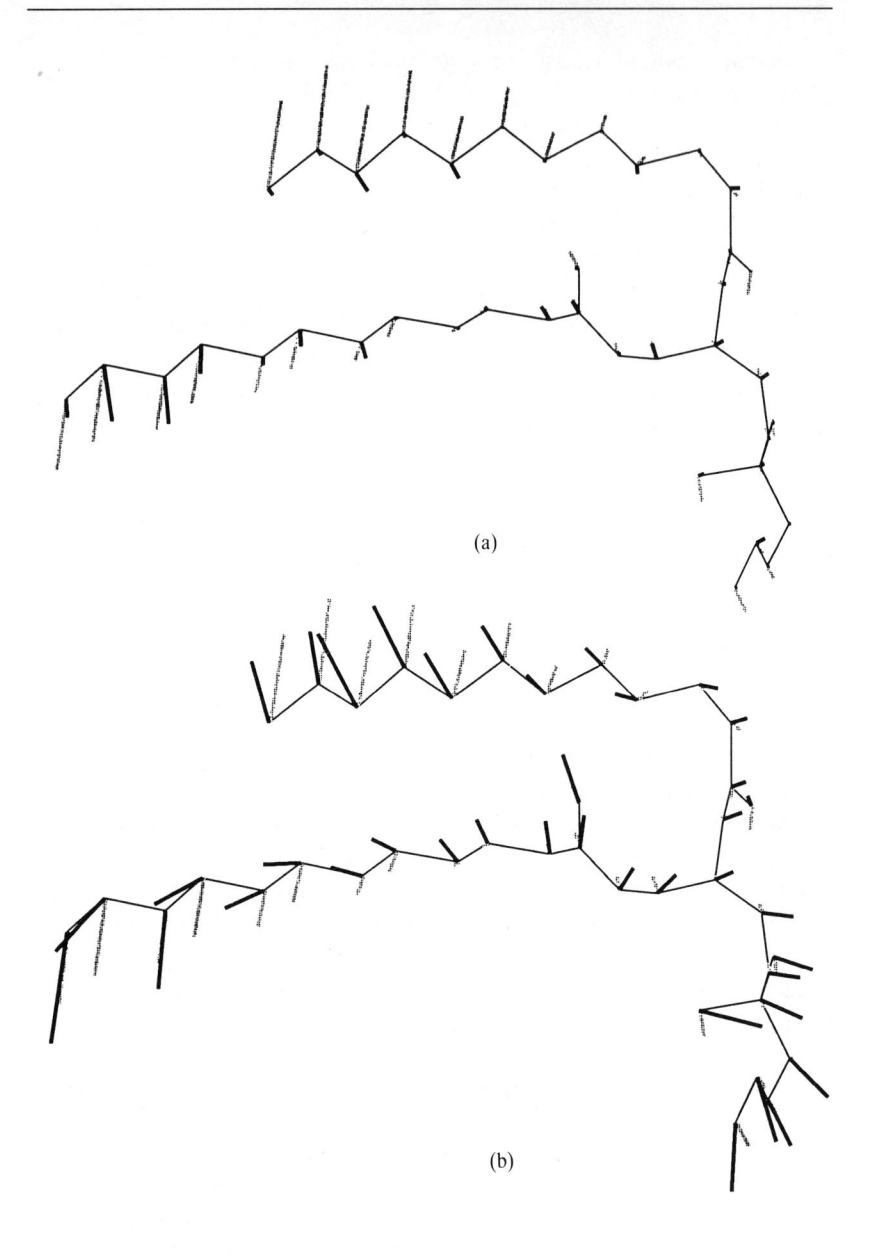

Figure 9.5 Modes of motion of the substrate molecule. The molecule is shown by thin black lines. The lowest frequency mode obtained by NM analysis of the isolated molecule is shown in thick grey lines. The low frequency modes with highest amplitudes extracted from MD simulatons are shown in black for (a) substrate–enzyme complex and (b) isolated substrate.

9.3.3 Modelling the structure of a transmembrane receptor

The final example we will look at shows a system which involves proteins which exist in the environment of the lipid bilayer which surrounds cells. Further, it is a system for which there is extensive amino acid sequence data and biological data available but little structural data, as is common for many membrane proteins. This demonstrates the application of less rigorous techniques to the creation of structural models which are to be tested by further experimentation and revision of the model in the light of the experimental results.

The nicotinic acetylcholine (nACh) receptor allows cations to flow across the membrane into the cytoplasm of a neurone after binding acetylcholine or its agonists. Thus, it is crucial to the function of the nervous system of many animals, including insects, and therefore provides a target for insecticidal compounds. It is a member of a large homologous and functionally similar set of proteins, the ligand-gated ion-channel (LGIC) superfamily. The superfamily includes the nACh receptor, γ-aminobutyric acid receptors, and the glycine receptors. The first type is selective for cations whereas the latter two are selective for anions.

Although there is a wealth of sequence data available, the current structural data are limited to a resolution of between 8 and 15 Å,[73,80] which is not enough to determine the position of the atoms. The objective of the study described in this section has been to provide a structural rationalization of the biochemical, pharmaceutical and biophysical information available for the nACh receptor by developing explicit 3-D models of the receptor at atomic level detail. The overall shape of the receptor has been defined by the structural studies based on electron microscopy[80] which show the receptor is composed of five subunits arranged symmetrically to give a barrel-shape with a cylindrical extracellular funnel and a centrally located ion-channel. The modelling study was carried out in three parallel stages – modelling a putative agonist binding region, modelling the transmembrane region and modelling the extracellular region.

9.3.3.1 The Cys-loop model

Two regions of the extracellular portion of the *Torpedo* nACh receptor have been suggested by biochemical studies to be involved in agonist binding. The first is the region around cysteines 192–193, and the second region is residues 125–147, termed the Cys-loop, because a disulphide bridge links cysteine residues at positions 128 and 142. This study concentrated on the Cys-loop region, which is a highly conserved stretch of residues. This loop contains an invariant aspartate residue which could interact with the positively charged nitrogen that occurs in all LGIC agonists and most competitive antagonists. Thus, the presence of this invariant aspartate suggests that this loop is a good candidate for the proposed anionic binding site.

Sequence alignment revealed a conserved two residue periodicity in the hydrophobicity of the residues, which suggested a β-strand structure, with an exposed hydrophilic face and buried hydrophobic face. The disulphide bridge between residues 1 and 15 of the loop requires a turn to occur around residues 8–9. By examining turns occurring in known protein structures with similar sequences, in particular a Phe or Tyr followed by a Pro, the conformation for this turn was chosen to be a type VIa β-turn. The side chain angles were taken from common values observed for β-sheet in experimental structure.[6] Energy minimization from this starting structure provided the basic structural model of the Cys-loop. Additional tests were carried out to determine the structural stability of this loop, by performing MD simulations at high temperature to allow the molecule to find any nearby accessible conformations. The Cys-loop remained close to the β-sheet structure as initially defined.

Similar models were built for the corresponding loops of other members of the LGIC superfamily. These models were constructed by mutating the residues of the initial nACh Cys-loop structure according to the sequence of the new receptor, followed by energy minimization. All of the LGIC Cys-loop sequences had the same hydrophobic/hydrophilic pattern, with an invariant Pro at position 9 preceded by a Phe or Tyr and were thus consistent with the β strands and turn suggested above. Interestingly, it was also possible to account for affinities to different ligands in the different receptors, by differences in sequence on the hydrophilic face of the loop. The residue at position 6 changes for each member of the LGIC superfamily. The different size, charge and hydrogen bonding capability of these residues can be associated with a specific interaction with the corresponding ligands that bind to that receptor.[80]

9.3.3.2 Transmembrane helices modelling

The ion channel in the LGIC receptors is thought to be created by the combination of subunits, with each subunit contributing one helix to the ion-channel. From hydrophobicity profiles calculated from the sequence data, four regions of the nACh receptor have been identified as possible trans-membrane helices (M1–M4),[81] and one region (MA) has been identified as a highly charged amphiphilic helix.[82] Recent work[83,84] demonstrated that partial ion-channel activity still existed even after the deletion of the MA helix region, thus removing the MA helix as a contender for the ion channel lining helix. On the other hand, novel peptides have been synthesized based on the M2 sequence which have been shown to act as ion-channels.[85] This helix contains a number of Ser and Thr residues which can be lined up along one side of the helix surface, and thus the hydroxyl groups can interact favourably with the ion as it passes through the channel.

The packing of the transmembrane helices was based on the known structure of the protein myohemerythrin, which has a 4-helical bundle of similar length to that required for spanning the membrane. The antiparallel packing in this bundle is appropriate as well since the connecting loop of M1–M2 and M2–M3 are too short to allow parallel packing. Information from molecular biology was used at the superimposition step to fit the pattern seen in the LGIC sequences onto the patterns found in an alignment of myohemerythrin sequences. In addition, an MD simulation was carried out to test the overall features of the packing of these helices. The initial structure for this simulation was generated by assigning the helical conformation to the residues corresponding to helices M1–M3 with the intervening linker regions being set to the extended conformation. Over the course of this simulation the three helical segments did indeed fold into an antiparallel bundle. An analysis of the final structure from the simulation showed several features that were consistent with experimental data. In particular, the partly hydrophilic surface of the M2 segment was exposed and could contribute to the wall of the ion-channel. In addition, there was a sufficient gap in the final structure for the M4 helix to be packed to give a four-helix bundle arrangement. A model of the complete ion-channel was created by assembling five copies of the 4-helix bundle. (See Fig. 9.6.) The M2 helix forms the ion-channel lining, M1 is the most tightly packed helix, making both intra-subunit contacts with M2 and M4, and inter-subunit contacts with M2 and M3 of the next subunit. The M4 helix is on the outside of the channel and makes no contacts with the helices of adjacent subunits. Thus, from the model there seem to be few structural constraints on this helix and the sequence alignments show that this is the least conserved helix. Further, this helix was replaced with a hydrophobic helix from the insulin receptor and retained activity.[84] Several other experimental features could be accounted for by this model.[86]

9.3.3.3 Extra-cellular domain modelling

Electron micrographs indicated that the first ≈ 250 residues are entirely external to the membrane. It is therefore highly likely that its structure is determined by the same rules as for globular proteins, for which there is a large body of structural data. However, no sequence homology has

Membrane

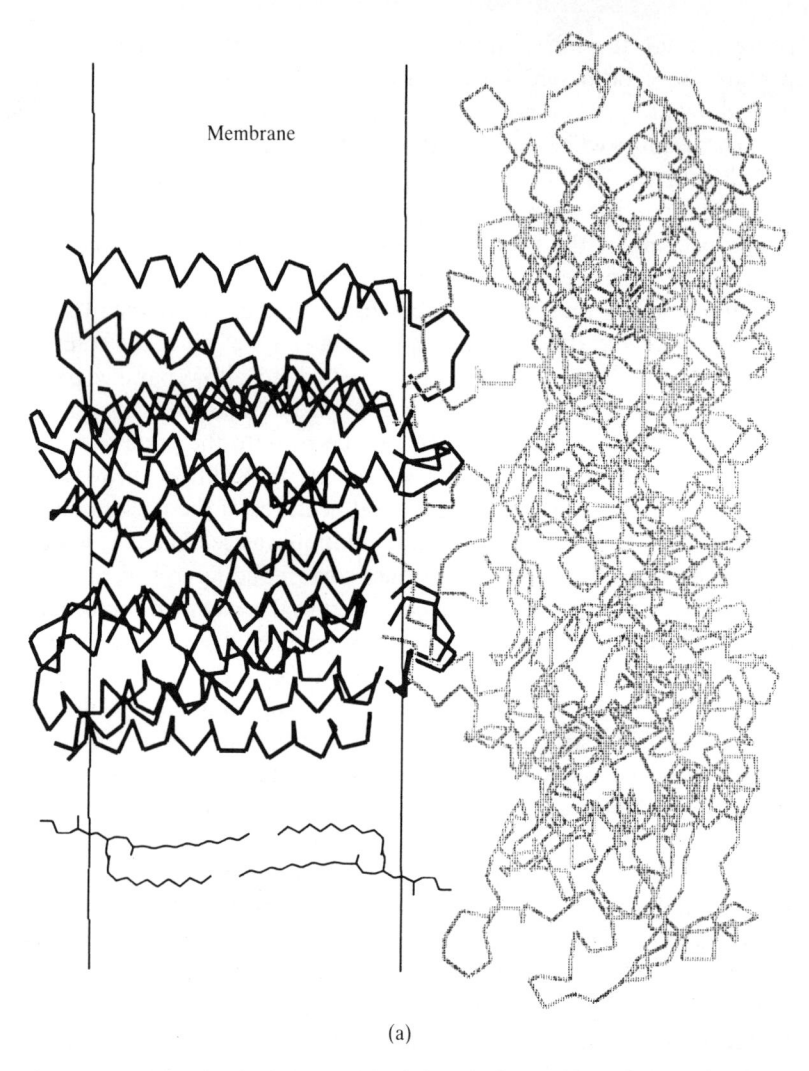

(a)

Figure 9.6 Two views of the model of the nACh receptor. The membrane spanning domain is shown in black and the extracellular domain is shown in grey. (a) View along the membrane (perpendicular to the channel). *(continued)*

been found between the extra-cellular domain and any protein of known structure using standard techniques. Instead, a method based on matching patterns of surface accessibilities was used to find relevant structures.[87] An aligned set of LGIC sequences was used to calculate a predicted surface accessibility at each amino acid position in the sequence. These accessibilities were then used to scan the computed accessibilities of known

(b)

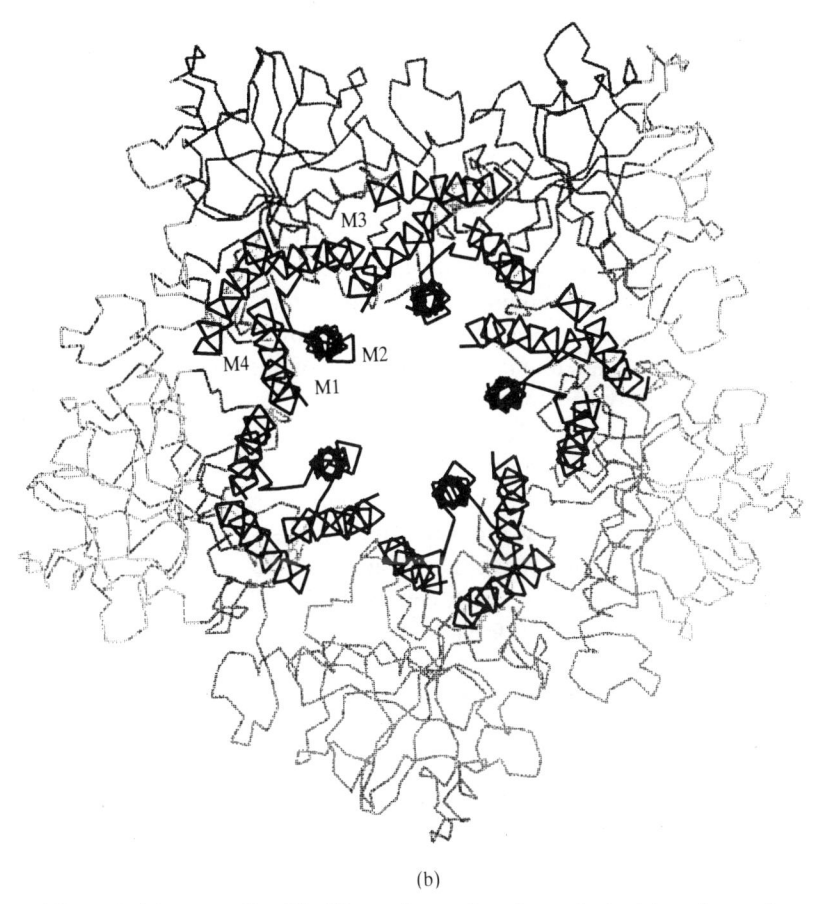

Figure 9.6 (*continued*) (b) View along the channel, looking from the intra-cellular side of the membrane.

structures to find segments of known structures which correspond in accessibility to the unknown structure. This resulted in the protein pyrophosphatase (PPase) being identified as having a high degree of similarity to the extra-cellular domain of the nACh receptor. A model for the entire extra-cellular domain of the receptor was then generated by amino acid mutation of the PPase structure. Examination of the resulting structure revealed features which are in agreement with the biochemical data. For example, the main immunogenic region occurs on the surface of this protein as does the Cys-loop. In this model, the two regions implicated in ligand binding, the region around Cys 192–193 and the Cys-loop of the same subunit, are not in close proximity. However, the Cys 192–193 and Cys-loop from adjacent subunits could form a binding site. None of the currently available experimental data can rule out a binding site between two subunits and experiments need to be devised to probe this possibility.

9.4 Conclusions and future directions

The rapid advances in computer power and simulation techniques in recent years have led to a steady increase in the number of theoretical studies of biopolymers. Simulations are used in basic research aimed at the understanding of fundamental biological phenomena, as well as in applied research in drug design and biotechnology. The trend of applying simulation techniques to biological systems of ever increasing size and complexity is likely to continue in the future. Whereas in the early days of computer simulations of biopolymers only peptides and small proteins such as PTI could be studied, today simulations involving thousands of atoms are common. The first MD simulations were carried out for only a few picoseconds (i.e. a few thousand time-steps), current simulations are reaching the nanosecond range (i.e. millions of time-steps). In addition to improvements in computer power, new methods and techniques have been developed which have made feasible simulations previously thought to be prohibitively expensive. For example, although a system which includes a biopolymer and solvent with boundary conditions may be a very large system, quite frequently the region of interest might be relatively small (e.g. an active site of an enzyme). In this case the region of interest can be represented in atomic detail, while the surrounding region is modelled by stochastic dynamics.[88]

The field of theoretical simulation has also benefited from a large increase in experimental data, including sequences and crystal structures of proteins and nucleic acids. The information gathered from the known structures is utilized in techniques for predicting structures of biopolymers with known sequence. Information theory,[89] machine learning,[90] neural networks[91] and genetic algorithms[92] are used to predict protein structure. Further developments are taking place also in 'homology modelling' in which a three-dimensional model of a protein is generated by analogy to a known structure of an homologous (i.e. with similar sequence) protein.[93-95]

Since most of the simulations depend on an analytical representation of the energy surface as a function of atomic coordinates (Eqn [1]), effort is being made,[96,97] and will be made in the future, to improve its accuracy. Of particular importance are the electrostatic forces and their treatment in systems including biopolymer molecules. These treatments have to account for the effects of charges on the biopolymer or in the solution, and the mutual effects of charge polarization and solvent orientation. In recent years a few approaches to the problem of electrostatics in biomolecular systems are emerging.[98,99] One approach treats the protein and solvent as two dielectric continuums, and uses finite difference grid methods to solve the Poisson–Boltzmann equation.[100] In another approach induced dipoles are put on a grid for the solvent and centred on atoms for the protein (protein dipoles/Langevin dipoles method).[101]

Another field of theoretical investigations which has opened recently is the modelling of enzymatic reactions. Classical molecular mechanics calculations are not sufficient to give a full picture of these events, since these reactions involve changes in the electronic state of the system and/or making and breaking of chemical bonds. Quantum mechanical calculations have to be introduced to study this aspect of biopolymer function. Quantum mechanics calculations on the small fragments involved in the reaction do not provide a correct picture of the reaction in the enzyme's active site. On the other hand, since quantum mechanics calculations are very expensive computationally, it is not possible to include the whole enzyme in the calculations. The emerging solution seems to be a hybrid approach. The reaction centre is treated quantum mechanically, while the remainder of the system is treated by molecular mechanics approximations.[102]

In conclusion, the advances in computer technology, experimental databases and new methodologies and simulation techniques are expected to result in powerful tools for the analysis and design of biopolymer structure and function, which will be applied in many topics of biological interest.

References

1. Fine R M, Wang H, Shenkin P S, Yarmush D L, Levinthal C *Proteins: Struct., Funct., Genet.* **1**: 342–362 (1986).
2. Moult J, James M N G *Proteins: Struct., Funct., Genet.* **1**: 146–163 (1986).
3. Bruccoleri R E, Karplus M *Biopolymers* **26**: 137 (1987).
4. Chou P Y, Fasman G D *Adv. Enzymol. Relat. Areas Mol. Biol.* **47**: 45–148 (1978).
5. Garnier J, Osguthorpe D J, Robson B *J. Mol. Biol.* **120**: 97 (1979).
6. Sutcliffe M J, Hayes F R F, Blundell T L *Protein Engineering* **1**: 385–392 (1987).
7. Sutcliffe M J, Haneef I, Carney D, Blundell T L *Protein Engineering* **1**: 377–384 (1987).
8. Hehre W J, Radom L, Schleyer P v R, Pople J A *Ab Initio Molecular Orbital Theory*, Wiley, New York, 1986.
9. Wilson E B, Decius J C, Cross P C *Molecular Vibrations*, McGraw-Hill, New York, 1955.
10. Dauber-Osguthorpe P, Roberts V A, Osguthorpe D J, Wolff J, Genest M, Hagler A T *Proteins: Struct., Funct., Genet.* **4**: 31–47 (1988).

11. Burkert U, Allinger N L *Molecular Mechanics*, American Chemical Society, Washington, DC, 1982.
12. Brooks B R, Bruccoleri R E, Olafson B D, States D J, Swaminathan S, Karplus M *J. Comp. Chem.* **4**: 187 (1983).
13. Weiner S J, Kollman P A, Case D A, Singh U C, Ghio C, Alagona G, Profeta S Jr, Weiner P *J. Am. Chem. Soc.* **106**: 765–784 (1984).
14. Berendsen H J C, Postma J P M, van Gunsteren W F, DiNola A, Haak J R *J. Chem. Phys.* **81**: 3684–3690 (1984).
15. Biosym Technologies, San Diego, USA.
16. Viner R C, PhD Thesis, University of Bath, 1990.
17. Lifson S, Warshel A *J. Chem. Phys.* **49**: 5116 (1968).
18. Hagler A T, Huler E, Lifson S *J. Am. Chem. Soc.* **96**: 5319–5326 (1974).
19. Lifson S, Hagler A T, Dauber P *J. Am. Chem. Soc.* **101**: 5111–5121 (1979).
20. Dauber P, Hagler A T *Accts. of Chem. Res.* **13**: 105 (1980).
21. McCammon J A, Harvey S C, *Dynamics of Proteins and Nucleic Acids*, Cambridge University Press, Cambridge, 1987.
22. Wood W W, in *Physics of Simple Liquids*, Temperely H N V, Rowlinson J S, Rushbrooke G S (eds), p. 117, North-Holland, 1968.
23. Hagler A T, Moult J *Nature* **272**: 222 (1978).
24. Hagler A T, Osguthorpe D, Robson B *Science* **208**: 599 (1980).
25. Mezei M, Beveridge D L, Berman H M, Goodfellow J M, Finney J L, Neidle S *J. Biomolecular Structure and Dynamics* **1**: 111 (1983).
26. Hagler A T, Stern P S, Sharon R, Becker J M, Naider F *J. Am. Chem. Soc.* **101**: 6842 (1979).
27. McCammon J A, Gelin B R, Karplus M *Nature* **267**: 585 (1977).
28. van Gunsteren W F, Berendsen H J C, Hermans J, Hol W G J, Postma J P M *Proc. Nat. Acad. Sci. USA* **80**: 4315–4319 (1983).
29. Fletcher R *Practical Methods of Optimization*, Vol. 1, Wiley, New York, 1980.
30. Dauber-Osguthorpe P, Campbell M M, Osguthorpe D J *Int. J. Peptide Protein Res.* **38**: 357–377 (1991).
31. Metropolis N, Rosenbluth A W, Rosenbluth M N, Teller A, Teller E *J. Chem. Phys.* **21**: 1087 (1953).
32. Brunger A T, Karplus M *Acc. Chem. Res.* **24**: 54–61 (1991).
33. Clore G M, Gronenborn A M, Brunger A T, Karplus M *J. Mol. Biol.* **186**: 435–455 (1985).
34. de Vlieg J, van Gunsteren W F *Methods in Enzymology* **202**: 268–300 (1991).
35. Brunger A T, Kuriyan J, Karplus M *Science* **235**: 45 (1987).
36. Ichiye T, Karplus M *Proteins: Struct., Funct., Genet.* **11**: 205–217 (1991).
37. Horiuchi T, Gō N *Proteins: Struct., Funct., Genet.* **10**: 106–116 (1991).
38. Kitao A, Hirata F, Gō N *Chem. Phys.* **158**: 447–472 (1991).
39. Osguthorpe D J, Dauber-Osguthorpe P *J. Mol. Graphics* **10**: 178–184 (1992).
40. Dauber-Osguthorpe P, Osguthorpe D J *J. Am. Chem. Soc.* **112**: 7921–7935 (1990).
41. Dauber-Osguthorpe P, Osguthorpe D J *Biochemistry* **29**: 8223–8228 (1990).
42. Sessions R B, Dauber-Osguthorpe P, Osguthorpe D J *J. Mol. Biol.* **210**: 617–634 (1989).
43. Dauber-Osguthorpe P, Osguthorpe D J *J. Comp. Chem.*, in press.
44. Gō N, Scheraga H A *J. Chem. Phys.* **31**: 4751 (1969).
45. Gō N, Scheraga H A *Macromolecules* **9**: 535 (1976).

46. Hill T L *An Introduction to Statistical Thermodynamics*, Addison-Wesley, Reading, MA, 1960.
47. Dauber P, Goodman M, Hagler A T, Osguthorpe D J, Sharon R, Stern P S *Proc. of the ACS Symposium on Supercomputers in Chemistry* **173**: 161 (1981).
48. Karplus M, Kushick J N *Macromolecules* **14**: 325–332 (1981).
49. Tembe B L, McCammon J A *Comput. Chem.* **8**: 281 (1984).
50. Beveridge D L, Dicapua F M *Ann. Rev. Biophys. Chem.* **18**: 431–492 (1989).
51. van Gunsteren W F, Berendsen H J C *Angew. Chem.* **29**: 992–1023 (1990).
52. Straatsma T P, McCammon J A *Methods in Enzymology* **202**: 497–311 (1991).
53. McCammon J A *Curr. Opin. Struct. Biol.* **1**: 196–200 (1991).
54. Lee C *Curr. Opin. Struct. Biol.* **2**: 217–222 (1992).
55. Meirovitch H, Kitson D H, Hagler A T *J. Am. Chem. Soc.* **114**: 5386–5399 (1992).
56. Momany F A *J. Am. Chem. Soc.* **98**: 2990 (1976).
57. Momany F A *J. Am. Chem. Soc.* **98**: 2996 (1976).
58. Struthers R S, Rivier J, Hagler A T *Ann. N.Y. Acad. Sci.* **439**: 81 (1985).
59. Rizo J, Gierasch L M *Annu. Rev. Biochem.* **61**: 387–418 (1992).
60. Dutta A S, Furr B J A *Annu. Rep. Med. Chem.* **20**: 203–214 (1985).
61. Paul P K C, Dauber-Osguthorpe P, Campbell M M, Osguthorpe D J *Biochem. Biophys. Res. Comm.* **165**: 1051–1058 (1989).
62. Dutta A S, Mclachlan P, Woodburn J R, Paul P K C, Campbell M M, Osguthorpe D J Submitted.
63. Dutta A S, Gormley J J, Woodburn J R, Paul P K C, Osguthorpe D J, Campbell M M *Bioorg. Med. Chem. Lett.* **3**: 943 (1993).
64. Drenth J, Enzing C M, Kalk K H, Vessies J C A *Nature* **264**: 373–377 (1976).
65. Sessions R B, Dauber-Osguthorpe P, Osguthorpe D J *Proteins: Struct., Funct., Genet.* **14**: 45–64 (1992).
66. Kuipers O P, Dijkman R, Pals C E G M, Verheij H M, de Haas G H *Protein Engineering* **2**: 467–471 (1989).
67. Campbell M M, Long-Fox J, Osguthorpe D J, Sainsbury M, Sessions R B *J. Chem. Soc., Chem. Commun.* 1560–1562 (1988).
68. Achari A, Scott D, Barlow P, Vidal J C, Otwinowski Z, Brunie S, Sigler P B *Cold Spring Harb. Symp. Quant. Biol.* **LII**: 441–452 (1987).
69. Barlow P N, Lister M D, Sigler P B, Dennis E A *J. Biol. Chem.* **263**: 12954–12958 (1988).
70. Iin G, Noel J, Loffredo W, Stable H Z, Tsai M-D *J. Biol. Chem.* **263**: 13208–13214 (1988).
71. Lister M D, Hancock A J *J. Lipid Res.* **29**: 1297–1308 (1988).
72. Yu L, Deems R A, Hajdu J, Dennis E A *J. Biol. Chem.* **265**: 2657–2664 (1990).
73. Thunnissen M M G M, Ab E, Kalk K H, Drenth J, Dijkstra B W, Kuipers O P, Dijkman R, de Haas G H, Verheij H M *Nature* **347**: 689–691 (1990).
74. Yuan W, Gelb M H *J. Am. Chem. Soc.* **110**: 2665–2666 (1988).
75. Yuan W, Berman R J, Gelb M H *J. Am. Chem. Soc.* **109**: 8071–8081 (1987).
76. White S P, Scott D L, Otwinowski Z, Gelb M H, Sigler P B *Science* **250**: 1560–1563 (1990).
77. Unwin N *Neuron* **3**: 655 (1989).
78. Mitra M, McCarthy M P, Stroud R M *J. Cell. Biol.* **109**: 755 (1989).
79. Toyoshima C, Unwin N *Nature* **336**: 247–250 (1988).

80. Cockroft V B, Osguthorpe D J, Barnard D J, Lunt G G *Proteins: Struct., Funct., Genet.* **8**: 386 (1990).
81. Claudio T, Ballivet M, Patrick J, Heinemann S *Proc. Natl. Acad. Sci. USA* **80**: 1111 (1983).
82. Stroud R M, Finer-Moore J *Proc. Natl. Acad. Sci. USA* **81**: 155 (1984).
83. Mishina M, Tobimatsu T, Imoto K, Tanaka K, Fulita Y, Fukuda Y, Hirose T, Inayama S, Takahashi T, Kuno M, Numa S *Nature* **313**: 364–369 (1985).
84. Tobimatsu T, Fujita Y, Fukuda K, Tanaka K, Mori Y, Konno T, Mishina M, Numa S *FEBS Lett.* **222**: 56 (1987).
85. Lear J D, Wasserman Z R, DeGrado W F *Science* **240**: 1177 (1988).
86. Osguthorpe D J, Lunt G G, Cockroft V B, in *Neurotox '91*, Duce I R (ed.), pp. 241–253, Elsevier Science Publishers, London, 1992.
87. Cockroft V B, Osguthorpe D J *FEBS* **293**: 149–152 (1991).
88. Brooks C L, Karplus M *J. Mol. Biol.* **208**: 159–181 (1989).
89. Gibart J F, Robson B, Garnier J *Biochemistry* **230**: 1578–1586 (1991).
90. King R D, Sternberg M J E *J. Mol. Biol.* **216**: 441–457 (1990).
91. Holley H, Karplus M *Methods in Enzymology* **202**: 204–224 (1991).
92. Unger R, Moult J *J. Mol. Biol.* **231**: 75–81 (1993).
93. Greer J *Methods in Enzymology* **202**: 239–252 (1991).
94. Lesk A M, Boswell D R *Curr. Opin. Struct. Biol.* **2**: 242–247 (1992).
95. Summers N L, Karplus M *Methods in Enzymology* **202**: 156–205 (1991).
96. Hwang M J, Maple J R, Stocfisch T, Hagler A T *Abst. of Papers Am. Chem. Soc.* **204**: 37 (1992).
97. Allinger N L, Zhu Z Q, Chen Z H *J. Am. Chem. Soc.* **114**: 6120–6133 (1992).
98. Bashford D *Curr. Opin. Struct. Biol.* **1**: 175–184 (1991).
99. Moult J *Curr. Opin. Struct. Biol.* **2**: 223–229 (1992).
100. Sharp K A, Honig B *Ann. Rev. Biophys. Biophys. Chem.* **19**: 301–332 (1990).
101. Warshel A, Aaqvist J *Ann. Rev. Biophys. Biophys. Chem.* **20**: 267–298 (1991).
102. Warshel A *Curr. Opin. Struct. Biol.* **2**: 230–235 (1992).

INDEX

acrylates 27
adhesive 263
aggregation 293
alkanes, normal 192, 254
AM1 semi-empirical calculations 159
AMBER force field 133, 158, 308
amorphous layer 165, 189
  polymers, sample preparation 57
  region 190, 193
analysis, regression *see* regression analysis
apportioning 171, 172, 182
aromatic polyesters 140
  polymers 137
athermal mode of nucleation 194
atomic charges, determination of 138, 140
Avrami equation 193

balance, detailed *see* detailed balance
barrier properties 47, 130
bead model 46, 250
  -spring model 100, 109, 229
behaviour, elastic *see* elastic behaviour
  phase *see* phase behaviour
  stress–strain *see* stress–strain behaviour
  viscous *see* viscous behaviour
Bethe lattice 269
biaxial deformation of network 221

bimodal network 263
bioactive peptide analogues 317
biopolymers 303–36
bisphenol-A–polycarbonate 104, 105, 126
blending 36
blends *see* polymer blends
blob 106, 107
block copolymer 91, 92, 101, 104, 121, 125
bond bending terms in force fields 134
  fluctuation model 96, 97, 98, 101, 102, 104, 107, 109, 110, 113, 115, 116, 126
  percolation, in network formation 281
  stretching terms in force fields 133
BPA–PC *see* bisphenol-A–polycarbonate
branched molecule 264
branching ratio 269
Brill transition 158
bulk moduli 17, 292

canonical ensemble, grand *see* grand canonical ensemble
cascade theory 273
chain collapse 115, 242, 253
  dynamics 169, 192
  extension 270
  fold lengths 188

337

chain (*cont.*)
  folding 132, 165, 185, 189
  growth 58
  growth, kinetics of 169
  growth, phantom *see* phantom chain growth
  mobility 168
  packing, in crystalline polymers 146
  re-entry 166, 182, 190
  re-entry, adjacent 181
CHARGE2 algorith 143
CHARMM 133, 308
chemical potential 229, 233, 237, 240, 249, 250
chi parameter, Flory–Huggins 37, 111, 112, 114, 116, 229, 244
circuits, in network formation 265
clotting of blood, as example of network formation 263
clusters, in percolation theory 281
CNDO/2 calculations 10, 16, 18
coarse-grained models 91, 94, 97, 99, 108, 125, 126
coatings 263
cohesive energy density 17
collapse, chain *see* chain collapse
collective phenomena 91
combs 240
condensation reactions, in network formation 263
condis crystals 156, 157
configurational bias Monte Carlo scheme 233
conformational energy 10, 17
  entropy 9, 15, 41, 180
  repeat unit 150
  search 318
  search, systematic 310
conformtionally disordered crystals, *see* condis crystals
connectivity, in network formation 281
controlled pressure dynamics 51
copolymer 94, 189, 191
  block *see* block copolymer
correlation hole 113
COSMIC force field 133
critical amplitudes, in network formation 282
  condition, in network formation 265

  exponent 118, 281
  phenomena, 263
  point 92, 118, 229
  scattering 118
  temperature 93, 112, 113, 119
crosslinking 211, 263, 305
crossover scaling theory 91, 106, 107, 108, 109, 110, 111
crystal packing, effect of non-bonded terms 143
crystalline polymers, simulation of 130–62
  volume fraction 193
crystallization, lamellar *see* lamellar crystallization
  non-isothermal *see* non-isothermal crystallization
cycle rank, in theory of elasticity 288
  rank density 288

defects, in polyethylene crystals 155
deformation 200
  biaxial 221
  multiaxial 219
deformations, mechanical *see* mechanical deformations
dendrimer 268
density, cohesive energy, *see* cohesive energy density
detailed balance 171
diblock copolymer 126, 229
dielectric constant 135
diffusion 130
digital signal processing techniques 314
dilute solution hydrodynamic forces 106
  solution 107, 228
dipole moment 16, 139
DISCOVER computer program 308
distributed multipoles 139
distribution function, radial *see* radial distribution function
DNA 304
drawability 205
Dreiding force field 133, 161
dynamic Monte Carlo *see* Monte Carlo, dynamic
  relaxation 60
dynamics, chain *see* chain dynamics

edge, in graph theory 267
elastic behaviour 66, 218
elasticity active chain 277
elasticity 263
elastomeric network 213
electrostatic interactions 18, 133, 134, 138, 307
endlinked silicone 288
energies, conformational *see* conformational energies
energies, intermolecular probe, *see* intermolecular probe energies
energy minimization 309, 310, 318, 322, 328
energy, of unit cell 143
enrichment, surface *see* surface enrichment
ensemble, grand canonical *see* grand canonical ensemble
entanglement 98, 104, 109, 208, 216, 300
  length 48
entropic forces 99, 190
entropy 168, 180, 182, 184, 189, 191, 314
environmental effects 316
enzyme–ligand interactions 320
Ewald summation 150, 161
excluded volume forces 106
  volume 58, 59
expectation, in Miller–Macosko probability theory 278
extensional modulus 71
extent of reaction, in network formation 269

failure 200
filtering of MD results 314, 325
finite size scaling 118
flexibility, intermolecular *see* intermolecular flexibility
Flory–Huggins theory 37, 111, 112, 114, 115, 118, 126, 228
Flory–Stockmayer theory of gelation 263
fluctuating fold lengths 180
fluctuation 168, 172, 174, 282
fluids, 'fragile' *see* fragile fluids
fold lengths, fluctuating *see* fluctuating fold lengths
folding, chain *see* chain folding

force field 18, 132, 307
  field, Dreiding *see* Dreiding force field
  field, AMBER *see* AMBER force field
  field, COSMIC *see* COSMIC force field
  field, MM2 *see* MM2
  field, MM3 *see* MM3
  field, determination of parameters 136
forces, entropic *see* entropic forces
  excluded volume *see* excluded volume forces
Fourier transform 313, 325
fracture 200, 201
fragile fluids 75
free energy 228, 314
frequency distribution 314
function, radial distribution *see* radial distribution function

GAP *see* Group Additive Property
gaussian chain statistics 277
gel point 263
gelation 263
generating function, in cascade theory 273
glass transition temperature 23, 70
  transition 2, 47, 73, 99, 104, 107, 108, 126
glasses, polymer *see* polymer glasses
grafted polymer layers 125
Grand Canonical Monte Carlo methods 191
grand canonical ensemble 100, 106
graph theory 266
GROMOS computer program 308
group additive property 2
growth rate of crystals 167, 174, 183, 184, 186, 187, 193, 197
  rate minima 188
  chain *see* chain growth
  phantom chain *see* phantom chain growth

hardening, strain 76
helical repeat unit, in crystalline polymers 149

helix formation in polymer chains 145
homology modelling 332
Hooke's Law 133
hydrodynamic forces, dilute solution
    see dilute solution hydrodynamic
    forces
  interactions 107

integral equations theories 112, 113, 115
interdiffusion 98, 99
interface, polymer 98, 99, 125
intermolecular probe energies 17
intramolecular flexibility 9
intramolecular reaction in network formation 265
Ising model 118

junction, in graph theory 267

KEVLAR 157
Kuhn length 76

lamellae, crystal 131, 165, 193, 230
lamellar crystallization 166
  mesophase 121
  structure 94
  thickness 165, 167, 174, 177, 183, 186, 191
lattice models 47, 100, 179, 201, 228
  model, Flory–Huggins *see*
    Flory–Huggins lattice model
length, Kuhn *see* Kuhn length
  persistence *see* persistence length
  screening *see* screening length
linearized theory 118, 121, 126
loop, in network formation 265
  probability 277
loose-coupling algorithm 82–86
  controlled pressure dynamics 51, 52
low frequency collective motions 325
lower critical consolute point 119

MC *see* Monte Carlo
MD *see* molecular dynamics
mechanical deformations 63
melt transition temperature 30, 31
melting point 189
melts, polymer *see* polymer melts

mesophase formation 92
methacrylates 27
Metropolis Monte Carlo 190
microphase separation 94, 121, 124
microscopic reversibility 169, 171
minimization *see* energy minimization
miscibility 111
mixture 36, 92 *see also* symmetric polymer mixture
MM2 133, 136, 308
MM3 133, 136
MNDO semi-empirical calculations 144
mobile phase, in crystalline region 192
mobility, chain *see* chain mobility
modulus, bulk 17, 292
modulus, extensional *see* extensional modulus
modulus, Young's *see* Young's modulus
molecular dynamics 99, 108, 109, 136, 151, 152, 192, 193, 309, 312, 318, 322, 325, 328, 329
molecular mechanics 132, 307
molecular weight 203
molecular weight distributions 267, 272
moment properties in QSPR 16
Monte Carlo methods 62, 91, 109, 125, 169, 173, 179, 182, 186, 193, 201, 228, 281, 309, 311
  sampling, unbiased 88
  sampling, simple 104
  configurational bias scheme *see* configurational bias Monte Carlo scheme
  dynamic 170
morphology 186, 194, 195
multipole expansion 138, 139
MUSIC method 156

n-alkanes 192, 254
natural rubber 263
network models for deformation 211
  bimodal *see* bimodal network
  biaxial deformation of *see* biaxial deformation
  polymer *see* polymer network
  random *see* random network
  elastomeric *see* elastomeric network
NM analysis *see* normal mode analysis

node, in graph theory 267
non-isothermal crystallization 189, 193
normal mode analysis, of MD results 312, 325
Nose–Hoover algorithm 51
nucleation 171, 175, 191
    distribution 193, 196
    mode, athermal 194
    mode, thermal 194
    primary *see* primary nucleation
    rate 173, 179, 183
nylon 66, 36, 158
    3Me6T 36

ODT *see* order–disorder transition
order, orientational *see* orientational order
order–disorder transition 124, 125, 229
orientational order 97, 104

paint 263
paraffins 189
PCG *see* phantom chain growth
PDMS *see* polydimethyl-siloxane
PEEK *see* poly(phenylene ether ether ketone)
PEK *see* poly(phenylene ether ketone)
PEKK *see* poly(phenylene ether ketone ketone)
PEO *see* poly(ethylene oxide)
percolation simulation 263
    bond *see* bond percolation
    site *see* site percolation
periodic boundary conditions 57, 135, 146, 150, 207, 213
persistence length 77, 81, 92, 94, 102, 106, 145
perturbation techniques for free energy 315
phantom chain growth 59
    sampling, unbiased 62
phantom modulus 288
phase behaviour 36, 97, 114, 118, 228
pinning models 185
Poisson's ratio 53, 65, 71
poly(ethylene-propylene)-poly(ethylene) diblock copolymer 122, 124

poly(ethylene oxide) 188, 195
poly(oxymethylene) 149
poly(p-phenyleneterephthalamide) 150, 159, 203
poly(phenylene ether ether ketone) 143, 144, 147, 148, 189
poly(phenylene ether ketone) 143, 144, 147
poly(phenylene ether ketone ketone) 147, 148
poly(phenylene oxide) 143, 144
poly(vinylchloride) 230
polyamides 157
polybenzamide 160
polybutadiene 93
polycarbonate 230
polydimethyl-siloxane 216
polyesters, aromatic 140
polyethylene 63, 67, 75, 92, 95, 97, 99, 103, 109, 116, 133, 149, 152, 169, 183, 184, 187, 192, 201, 230, 241
polymer design 2
    blends 91, 101, 104, 106, 111, 117, 244
    glasses 47
    melts 106
    miscibility, *see* miscibility
    mixture *see* mixture
    mixtures, symmetric *see* symmetric polymer mixtures
    network 263
    solutions 91, 126 *see also* dilute solutions, semi-dilute solutions
polymers, semi-crystalline 145
polymethylvinylether 119
polypropylene 230
polysaccharides 305
polystyrene 119, 230, 244
polysulphones 230
potential derived charges 139
    energy function in force fields 10, 307
PPTA *see* poly(p-phenylene-terephthalamide)
pressure loose-coupling algorithm *see* loose-coupling algorithm, pressure
primary nucleation 193, 195
principal moments of inertia 15
probe energies, intermolecular 17

properties, interfacial *see* interfacial
    properties
protein structure 332

QSPR *see* quantitative structure
    property relationship
quadrupole moment 139
Quantitative Structure Property
    Relationship 5, 23, 41
quantum chemical calculations 104
    chemical calculations, *ab initio* 137,
    138, 141, 308
quenching experiments 118

radial distribution function 61
radiation cured polymers 289
radius of gyration 145
Rahman–Parrinello algorithm 51
random walk 57
    network 263
    phase approximation 91, 114, 115,
    121, 125, 126, 248
rate equations in crystallization 169,
    170, 180, 182
re-entry, chain *see* chain re-entry
reaction, extent of *see* extent of reaction
recursion relation, for calculation of
    molecular weight distribution 273
    probability rules 278
Regime I crystallization 174, 191
Regime II crystallization 175, 191
Regime I–II transition 175, 187
Regime II–III transition 175
regime analysis 183
regimes 173
regression analysis 5, 8, 23
relaxation, dynamic 60
Reneker defect 155
renormalization group theory 263
repeat unit, helical *see* helical repeat
    unit
    in crystalline polymers 149
    conformational *see* conformational
    repeat unit
reptation 98, 107, 108, 110
reversibility, microscopic *see*
    microscopic reversibility
rings, in gelation theory 280

RIS *see* Rotational Isomeric State
    theory
RIS-MC 282
Rotational Isomeric State theory 4,
    59, 62, 145, 282
roughening 186
    transition 187
    surface *see* surface roughening
Rouse model 95, 106, 108, 109
RTV adhesive 263
rubber, natural 263

scaling theory, crossover *see* crossover
    scaling theory
    function 106, 107
scattering function 119
    light or neutrons 114
screening length 92, 106
secondary nucleus in crystallization 181
segment–segment interaction
    parameter, Flory–Huggins *see* chi
    parameter
self-averaging, of structure factor 118
self-avoiding walk, on lattice 10, 111,
    240, 241
self-diffusion constant 104, 106, 109
semi-crystalline polymers 145, 228
semi-dilute solution 91, 106, 111
separation, phase *see* phase separation
    microphase *see* microphase
    separation
SHAKE algorithm for molecular
    dynamics 158
shapes, crystallite 177
site percolation 281
slithering snake algorithm 100
solutions, polymer *see* polymer
    solutions
    semi-dilute *see* semi-dilute solutions
spherulites 132, 165, 193, 194, 195
spinodal decomposition of blends 92,
    93, 99, 109, 115, 118, 121, 126
SPR *see* structure-property
    relationship
SRU *see* structural repeat unit
star polymers 240
step-gas-model 173
strain hardening 76
    energy density 219

strength, tensile *see* tensile strength
stress, yield *see* yield stress
stress–strain behaviour 76
structural repeat unit 4
structure factor 114, 118, 121, 122
   lamellar *see* lamellar structure
structure–property relationship 5
sulphur crosslinkage 273
surface accessibilities, of biopolymers 330
   enrichment of polymer blends 125
   nucleation 167, 175
   roughening 177, 180
   segregation of polymer melt chains 228
swelling 131, 228
symmetric polymer mixture 113, 115

TAU *see* torsional angle unit theory
TAU dipole moments 16
TAU masses 11
TAU principal moments of inertia 15
T_c *see* critical temperature
telechelic polyoxyethylene 289
temperature, critical *see* critical temperature
   glass transition *see* glass transition temperature
   melt transition *see* melt transition temperature
tensile strength 158, 201
tension experiments, uniaxial 65
T_g *see* glass transition temperature
theory of elasticity 277
thermal mode of nucleation 194
thermodynamic integration for free energy 315
thermosetting polymers 263
theta solvent 59
three-dimensional static modelling 4
T_m *see* melt transition temperature
topology 281

torsion angle unit theory 4, 5, 41
torsional angle unit 5
   potential 50, 95, 133, 134, 144
total probability, law for 278
transcrystallization 196
transition, glass *see* glass transition
   temperature, glass, *see* glass transition temperature
   temperature, melt *see* melt transition temperature
transitions, conformational *see* conformational transitions
transmembrane receptor 327
tube 78, 79, 110, 213
   diameter 92

unbiased Monte Carlo sampling *see* Monte Carlo sampling, unbiased
   phantom chain sampling *see* phantom chain sampling, unbiased
uniaxial tension experiments *see* tension experiments, uniaxial
universality 282
   class 265
upper critical consolute point 119
urethane 289

Van Krevelen 3, 23, 31
Van Krevelen GAP models 3
van der Waals potentials 134, 144
Verdier–Stockmayer algorithm 100
vertex, in graph theory 267
viscous behaviour 66
Vogel–Fulcher law 104, 105
volume fraction, crystalline *see* crystalline volume fraction
vulcanizing 296

yield stress 53, 69
Young's modulus 53, 66, 71, 130